# Praise for *The Rational Optimist*

"Ridley does an impressively comprehensive job of looking at the entirety of human history through the lens of a single big idea: that of the awesome power of trade. . . . Trade in ideas and goods, and of resource management, and of the way in which humans, left to their own devices in contact with other humans, have a tendency to grow and prosper unless quashed by outside forces or rentiers. . . . A useful corrective to prevailing pessimism."
—Felix Salmon, *Barnes & Noble Review*

"Audacious. . . . Ridley is on the mark with the big things."
—*The Economist*

"It's a fantastic read essentially arguing that specialization and the trade that emerged from specialization was the essential driver of human cultural evolution. It's a great book."
—*Psychology Today*

"It does much more than debunk the doomsaying. Dr. Ridley provides a grand unified theory of history from the Stone Age to the better age awaiting us in 2100. It's an audacious task, but he has the intellectual breadth for it."
—John Tierney, *New York Times*

"Ridley eloquently weaves together ec--nomics, archaeology, history, and evolu*i*-- ... e argument [that we are livin~ ... ~ceful times in history] a step ... turn complicated eco- nomi~ ... tertaining, digestible nugge~ ... _ntalist or a WTO pro- tester, ~ ... , ~~ ~e rationally optimistic about here."
—Barrett Sheridan, *Newsweek*

"A mesmerizing book."
—*Los Angeles Times*

"Invigorating. . . . Driven by an intelligence that seems to have mated profitably with a wide array of ideas . . . Mr. Ridley shows a commitment to rational optimism . . . [and to] the idea of openness as fundamental to progress. For Mr. Ridley, the market for ideas needs to be as open as possible in order to breed ingenuity from collaboration."
—Trevor Butterworth, *Wall Street Journal*

"A superb book. . . . Elegant, learned, and cogent. . . . Combining Adam Smith's division of labor with Charles Darwin's natural selection, Ridley frames a far-reaching synthesis of economics and ecology, a triumphant new démarche in the understanding of wealth and poverty. . . . This inspiring Ridley vision is full of fascinating insights. . . . Unexpected . . . ambitious . . . and contrarian."
—George Gilder, *National Review*

"Without sounding like a cockeyed optimist, *The Rational Optimist* will give a reader solid reasons for believing that the human species will overcome its economic, political, and environmental woes during this century."
—*Fort Worth Star-Telegram*

"The indefatigable Matt Ridley can help—and just in time. Ridley has a new book called *The Rational Optimist* for a dose of just the kind of glass-half-full information we need right now. . . . Ridley's book is a powerful antidote to the gloom-n-doom-mongering of the hard left." —*Washington Examiner*

"The chapters tracing the human story from 50,000 years ago through the seventeenth century are themselves worth the price of admission, with vivid storytelling illuminating the huge role of markets and trade in material progress. . . . Read *The Rational Optimist* for its fascinating history of trade and innovation." —William Easterly, *New York Times Book Review*

"Ridley's dazzling, insightful, and entertaining book on the unstoppable march of innovation is a refresher course in human history."
—Kyle Smith, *New York Post*

"A very good book. . . . A rich analysis. . . . Ridley bolsters his argument with an impressive tour of evolutionary biology, anthropology, economics, philosophy, and world history. . . . He is a cogent and erudite social critic."
—Wray Herbert, *Washington Post*

"*The Rational Optimist* is chock-full of in-your-face challenges to conventional wisdom. . . . Ridley is a sworn enemy of Cassandras and Chicken Littles. In *The Rational Optimist*, he covers 200,000 years of human history to make a compelling case that over the millennia poverty declined, disease retreated, violence atrophied, freedom grew, and happiness increased."
—Glenn C. Altschuler, *Oregonian*

"Ridley provides a selective economic history of the human race from before its beginning. . . . The author has the great advantage over most of those who pontificate on 'the science' of [climate change] in having a scientific background himself and is able to show that much of this is not science at all and, therefore, discounts much of the fashionable alarmism."
—*Financial Times*

"Ridley's important book shoots down the culture of doom that stands in such stark contrast to the generally optimistic arc of human history. . . . *The Rational Optimist* teases out the contradiction between our increasingly comfortable lives and the intellectual climate of deep, dark pessimism. With simplicity, clarity, and verve, he stands up for 'the bright side of human endeavor' in a book that feels like an act of intellectual rebellion against the tyranny of misery gripping this young century."
—Brendan O'Neill, *American Conservative*

"A cleverly written, deeply researched effort meant to be savored slowly. My guess is that readers will find themselves often quoting its tasty bits of evidence, or even reading it aloud to others. . . . His book should seduce and charm thinking readers everywhere."
—DailyFinance.com

"This inspiring book, a glorious defense of our species, explains why: it is a devastating rebuke to humanity's self-haters."
—Dominic Lawson, *Sunday Times* (London)

"Original, clever, and . . . controversial."
—Jon Henley, *The Guardian* (London)

Also by Matt Ridley

*The Red Queen: Sex and the Evolution of Human Nature*

*The Origins of Virtue:*
*Human Instincts and the Evolution of Cooperation*

*Genome: The Autobiography of a Species in 23 Chapters*

*The Agile Gene: How Nature Turns on Nurture*
(Originally published as *Nature Via Nurture: Genes,*
*Experience, and What Makes Us Human*)

*Francis Crick: Discoverer of the Genetic Code*

# The Rational Optimist

## How Prosperity Evolves

# Matt Ridley

HARPER PERENNIAL

NEW YORK • LONDON • TORONTO • SYDNEY • NEW DELHI • AUCKLAND

HARPER ● PERENNIAL

A U.S. hardcover edition was published, in slightly different form, by Harper-Collins Publishers in 2010.

A paperback edition of this book was originally published by HarperCollins UK in 2011.

P.S.™ is a trademark of HarperCollins Publishers.

HarperCollins books may be purchased for educational, business, or sales promotional use. For information, please e-mail the Special Markets Department at SPsales@harpercollins.com.

FIRST HARPER PERENNIAL EDITION PUBLISHED 2011.

Library of Congress Cataloging-in-Publication Data is available upon request.

ISBN 978-0-06-145206-2 (pbk.)

23 24 25 26 27  LBC  31 30 29 28 27

*For Matthew and Iris*

This division of labour, from which so many advantages are derived, is not originally the effect of any human wisdom, which foresees and intends that general opulence to which it gives occasion. It is the necessary, though very slow and gradual, consequence of a certain propensity in human nature which has in view no such extensive utility; the propensity to truck, barter, and exchange one thing for another.

ADAM SMITH
*The Wealth of Nations*

# CONTENTS

# When ideas have sex

In other classes of animals, the individual advances from infancy to age or maturity; and he attains, in the compass of a single life, to all the perfection his nature can reach: but, in the human kind, the species has a progress as well as the individual; they build in every subsequent age on foundations formerly laid.

ADAM FERGUSON
*An Essay on the History of Civil Society*

On my desk as I write sit two artefacts of roughly the same size and shape: one is a cordless computer mouse; the other a hand axe from the Middle Stone Age, half a million years old. Both are designed to fit the human hand – to obey the constraints of being used by human beings. But they are vastly different. One is a complex confection of many substances with intricate internal design reflecting multiple strands of knowledge. The other is a single substance reflecting the skill of a single individual. The difference between them shows that the human experience of today is vastly different from the human experience of half a million years ago.

This book is about the rapid, continuous and incessant change that human society experiences in a way that no other animal does. To a biologist this is something that needs explaining. In the past two decades I have written four books about how similar human beings are to other animals. This book is about how different they are from other animals. What is it about human beings that enables them to keep changing their lives in this tumultuous way?

It is not as if human nature changes. Just as the hand that held the hand axe was the same shape as the hand that holds the mouse, so people always have and always will seek food, desire sex, care for offspring, compete for status and avoid pain just like any other animal. Many of the idiosyncrasies of the human species are unchanging, too. You can travel to the farthest corner of the earth and still expect to encounter singing, smiling, speech, sexual jealousy and a sense of humour – none of which you would find to be the same in a chimpanzee. You could travel back in time and empathise easily with the motives of Shakespeare, Homer, Confucius and the Buddha. If I could meet the man who painted exquisite images of rhinos on the wall of the Chauvet Cave in southern France 32,000 years ago, I have no doubt that I would find him fully human in every psychological way. There is a great deal of human life that does not change.

Yet to say that life is the same as it was 32,000 years ago would be absurd. In that time my species has multiplied by 100,000 per cent, from perhaps three million to nearly seven billion people. It has given itself comforts and luxuries to a level that no other species can even imagine. It has colonised every habitable corner of the planet and explored almost every uninhabitable one. It has altered the appearance, the genetics and the chemistry of the world and pinched perhaps 23 per cent of the productivity of all land plants for its own purposes. It has surrounded itself with peculiar, non-random arrangements of atoms called technologies, which it invents, reinvents and discards almost continuously. This is not true for other creatures, not even brainy ones like chimpanzees, bottlenose dolphins, parrots and octopi. They may occasionally use tools, they may occasionally shift their ecological niche, but they do not 'raise their standard of living', or experience 'economic growth'. They do not encounter 'poverty' either. They do not progress from one mode of living to another – nor do they deplore doing so. They do not experience agricultural, urban, commercial, industrial and information revolutions, let alone Renaissances, Reformations, Depressions, Demographic Transitions, civil wars, cold wars, culture wars and credit crunches. As I sit here at my desk, I am surrounded by things – telephones, books, computers, photographs, paper clips, coffee mugs – that no monkey has ever come close to making. I am spilling digital information on to a screen in a way that no dolphin has ever managed. I am aware of abstract concepts – the date, the weather forecast, the second law of thermodynamics – that no parrot could begin to grasp. I am definitely different. What is it that makes me so different?

It cannot just be that I have a bigger brain than other animals. After all, late Neanderthals had on average bigger brains than I do, yet did not experience this headlong cultural change. Moreover, big though my brain may be compared with another animal species, I have barely the foggiest inkling how to make coffee cups and paper clips, let alone weather forecasts. The psychologist Daniel Gilbert likes to joke that every member of his profession lives under the obligation at some time in his career to complete a sentence which begins: 'The human being is the only animal that ...' Language, cognitive reasoning, fire, cooking, tool making, self-awareness, deception, imitation, art, religion, opposable thumbs, throwing weapons, upright stance, grandparental care – the list of features suggested as unique to human beings is long indeed. But then the list of features unique to aardvarks or bare-faced go-away birds is also fairly long. All of these features are indeed uniquely human and are indeed very helpful in enabling modern life. But I will contend that, with the possible exception of language, none of them arrived at the right time, or had the right impact in human history to explain the sudden change from a merely successful ape-man to an ever-expanding progressive moderniser. Most of them came much too early in the story and had no such ecological effect. Having sufficient consciousness to want to paint your body or to reason the answer to a problem is nice, but it does not lead to ecological world conquest.

Clearly, big brains and language may be necessary for human beings to cope with a life of technological modernity. Clearly, human beings are very good at social learning, indeed compared with even chimpanzees humans are almost obsessively inter-ested in faithful imitation. But big brains and imitation and language are not themselves the explanation of prosperity and progress and poverty. They do not themselves deliver a chang-ing standard of living. Neanderthals had all of these: huge brains, probably complex languages, lots of technology. But they never burst out of their niche. It is my contention that in looking inside our heads, we would be looking in the wrong place to explain this extraordinary capacity for change in the species. It was not something that happened within a brain. It was something that happened between brains. It was a collective phenomenon.

Look again at the hand axe and the mouse. They are both 'man-made', but one was made by a single person, the other by hundreds of people, maybe even millions. That is what I mean by collective intelligence. No single person knows how to make a computer mouse. The person who assembled it in the factory did not know how to drill the oil well from which the plastic came, or vice versa. At some point, human intelligence became collective and cumulative in a way that happened to no other animal.

## Mating minds

To argue that human nature has not changed, but human culture has, does not mean rejecting evolution – quite the reverse. Humanity is experiencing an extraordinary burst of evolutionary change, driven by good old-fashioned Darwinian natural selection. But it is selection among ideas, not among genes. The habitat in which these ideas reside consists of human brains. This notion has been trying to surface in the social sciences for a long time. The French sociologist Gabriel Tarde wrote in 1888: 'We may call it social evolution when an invention quietly spreads through imitation.' The Austrian economist Friedrich Hayek wrote in the 1960s that in social evolution the decisive factor is 'selection by imitation of successful institutions and habits'. The evolutionary biologist Richard Dawkins in 1976 coined the term 'meme' for a unit of cultural imitation. The economist Richard Nelson in the 1980s proposed that whole economies evolve by natural selection.

This is what I mean when I talk of cultural evolution: at some point before 100,000 years ago culture itself began to evolve in a way that it never did in any other species – that is, to replicate, mutate, compete, select and accumulate – somewhat as genes had been doing for billions of years. Just like natural selection cumulatively building an eye bit by bit, so cultural evolution in human beings could cumulatively build a culture or a camera. Chimpanzees may teach each other how to spear bushbabies with sharpened sticks, and killer whales may teach each other how to snatch sea lions off beaches, but only human beings have the cumulative culture that goes into the design of a loaf of bread or a concerto.

Yes, but why? Why us and not killer whales? To say that people have cultural evolution is neither very original nor very helpful. Imitation and learning are not themselves enough, however richly and ingeniously they are practised, to explain why human beings began changing in this unique way. Something else is necessary; something that human beings have and killer whales do not. The answer, I believe, is that at some point in human history, ideas began to meet and mate, to have sex with each other.

Let me explain. Sex is what makes biological evolution cumulative, because it brings together the genes of different individuals. A mutation that occurs in one creature can therefore join forces with a mutation that occurs in another. The analogy is most explicit in bacteria, which trade genes without replicating at the same time – hence their ability to acquire immunity to antibiotics from other species. If microbes had not begun swapping genes a few billion years ago, and animals had not continued doing so through sex, all the genes that make eyes could never have got together in one animal; nor the genes to make legs or nerves or brains. Each mutation would have remained isolated in its own lineage, unable to discover the joys of synergy. Think, in cartoon terms, of one fish evolving a nascent lung, another nascent limbs and neither getting out on land. Evolution can happen without sex; but it is far, far slower.

And so it is with culture. If culture consisted simply of learning habits from others, it would soon stagnate. For culture to turn cumulative, ideas needed to meet and mate. The 'cross-fertilisation of ideas' is a cliché, but one with unintentional fecundity. 'To create is to recombine' said the molecular biologist François Jacob. Imagine if the man who invented the railway and the man who invented the locomotive could never meet or speak to each other, even through third parties. Paper and the printing press, the internet and the mobile phone, coal and turbines, copper and tin, the wheel and steel, software and hardware. I shall argue that there was a point in human pre-history when big-brained, cultural, learning people for the first time began to exchange things with each other, and that once they started doing so, culture suddenly became cumulative, and the great headlong experiment of human economic 'progress' began. Exchange is to cultural evolution as sex is to biological evolution.

By exchanging, human beings discovered 'the division of labour', the specialisation of efforts and talents for mutual gain. It would at first have seemed an insignificant thing, missed by passing primatologists had they driven their time machines to the moment when it was just starting. It would have seemed much less interesting than the ecology, hierarchy and super-stitions of the species. But some ape-men had begun exchanging food or tools with others in such a way that both partners to the exchange were better off, and both were becoming more specialised.

Specialisation encouraged innovation, because it encouraged the investment of time in a tool-making tool. That saved time, and prosperity is simply time saved, which is proportional to the division of labour. The more human beings diversified as consumers and specialised as producers, and the more they then exchanged, the better off they have been, are and will be. And the good news is that there is no inevitable end to this process. The more people are drawn into the global division of labour, the more people can specialise and exchange, the wealthier we will all be. Moreover, along the way there is no reason we cannot solve the problems that beset us, of economic crashes, population explosions, climate change and terrorism, of poverty, AIDS, depression and obesity. It will not be easy, but it is perfectly possible, indeed probable, that in the year 2110, a century after this book is published, humanity will be much, much better off than it is today, and so will the ecology of the planet it inhabits. This book dares the human race to embrace change, to be rationally optimistic and thereby to strive for the betterment of humankind and the world it inhabits.

Some will say that I am merely restating what Adam Smith said in 1776. But much has happened since Adam Smith to change, challenge, adjust and amplify his insight. He did not realise, for instance, that he was living through the early stages of an industrial revolution. I cannot hope to match Smith's genius as an individual, but I have one great advantage over him – I can read his book. Smith's own insight has mated with others since his day.

Moreover, I find myself continually surprised by how few people think about the problem of tumultuous cultural change. I find the world is full of people who think that their dependence on others is decreasing, or that they would be better off if they were more self-sufficient, or that technological progress has brought no improvement in the standard of living, or that the world is steadily deteriorating, or that the exchange of things and ideas is a superfluous irrelevance. And I find a deep incuriosity among trained economists – of which I am not one – about defining what prosperity is and why it happened to their species. So I thought I would satisfy my own curiosity by writing this book.

I am writing in times of unprecedented economic pessimism. The world banking system has lurched to the brink of collapse; an enormous bubble of debt has burst; world trade has contracted; unemployment is rising sharply all around the world as output falls. The immediate future looks bleak indeed, and some governments are planning further enormous public debt expansions that could hurt the next generation's ability to prosper. To my intense regret I played a part in one phase of this disaster as non-executive chairman of Northern Rock, one of many banks that ran short of liquidity during the crisis. This is not a book about that experience (under the terms of my employment there I am not at liberty to write about it). The experience has left me mistrustful of markets in capital and assets, yet passionately in favour of markets in goods and services. Had I only known it, experiments in laboratories by the economist Vernon Smith and his colleagues have long confirmed that markets in goods and services for immediate consumption – haircuts and hamburgers – work so well that it is hard to design them so they fail to deliver efficiency and innovation; while markets in assets are so automatically prone to bubbles and crashes that it is hard to design them so they work at all. Speculation, herd exuberance, *irrational* optimism, rent-seeking and the temptation of fraud drive asset markets to overshoot and plunge – which is why they need careful regulation, something I always supported. (Markets in goods and services need less regulation.) But what made the bubble of the 2000s so much worse than most was government housing and monetary policy, especially in the United States, which sluiced artificially cheap money towards bad risks as a matter of policy and thus also towards the middlemen of the capital markets. The crisis has at least as much political as economic causation, which is why I also mistrust too much government.

(In the interests of full disclosure, I here note that as well as banking I have over the years worked in or profited directly from scientific research, species conservation, journalism, farming, coal mining, venture capital and commercial property, among other things: experience may have influenced, and has certainly informed, my views of these sectors in the pages that follow. But I have never been paid to promulgate a particular view.)

Rational optimism holds that the world will pull out of the current crisis because of the way that markets in goods, services and ideas allow human beings to exchange and specialise honestly for the betterment of all. So this is not a book of unthinking praise or condemnation of all markets, but it is an inquiry into how the market process of exchange and specialisation is older and fairer than many think and gives a vast reason for optimism about the future of the human race. Above all, it is a book about the benefits of change. I find that my disagreement is mostly with reactionaries of all political colours: blue ones who dislike cultural change, red ones who dislike economic change and green ones who dislike technological change.

I am a rational optimist: rational, because I have arrived at optimism not through temperament or instinct, but by looking at the evidence. In the pages that follow I hope to make you a rational optimist too. First, I need to convince you that human progress has, on balance, been a good thing, and that, despite the constant temptation to moan, the world is as good a place to live as it has ever been for the average human being – even now in a deep recession. That it is richer, healthier, and kinder too, as much because of commerce as despite it. Then I intend to explain why and how it got that way. And finally, I intend to see whether it can go on getting better.

# CHAPTER 1

# A better today: the unprecedented present

On what principle is it, that when we see nothing but improvement
behind us, we are to expect nothing but deterioration before us?

THOMAS BABINGTON MACAULAY
*Review of Southey's Colloquies on Society*

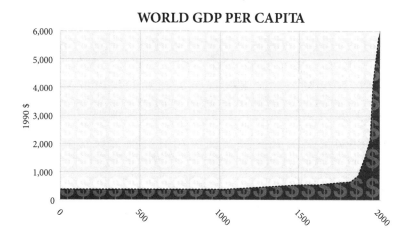

WORLD GDP PER CAPITA

By the middle of this century the human race will have expanded in ten thousand years from less than ten million to nearly ten billion people. Some of the billions alive today still live in misery and dearth even worse than the worst experienced in the Stone Age. Some are worse off than they were just a few months or years before. But the vast majority of people are much better fed, much better sheltered, much better entertained, much better protected against disease and much more likely to live to old age than their ancestors have ever been. The availability of almost everything a person could want or need has been going rapidly upwards for 200 years and erratically upwards for 10,000 years before that: years of lifespan, mouthfuls of clean water, lungfuls of clean air, hours of privacy, means of travelling faster than you can run, ways of communicating farther than you can shout. Even allowing for the hundreds of millions who still live in abject poverty, disease and want, this generation of human beings has access to more calories, watts, lumen-hours, square feet, gigabytes, megahertz, light-years, nanometres, bushels per acre, miles per gallon, food miles, air miles, and of course dollars than any that went before. They have more Velcro, vaccines, vitamins, shoes, singers, soap operas, mango slicers, sexual partners, tennis rackets, guided missiles and anything else they could even imagine needing. By one estimate, the number of different products that you can buy in New York or London tops ten billion.

This should not need saying, but it does. There are people today who think life was better in the past. They argue that there was not only a simplicity, tranquillity, sociability and spirituality about life in the distant past that has been lost, but a virtue too. This rose-tinted nostalgia, please note, is generally confined to the wealthy. It is easier to wax elegiac for the life of a peasant when you do not have to use a long-drop toilet. Imagine that it is 1800, somewhere in Western Europe or eastern North America. The family is gathering around the hearth in the simple timber-framed house. Father reads aloud from the Bible while mother prepares to dish out a stew of beef and onions. The baby boy is being comforted by one of his sisters and the eldest lad is pouring water from a pitcher into the earthenware mugs on the table. His elder sister is feeding the horse in the stable. Outside there is no noise of traffic, there are no drug dealers and neither dioxins nor radioactive fall-out have been found in the cow's milk. All is tranquil; a bird sings outside the window.

Oh please! Though this is one of the better-off families in the village, father's Scripture reading is interrupted by a bronchitic cough that presages the pneumonia that will kill him at 53 – not helped by the wood smoke of the fire. (He is lucky: life expectancy even in England was less than 40 in 1800.) The baby will die of the smallpox that is now causing him to cry; his sister will soon be the chattel of a drunken husband. The water the son is pouring tastes of the cows that drink from the brook. Toothache tortures the mother. The neighbour's lodger is getting the other girl pregnant in the hayshed even now and her child will be sent to an orphanage. The stew is grey and gristly yet meat is a rare change from gruel; there is no fruit or salad at this season. It is eaten with a wooden spoon from a wooden bowl. Candles cost too much, so firelight is all there is to see by. Nobody in the family has ever seen a play, painted a picture or heard a piano. School is a few years of dull Latin taught by a bigoted martinet at the vicarage. Father visited the city once, but the travel cost him a week's wages and the others have never travelled more than fifteen miles from home. Each daughter owns two wool dresses, two linen shirts and one pair of shoes. Father's jacket cost him a month's wages but is now infested with lice. The children sleep two to a bed on straw mattresses on the floor. As for the bird outside the window, tomorrow it will be trapped and eaten by the boy.

If my fictional family is not to your taste, perhaps you prefer statistics. Since 1800, the population of the world has multiplied six times, yet average life expectancy has more than doubled and real income has risen more than nine times. Taking a shorter perspective, in 2005, compared with 1955, the average human being on Planet Earth earned nearly three times as much money (corrected for inflation), ate one-third more calories of food, buried one-third as many of her children and could expect to live one-third longer. She was less likely to die as a result of war, murder, childbirth, accidents, tornadoes, flooding, famine, whooping cough, tuberculosis, malaria, diphtheria, typhus, typhoid, measles, smallpox, scurvy or polio. She was less likely, at any given age, to get cancer, heart disease or stroke. She was more likely to be literate and to have finished school. She was more likely to own a telephone, a flush toilet, a refrigerator and a bicycle. All this during a half-century when the world population has more than doubled, so that far from being rationed by population pressure, the goods and services available to the people of the world have expanded. It is, by any standard, an astonishing human achievement.

Averages conceal a lot. But even if you break down the world into bits, it is hard to find any region that was worse off in 2005 than it was in 1955. Over that half-century, real income per head ended a little lower in only six countries (Afghanistan, Haiti, Congo, Liberia, Sierra Leone and Somalia), life expectancy in three (Russia, Swaziland and Zimbabwe), and infant survival in none. In the rest they have rocketed upward. Africa's rate of improvement has been distressingly slow and patchy compared with the rest of the world, and many southern African countries saw life expectancy plunge in the 1990s as the AIDS epidemic took hold (before recovering in recent years). There were also moments in the half-century when you could have caught countries in episodes of dreadful deterioration of living standards or life chances – China in the 1960s, Cambodia in the 1970s, Ethiopia in the 1980s, Rwanda in the 1990s, Congo in the 2000s, North Korea throughout. Argentina had a disappointingly stagnant twentieth century. But overall, after fifty years, the outcome for the world is remarkably, astonishingly, dramatically positive. The average South Korean lives twenty-six more years and earns fifteen times as much income each year as he did in 1955 (and earns fifteen times as much as his North Korean counterpart). The average Mexican lives longer now than the average Briton did in 1955. The average Botswanan earns more than the average Finn did in 1955. Infant mortality is lower today in Nepal than it was in Italy in 1951. The proportion of Vietnamese living on less than $2 a day has dropped from 90 per cent to 30 per cent in twenty years.

The rich have got richer, but the poor have done even better. The poor in the developing world grew their consumption twice as fast as the world as a whole between 1980 and 2000. The Chinese are ten times as rich, one-third as fecund and twenty-eight years longer-lived than they were fifty years ago. Even Nigerians are twice as rich, 25 per cent less fecund and nine years longer-lived than they were in 1955. Despite a doubling of the world population, even the raw number of people living in absolute poverty (defined as less than a 1985 dollar a day) has fallen since the 1950s. The percentage living in such absolute poverty has dropped by more than half – to less than 18 per cent. That number is, of course, still all too horribly high, but the trend is hardly a cause for despair: at the current rate of decline, it would hit zero around 2035 – though it probably won't. The United Nations estimates that poverty was reduced more in the last fifty years than in the previous 500.

## Affluence for all

Nor was 1955 a time of deprivation. It was in itself a record – a moment when the world was richer, more populous and more comfortable than it had ever been, despite the recent efforts of

Hitler, Stalin and Mao (who was then just starting to starve his people so that he could use their grain to buy nuclear weapons from Russia). The 1950s were a decade of extraordinary abundance and luxury compared with any preceding age. Infant mortality in India was already lower than it had been in France and Germany in 1900. Japanese children had almost twice as many years in education in 1950 as at the turn of the century. World income per head had almost doubled in the first half of the twentieth century. In 1958 J.K. Galbraith declared that the 'affluent society' had reached such a pitch that many unnecessary goods were now being 'overprovided' to consumers by persuasive advertisers.

He was right that Americans were especially well off compared with others: they were three inches taller in 1950 than they had been at the turn of the century and spent twice as much on medicine as funerals – the reverse of the ratio in 1900. Roughly eight out of ten American households had running water, central heating, electric light, washing machines and refrigerators by 1955. Almost none had these luxuries in 1900. In his 1890 classic *How the Other Half Lives*, Jacob Riis encountered a family of nine in New York living in a ten-foot-square room plus a tiny kitchen, and women earning 60 cents a day for sixteen hours' work in sweatshops and unable to afford more than one meal a day. This would have been unthinkable by mid-century.

Yet looking back now, another fifty years later, the middle class of 1955, luxuriating in their cars, comforts and gadgets, would today be described as 'below the poverty line'. The average British working man in 1957, when Harold Macmillan told him he had 'never had it so good', was earning less in real terms than his modern equivalent could now get in state benefit if unemployed with three children. Today, of Americans officially designated as 'poor', 99 per cent have electricity, running water, flush toilets, and a refrigerator; 95 per cent have a television, 88 per cent a telephone, 71 per cent a car and 70 per cent air conditioning. Cornelius Vanderbilt had none of these. Even in 1970 only 36 per cent of all Americans had air conditioning: in 2005 79 per cent of *poor* households did. Even in urban China 90 per cent of people now have electric light, refrigerators and running water. Many of them also have mobile phones, internet access and satellite television, not to mention all sorts of improved and cheaper versions of everything from cars and toys to vaccines and restaurants.

Well all right, says the pessimist, but at what cost? The environment is surely deteriorating. In somewhere like Beijing, maybe. But in many other places, no. In Europe and America rivers, lakes, seas and the air are getting cleaner all the time. The Thames has less sewage and more fish. Lake Erie's water snakes, on the brink of extinction in the 1960s, are now abundant. Bald eagles have boomed. Pasadena has few smogs. Swedish birds' eggs have 75 per cent fewer pollutants in them than in the 1960s. American carbon monoxide emissions from transport are down 75 per cent in twenty-five years. Today, a car emits less pollution travelling at full speed than a parked car did in 1970 from leaks.

Meanwhile, average life expectancy in the longest-lived country (Sweden in 1850, New Zealand in 1920, Japan today) continues to march upwards at a steady rate of a quarter of a year per year, a rate of change that has altered little in 200 years. It still shows no sign of reaching a limit, though surely it must one day. In the 1920s demographers confidently asserted that average life span would peak at 65 'without intervention of radical innovations or fantastic evolutionary change in our physiological make-up'. In 1990 they predicted life expectancy 'should not exceed … 35 years at age 50 unless major break-throughs occur in controlling the fundamental rate of ageing'. Within just five years both predictions were proved wrong in at least one country.

Consequently the number of years of retirement is rocketing upwards. Starting from 1901, it took sixty-eight years for the mortality of British men between 65 and 74 to fall by 20 per cent. Subsequent 20 per cent falls took seventeen years, ten years and six years – the improvement has accelerated. That is all very well, say pessimists, but what about quality of life in old age? Sure, people live longer, but only by having years of suffering and disability added to their lives. Not so. In one American study, disability rates in people over 65 fell from 26.2 per cent to 19.7 per cent between 1982 and 1999 – at twice the pace of the decrease in the mortality rate. Chronic illness before death is if anything shortening slightly, not lengthening, despite better diagnosis and more treatments – 'the compression of morbidity' is the technical term. People are not only spending a longer time living, but a shorter time dying.

Take stroke, a big cause of disability in old age. Deaths from stroke fell by 70 per cent between 1950 and 2000 in America and Europe. In the early 1980s a study of stroke victims in Oxford concluded that the incidence of stroke would increase by nearly 30 per cent over the next two decades, mainly because stroke incidence increases with age and people were predicted to live longer. They did live longer but the incidence of stroke in fact fell by 30 per cent. (The age-related increase is still present, but it is coming later and later.) The same is true of cancer, heart disease and respiratory disease: they all still increase with age, but they do so later and later, by about ten years since the 1950s.

Even inequality is declining worldwide. It is true that in Britain and America income equality, which had been improving for most of the past two centuries (British aristocrats were six inches taller than the average in 1800; today they are less than two inches taller), has stalled since the 1970s. The reasons for this are many, but they are not all causes for regret. For example, high earners now marry each other more than they used to (which concentrates income), immigration has increased, trade has been freed, cartels have been opened up to entrepreneurial competition and the skill premium has grown in the work place. All these are inequality-boosting, but they stem from liberalising trends. Besides, by a strange statistical paradox, while inequality has increased within some countries, globally it has been falling. The recent enrichment of China and India has increased inequality within those countries by making the income of the rich grow faster than that of the poor – an income gap is an inevitable consequence of an expanding economy. Yet the global effect of the growth of China and India has been to reduce the difference between rich and poor worldwide. As Hayek put it, 'once the rise in the position of the lower classes gathers speed, catering to the rich ceases to be the main source of great gain and gives place to efforts directed towards the needs of the masses. Those forces which at first make inequality self-accentuating thus later tend to diminish it.'

In another respect, too, inequality has been retreating. The spread of IQ scores has been shrinking steadily – because the low scores have been catching up with the high ones. This explains the steady, progressive and ubiquitous improvement in the average IQ scores people achieve at a given age – at a rate of 3 per cent per decade. In two Spanish studies, IQ proved to be 9.7 points higher after thirty years, most of it among the least intelligent half of the group. Known as the Flynn effect, after James Flynn who first drew attention to it, this phenomenon was at first dismissed as an artefact of changes in tests, or a simple reflection of longer or better schooling. But the facts do not fit such explanations because the effect is consistently weakest in the cleverest children and in the tests that relate most to educational content. It is a levelling-up caused by an equalisation of nutrition, stimulation or diversity of childhood experience. You can, of course, argue that IQ may not be truly representative of intelligence, but you cannot argue that something is getting better – and more equal at the same time.

Even justice has improved thanks to new technology exposing false convictions and identifying true criminals. To date

234 innocent Americans have been freed as a result of DNA fingerprinting after serving an average of twelve years in prison; seventeen of them were on death row. The very first forensic use of DNA in 1986 exonerated an innocent man and then helped to catch the real murderer, a pattern that has been repeated many times since.

## Cheap light

These richer, healthier, taller, cleverer, longer-lived, freer people – you lot – have been enjoying such abundance that most of the things they need have been getting steadily cheaper. The four most basic human needs – food, clothing, fuel and shelter – have grown markedly cheaper during the past two centuries. Food and clothing especially so (a brief rise in food prices in 2008 notwithstanding), fuel more erratically and even housing has probably got cheaper too: surprising as it may seem, the average family house probably costs slightly less today than it did in 1900 or even 1700, despite including far more modern conveniences like electricity, telephone and plumbing. If basic needs have got cheaper, then there is more disposable income to spend on luxuries. Artificial light lies on the border between necessity and luxury. In monetary terms, the same amount of artificial lighting cost 20,000 times as much in England in the year 1300 as it does today.

Enormous as that difference is, in labour terms the change is even more dramatic and the improvement is even more recent. Ask how much artificial light you can earn with an hour of work at the average wage. The amount has increased from twenty-four lumen-hours in 1750 BC (sesame oil lamp) to 186 in 1800 (tallow candle) to 4,400 in 1880 (kerosene lamp) to 531,000 in 1950 (incandescent light bulb) to 8.4 million lumen-hours today (compact fluorescent bulb). Put it another way, an hour of work today earns you 300 days' worth of reading light; an hour of work in 1800 earned you ten minutes of reading light. Or turn it round and ask how long you would have to work to earn an hour of reading light – say, the light of an 18-watt compact-fluorescent light bulb burning for an hour. Today it will have cost you less than half a second of your working time if you are on the average wage: half a second of work for an hour of light. In 1950, with a conventional filament lamp and the then wage, you would have had to work for eight seconds to get the same amount of light. Had you been using a kerosene lamp in the 1880s, you would have had to work for about fifteen minutes to get the same amount of light. A tallow candle in the 1800s: over six hours' work. And to get that much light from a sesame-oil lamp in Babylon in 1750 BC would have cost you more than fifty hours' of work. From six hours to half a second – a 43,200-fold improvement – for an hour of lighting: that is how much better off you are than your ancestor was in 1800, using the currency that counts, your time. Do you see why my fictional family ate by firelight?

Much of this improvement is not included in the cost-of-living calculations, which struggle to compare like with unlike. The economist Don Boudreaux imagined the average American time-travelling back to 1967 with his modern income. He might be the richest person in town, but no amount of money could buy him the delights of eBay, Amazon, Starbucks, Wal-Mart, Prozac, Google or BlackBerry. The lighting numbers cited above do not even take into account the greater convenience and cleanliness of modern electric light compared with candles or kerosene – its simple switching, its lack of smoke, smell and flicker, its lesser fire hazard. Nor is the improvement in lighting finished yet. Compact fluorescent bulbs may be three times as efficient as filament bulbs in turning electrons' energy into photons' energy, but light-emitting diodes (LEDs) are rapidly overtaking them (as of this writing LEDs with ten times the efficiency of incandescent bulbs have been demonstrated) and have the added benefit of working at a portable scale. A cheap LED flashlight, powered by a solar-charged battery, will surely soon transform the life of some of the 1.6 billion people who do not have mains electricity, African peasants prominent among them. Admittedly, LEDs are still far too expensive to replace most light bulbs, but that might change.

Think what these improvements in lighting efficiency mean. You can either have a lot more light, or do a lot less work, or acquire something else. Devoting less of your working week to earning your lighting means devoting more of it to doing something else. That something else can mean employment for somebody else. The improved technology of lighting has liberated you to make or buy another product or service, or do a charitable act. That is what economic growth means.

## Saving time

Time: that is the key. Forget dollars, cowrie shells or gold. The true measure of something's worth is the hours it takes to acquire it. If you have to acquire it for yourself, it usually takes longer than if you get it ready-made by other people. And if you can get it made efficiently by others, then you can afford more of it. As light became cheaper so people used more of it. The average Briton today consumes roughly 40,000 times as much artificial light as he did in 1750. He consumes fifty times as much power and 250 times as much transport (measured in passenger-miles travelled), too.

This is what prosperity is: the increase in the amount of goods or services you can earn with the same amount of work. As late as the mid-1800s, a stagecoach journey from Paris to Bordeaux cost the equivalent of a clerk's monthly wages; today the journey costs a day or so and is fifty times as fast. A half-gallon of milk cost the average American ten minutes of work in 1970, but only seven minutes in 1997. A three-minute phone call from New York to Los Angeles cost ninety hours of work at the average wage in 1910; today it costs less than two minutes. A kilowatt-hour of electricity cost an hour of work in 1900 and five minutes today. In the 1950s it took thirty minutes work to earn the price of a McDonald's cheeseburger; today it takes three minutes. Healthcare and education are among the few things that cost more in terms of hours worked now than they did in the 1950s.

Even the most notorious of capitalists, the robber barons of the late nineteenth century, usually got rich by making things cheaper. Cornelius Vanderbilt is the man for whom the *New York Times* first used the word 'robber baron'. He is the very epitome of the phrase. Yet observe what *Harper's Weekly* had to say about his railways in 1859:

> The results in every case of the establishment of opposition lines by Vanderbilt has been the permanent reduction of fares. Wherever he 'laid on' an opposition line, the fares were instantly reduced, and however the contest terminated, whether he bought out his opponents, as he often did, or they bought him out, the fares were never again raised to the old standard. This great boon – cheap travel – this community owes mainly to Cornelius Vanderbilt.

Rail freight charges fell by 90 per cent between 1870 and 1900. There is little doubt that Vanderbilt sometime bribed and bullied his way to success, and that he sometimes paid his workers lower wages than others – I am not trying to make him into a saint – but there is also no doubt that along the way he delivered to consumers an enormous benefit that would otherwise have eluded them – affordable transport. Likewise, Andrew Carnegie, while enormously enriching himself, cut the price of a steel rail by 75 per cent in the same period; John D. Rockefeller cut the price of oil by 80 per cent. During those thirty years, the per capita GDP of Americans rose by 66 per cent. They were enricher-barons, too.

Henry Ford got rich by making cars cheap. His first Model T sold for $825, unprecedentedly cheap at the time, and four years later he had cut the price to $575. It took about 4,700 hours of work to afford a Model T in 1908. It takes about 1,000 hours today to afford an ordinary car – though one that is brimming with features that Model Ts never had. The price of aluminium fell from $545 a pound in the 1880s to 20 cents a pound in the 1930s, thanks to the innovations of Charles Martin Hall and his successors at Alcoa. (Alcoa's reward for this price cut was to be sued by the government on 140 counts of criminal monopoly: the rapid decrease in the price of its product being used as evidence of a determination to deter competition. Microsoft suffered the same allegation later in the century.) When Juan Trippe sold cheap tourist-class seats on his Pan Am airline in 1945, the other airlines were so insulted that they petitioned their governments to ban Pan Am: Britain, shamefully, agreed, so Pan Am flew to Ireland instead. The price of computing power fell so fast in the last quarter of the twentieth century that the capacity of a tiny pocket calculator in 2000 would have cost you a lifetime's wages in 1975. The price of a DVD player in Britain fell from £400 in 1999 to £40 just five years later, a decline that exactly matched the earlier one of the video recorder, but happened much faster.

Falling consumer prices is what enriches people (deflation of asset prices can ruin them, but that is because they are using asset prices to get them the wherewithal to purchase consumer items). And, once again, notice that the true metric of prosperity is time. If Cornelius Vanderbilt or Henry Ford not only moves you faster to where you want to go, but requires you to work fewer hours to earn the ticket price, then he has enriched you by granting you a dollop of free time. If you choose to spend that spare time consuming somebody else's production then you can enrich him in turn; if you choose to spend it producing for his consumption then you have also further enriched yourself.

Housing, too, is itching to get cheaper, but for confused reasons governments go to great lengths to prevent it. Where it took sixteen weeks to earn the price of 100 square feet of housing in 1956, now it takes fourteen weeks and the housing is of better quality. But given the ease with which modern machinery can assemble a house, the price should have come down much faster than that. Governments prevent this by, first, using planning or zoning laws to restrict supply (especially in Britain); second, using the tax system to encourage mortgage borrowing (in the United States at least – no longer in Britain); and third, doing all they can to stop property prices falling after a bubble. The effect of these measures is to make life harder for those who do not yet have a house and massively reward those who do. To remedy this, governments then have to enforce the building of more affordable housing, or subsidise mortgage lending to the poor.

## Happiness

As necessities and luxuries get cheaper, do people get happier? A small cottage industry grew up at the turn of the twenty-first century devoted to the subject of the economics of happiness. It started with the paradox that richer people are not necessarily happier people. Beyond a certain level of per capita income ($15,000 a year, according to Richard Layard), money did not seem to buy subjective well-being. As books and papers cascaded out of the academy, *Schadenfreude* set in on a grand scale among commentators happy to see the unhappiness of the rich confirmed. Politicians latched on and governments from Thailand to Britain began to think about how to maximise gross national happiness instead of gross national product. British government departments now have 'well-being divisions' as a result. King Jigme Singye Wangchuck of Bhutan is credited with having been the first to get there in 1972 when he declared economic growth a secondary goal to national well-being. If economic growth does not produce happiness, said the new wisdom, then there was no point in striving for prosperity and the world economy should be brought to a soft landing at a reasonable level of income. Or, as one economist put it: 'The hippies were right all along'.

If true, this rather punctures the rational optimist's balloon. What is the point of celebrating the continuing defeat of death, dearth, disease and drudgery, if it does not make people happier? But it is not true. The debate began with a study by Richard Easterlin in 1974, which found that although within a country rich people were generally happier than poor people, richer countries did not have happier citizens than poor countries. Since then the 'Easterlin paradox' has become the central dogma of the debate. Trouble is, it is wrong. Two papers were published in 2008 analysing all the data, and the unambiguous conclusion of both is that the Easterlin paradox does not exist. Rich people are happier than poor people; rich countries have happier people than poor countries; and people get happier as they get richer. The earlier study simply had samples too small to find significant differences. In all three categories of comparison – within countries, between countries and between times – extra income does indeed buy general well-being. That is to say, on average, across the board, on the whole, other things being equal, more money does make you happier. In the words of one of the studies, 'All told, our time-series comparisons, as well as evidence from repeated international cross-sections, appear to point to an important relationship between economic growth and growth in subjective well-being'.

There are some exceptions. Americans currently show no trend towards increasing happiness. Is this because the rich had got richer but ordinary Americans had not prospered much in recent years? Or because America continually draws in poor (unhappy) immigrants, which keeps the happiness quotient low? Who knows? It was not because the Americans are too rich to get any happier: Japanese and Europeans grew steadily happier as they grew richer despite being often just as rich as Americans. Moreover, surprisingly, American women have become less happy in recent decades despite getting richer.

Of course, it is possible to be rich and unhappy, as many a celebrity gloriously reminds us. Of course, it is possible to get rich and find that you are unhappy not to be richer still, if only because the neighbour – or the people on television – are richer than you are. Economists call this the 'hedonic treadmill'; the rest of us call it 'keeping up with the Joneses'. And it is probably true that the rich do lots of unnecessary damage to the planet as they go on striving to get richer long after the point where it is having much effect on their happiness – they are after all endowed with instincts for 'rivalrous competition' descended from hunter-gatherers whose relative, not absolute, status determined their sexual rewards. For this reason a tax on consumption to encourage saving for investment instead is not necessarily a bad idea. However, this does not mean that anybody would be necessarily happier if poorer – to be well off and unhappy is surely better than to be poor and unhappy. Of course, some people will be unhappy however rich they are, while others manage to bounce back cheerful even in poverty: psychologists find people to have fairly constant levels of happiness to which they return after elation or disaster. Besides, a million years of natural selection shaped human nature to be ambitious to rear successful children, not to settle for contentment: people are programmed to desire, not to appreciate.

Getting richer is not the only or even the best way of getting happier. Social and political liberation is far more effective, says the political scientist Ronald Ingleheart: the big gains in happiness come from living in a society that frees you to make choices about your lifestyle – about where to live, who to marry, how to express your sexuality and so on. It is the increase in free choice since 1981 that has been responsible for the increase in happiness recorded since then in forty-five out of fifty-two countries. Ruut Veenhoven finds that 'the more individualized the nation, the more citizens enjoy their life.'

## Crunch

And yet, good as life is, today life is not good. Happy statistics of recent improvement sound as hollow to a laid-off car worker in Detroit or an evicted house owner in Reykjavik as they would to a cholera victim in Zimbabwe or a genocide refugee in Congo. War, disease, corruption and hate still disfigure the lives of millions; nuclear terrorism, rising sea levels and pandemic flu may yet make the twenty-first century a dreadful place. True, but assuming the worst will not avert these fates; striving to continue improving the human lot may. It is precisely because so much human betterment has been shown to be possible in recent centuries that the continuing imperfection of the world places a moral duty on humanity to allow economic evolution to continue. To prevent change, innovation and growth is to stand in the way of potential compassion. Let it never be forgotten that, by propagating excessive caution about genetically modified food aid, some pressure groups may have exacerbated real hunger in Zambia in the early 2000s. The precautionary principle – better safe than sorry – condemns itself: in a sorry world there is no safety to be found in standing still.

More immediately, the financial crash of 2008 has caused a deep and painful recession that will generate mass unemployment and real hardship in many parts of the world. The reality of rising living standards feels to many today to be a trick, a pyramid scheme achieved by borrowing from the future.

Until he was rumbled in 2008, Bernard Madoff offered his investors high and steady returns of more than 1 per cent a month on their money for thirty years. He did so by paying new investors' capital out to old investors as revenue, a chain-letter con trick that could not last. When the music stopped, $65 billion of investors' funds had been looted. It was roughly what John Law did in Paris with the Mississippi Company in 1719, what John Blunt did in London with the South Sea company in 1720, what Charles Ponzi did in Boston in 1920 with reply coupons for postage stamps, what Ken Lay did with Enron's stock in 2001.

Is it possible that not just the recent credit boom, but the entire postwar rise in living standards was a Ponzi scheme, made possible by the gradual expansion of credit? That we have in effect grown rich by borrowing the means from our children and that a day of reckoning is now at hand? It is certainly true that your mortgage is borrowed (via a saver somewhere else, perhaps in China) from your future self, who will pay it off. It is also true on both sides of the Atlantic that your state pension will be funded by your children's taxes, not by your payroll contributions as so many think.

But there is nothing unnatural about this. In fact, it is a very typical human pattern. By the age of 15 chimpanzees have produced about 40 per cent and consumed about 40 per cent of the calories they will need during their entire lives. By the same age, human hunter-gatherers have consumed about 20 per cent of their lifetime calories, but produced just 4 per cent. More than any other animal, human beings borrow against their future capabilities by depending on others in their early years. A big reason for this is that hunter-gatherers have always specialised in foods that need extraction and processing – roots that need to be dug and cooked, clams that need to be opened, nuts that need to be cracked, carcasses that need to be butchered – whereas chimpanzees eat things that simply need to be found and gathered, like fruit or termites. Learning to do this extraction and processing takes time, practice and a big brain, but once a human being has learnt, he or she can produce a huge surplus of calories to share with the children. Intriguingly, this pattern of production over the lifespan in hunter-gatherers is more like the modern Western lifestyle than it is like the farming, feudal or early industrial lifestyles. That is to say, the notion of children taking twenty years even to start to bring in more than they consume, and then having forty years of very high productivity, is common to hunter-gatherers and modern societies, but was less true in the period in between, when children could and did go to work to support their own consumption.

The difference today is that intergenerational transfers take a more collective form – income tax on all productive people in their prime pays for education for all, for example. In that sense, the economy (like a chain letter, but unlike a shark, actually) must keep moving forward or it collapses. The banking system makes it possible for people to borrow and consume when they are young and to save and lend when they are old, smoothing their family living standards over the decades. Posterity can pay for its ancestors' lives because posterity can be richer through innovation. If somebody somewhere takes out a mortgage, which he will repay in three decades' time, to invest in a business that invents a gadget that saves his customers time, then that money, brought forward from the future, will enrich both him and those customers to the point where the loan can be repaid to posterity. That is growth. If, on the other hand, somebody takes out a loan just to support his luxury lifestyle, or to speculate on asset markets by buying a second home, then posterity will be the loser. That is what, it is now clear, far too many people and businesses did in the 2000s – they borrowed more from posterity than their innovation rate would support. They misallocated the resources to unproductive ends. Most past bursts of human prosperity have come to naught because they allocated too little money to innovation and too much to asset price inflation or to war, corruption, luxury and theft.

In the Spain of Charles V and Philip II, the gigantic wealth of the Peruvian silver mines was wasted. The same 'curse of resources' has afflicted countries with windfalls ever since, especially those with oil (Russia, Venezuela, Iraq, Nigeria) that end up run by rent-seeking autocrats. Despite their windfalls, such countries experience lower economic growth than countries that entirely lack resources but get busy trading and selling – Holland, Japan, Hong Kong, Singapore, Taiwan, South Korea. Even the Dutch, those epitomes of seventeenth-century enterprise, fell under the curse of resources in the late twentieth century when they found too much natural gas: the Dutch disease, they called it, as their inflated currency hurt their exporters. Japan spent the first half of the twentieth century jealously seeking to grab resources and ended up in ruins; it spent the second half of the century trading and selling without resources and ended up topping the lifespan league. In the 2000s the West misspent much of the cheap windfall of Chinese savings that the United States Federal Reserve sluiced our way.

So long as somebody allocates sufficient capital to innovation, then the credit crunch will not in the long run prevent the relentless upward march of human living standards. If you look at a graph of world per capita GDP, the Great Depression of the 1930s is just a dip in the slope. By 1939 even the worst-affected countries, America and Germany, were richer than they were in 1930. All sorts of new products and industries were born during the Depression: by 1937, 40 per cent of DuPont's sales came from products that had not even existed before 1929, such as rayon, enamels and cellulose film. So growth will resume – unless prevented by the wrong policies. Somebody, somewhere, is still tweaking a piece of software, testing a new material, or transferring a gene that will make your and my life easier in the future. I cannot know who or where he is for sure, but let me give you a candidate. In the week I wrote this paragraph, a small company called Arcadia Biosciences in northern California signed an agreement with a charity working in Africa to license, royalty-free to smallholders, new varieties of rice that can be grown with less nitrogen fertiliser for the same yield, thanks to the over-expression in the roots of a version of a gene called alanine aminotransferase borrowed from barley. Assuming the varieties work in Africa as well as they do in California, some African will one day grow and sell more food (for less pollution), which in turn means that he will have more money to spend, earning the cost of, say, a mobile phone, which he will buy from a Western company, and which will help him find a better market for his rice. An employee of that Western company will get a pay rise, which she will spend on a new pair of jeans, which were made from cotton woven in a factory that employs the smallholder's neighbour. And so on.

As long as new ideas can breed in this way, then human economic progress can continue. It may be only a year or two till world growth resumes after the current crisis, or it may for some countries be a lost decade. It may even be that parts of the world will be convulsed by a descent into autarky, authoritarianism and violence, as happened in the 1930s, and that a depression will cause a great war. But so long as somewhere somebody is incentivised to invent ways of serving others' needs better, then the rational optimist must conclude that the betterment of human lives will eventually resume.

## The declaration of interdependence

Imagine you are a deer. You have essentially only four things to do during the day: sleep, eat, avoid being eaten and socialise (by which I mean mark a territory, pursue a member of the opposite sex, nurse a fawn, whatever). There is no real need to do much else. Now imagine you are a human being. Even if you only count the basic things, you have rather more than four things to do: sleep, eat, cook, dress, keep house, travel, wash, shop, work … the list is virtually endless. Deer should therefore have more free time than human beings, yet it is people, not deer, who find the time to read, write, invent, sing and surf the net. Where does all this free time come from? It comes from exchange and specialisation and from the resulting division of labour. A deer must gather its own food. A human being gets somebody else to do it for him, while he or she is doing something for them – and both win time that way.

Self-sufficiency is therefore not the route to prosperity. 'Which would have advanced the most at the end of a month,' Henry David Thoreau asked: 'the boy who had made his own jack-knife from the ore which he had dug and smelted, reading as much as would be necessary for this – or the boy who had attended the lectures on metallurgy at the Institute in the meanwhile, and had received a Rodgers' penknife from his father?' *Contra* Thoreau, it is the latter, by a mile, because he has far more spare time to learn other things. Imagine if you had to be completely self-sufficient (not just pretending, like Thoreau). Every day you must get up in the morning and supply yourself entirely from your own resources. How would you spend your day? The top four priorities would be food, fuel, clothing and shelter. Dig the garden, feed the pig, fetch water from the brook, gather wood from the forest, wash some potatoes, light a fire (no matches), cook lunch, repair the roof, fetch fresh bracken for clean bedding, whittle a needle, spin some thread, sew leather for shoes, wash in the stream, fashion a pot out of clay, catch and cook a chicken for dinner. No candle or book for reading. No time for smelting metal, drilling oil, or travel. By definition, you are at subsistence level and frankly, though at first you mutter, Thoreau-like, 'how marvellous to get away from all the appalling hustle and bustle', after a few days the routine is pretty grim. If you wish to have even the most minimal improvement in your life – say metal tools, toothpaste or lighting – you are going to have to get some of your chores done by somebody else, because there just is not time to do them yourself. So one way to raise your standard of living would be to lower somebody else's: buy a slave. That was indeed how people got rich for thousands of years.

Yet, though you have no slaves, today when you got out of bed you knew that somebody would provide you with food, fibre and fuel in a most convenient form. In 1900, the average American spent $76 of every $100 on food, clothing and shelter. Today he spends $37. If you are on an average wage you knew that it would take you a matter of tens of minutes to earn the cash to pay for your food, some more tens of minutes to earn the cash to buy whatever new clothing you need and maybe an hour or two to earn the cash to pay for the gas, electricity and oil you might need today. Earning the rent or mortgage payment that ensures you have a roof over your head might take rather more time. But still, by lunchtime, you could relax in the knowledge that food, fuel, fibre and shelter were taken care of for the day. So it was time to earn something more interesting: the satellite television subscription, the mobile phone bill, the holiday deposit, the cost of new toys for the children, the income tax. 'To produce implies that the producer desires to consume' said John Stuart Mill; 'why else should he give himself useless labour?'

In 2009, an artist named Thomas Thwaites set out to make his own toaster, of the sort that he could buy from a shop for about £4. He needed only a few raw materials: iron, copper, nickel, plastic and mica (an insulating mineral around which the heating elements are wrapped). But even to get these he found almost impossible. Iron is made from iron ore, which he could probably mine, but how was he to build a sufficiently hot furnace without electric bellows? (He cheated and used a microwave oven.) Plastic is made from oil, which he could not easily drill for himself, let alone refine. And so on. More to the point, the project took months, cost a lot of money and resulted in an inferior product. Yet to buy a £4 toaster would cost him less than an hour's work at the minimum wage. To Thwaites this illustrated his helplessness as a consumer so divorced from self-sufficiency. It also illustrates the magic of specialisation and exchange: thousands of people, none of them motivated by the desire to do Thwaites a favour, have come together to make it possible for him to acquire a toaster for a trivial sum of money. In the same vein, Kelly Cobb of Drexel University set out to make a man's suit entirely from materials produced within 100 miles of her home. It took twenty artisans a total of 500 man-hours to achieve it and even then they had to get 8 per cent of the materials from outside the 100-mile radius. If they worked for another year, they could get it all from within the limit, argued Cobb. To put it plainly, local sourcing multiplied the cost of a cheap suit roughly a hundred-fold.

As I write this, it is nine o'clock in the morning. In the two hours since I got out of bed I have showered in water heated by North Sea gas, shaved using an American razor running on electricity made from British coal, eaten a slice of bread made from French wheat, spread with New Zealand butter and Spanish marmalade, then brewed a cup of tea using leaves grown in Sri Lanka, dressed myself in clothes of Indian cotton and Australian wool, with shoes of Chinese leather and Malaysian rubber, and read a newspaper made from Finnish wood pulp and Chinese ink. I am now sitting at a desk typing on a Thai plastic keyboard (which perhaps began life in an Arab oil well) in order to move electrons through a Korean silicon chip and some wires of Chilean copper to display text on a computer designed and manufactured by an American firm. I have consumed goods and services from dozens of countries already this morning. Actually, I am guessing at the nationalities of some of these items, because it is almost impossible to define some of them as coming from any country, so diverse are their sources.

More to the point, I have also consumed minuscule fractions of the productive labour of many dozens of people. Somebody had to drill the gas well, install the plumbing, design the razor, grow the cotton, write the software. They were all, though they did not know it, working for me. In exchange for some fraction of my spending, each supplied me with some fraction of their work. They gave me what I wanted just when I wanted it – as if I were the *Roi Soleil*, Louis XIV, at Versailles in 1700.

The Sun King had dinner each night alone. He chose from forty dishes, served on gold and silver plate. It took a staggering 498 people to prepare each meal. He was rich because he consumed the work of other people, mainly in the form of their services. He was rich because other people did things for him. At that time, the average French family would have prepared and consumed its own meals as well as paid tax to support his servants in the palace. So it is not hard to conclude that Louis XIV was rich because others were poor.

But what about today? Consider that you are an average person, say a woman of 35, living in, for the sake of argument, Paris and earning the median wage, with a working husband and two children. You are far from poor, but in relative terms, you are immeasurably poorer than Louis was. Where he was the richest of the rich in the world's richest city, you have no servants, no palace, no carriage, no kingdom. As you toil home from work on the crowded Metro, stopping at the shop on the way to buy a ready meal for four, you might be thinking that Louis XIV's dining arrangements were way beyond your reach. And yet consider this. The cornucopia that greets you as you enter the supermarket dwarfs anything that Louis XIV ever experienced (and it is probably less likely to contain salmonella). You can buy a fresh, frozen, tinned, smoked or pre-prepared meal made with beef, chicken, pork, lamb, fish, prawns, scallops, eggs, potatoes, beans, carrots, cabbage, aubergine, kumquats, celeriac, okra, seven kinds of lettuce, cooked in olive, walnut, sunflower or peanut oil and flavoured with cilantro, turmeric, basil or rosemary … You may have no chefs, but you can decide on a whim to choose between scores of nearby bistros, or Italian, Chinese, Japanese or Indian restaurants, in each of which a team of skilled chefs is waiting to serve your family at less than an hour's notice. Think of this: never before this generation has the average person been able to afford to have somebody else prepare his meals.

You employ no tailor, but you can browse the internet and instantly order from an almost infinite range of excellent, affordable clothes of cotton, silk, linen, wool and nylon made up for you in factories all over Asia. You have no carriage, but you can buy a ticket which will summon the services of a skilled pilot of a budget airline to fly you to one of hundreds of destinations that Louis never dreamed of seeing. You have no woodcutters to bring you logs for the fire, but the operators of gas rigs in Russia are clamouring to bring you clean central heating. You have no wick-trimming footman, but your light switch gives you the instant and brilliant produce of hard-working people at a grid of distant nuclear power stations. You have no runner to send messages, but even now a repairman is climbing a mobile-phone mast somewhere in the world to make sure it is working properly just in case you need to call that cell. You have no private apothecary, but your local pharmacy supplies you with the handiwork of many thousands of chemists, engineers and logistics experts. You have no government ministers, but diligent reporters are even now standing ready to tell you about a film star's divorce if you will only switch to their channel or log on to their blogs.

My point is that you have far, far more than 498 servants at your immediate beck and call. Of course, unlike the Sun King's servants, these people work for many other people too, but from your perspective what is the difference? That is the magic that exchange and specialisation have wrought for the human species. 'In civilized society,' wrote Adam Smith, an individual 'stands at all times in need of the co-operation and assistance of great multitudes, while his whole life is scarce sufficient to gain the friendship of a few persons.' In Leonard Read's classic 1958 essay 'I, Pencil', an ordinary pencil describes how it came to be made by millions of people, from loggers in Oregon and graphite miners in Sri Lanka to coffee bean growers in Brazil (who supplied the coffee drunk by the loggers). 'There isn't a single person in all these millions,' the pencil concludes, 'including the president of the pencil company, who contributes more than a tiny, infinitesimal bit of know-how.' The pencil stands amazed at 'the absence of a master mind, of anyone dictating or forcibly directing these countless actions which bring me into being.'

This is what I mean by the collective brain. As Friedrich Hayek first clearly saw, knowledge 'never exists in concentrated or integrated form but solely as the dispersed bits of incomplete and frequently contradictory knowledge which all the separate individuals possess'.

## The multiplication of labour

You are not just consuming the labour and resources of others. You are consuming others' inventions, too. A thousand entrepreneurs and scientists devised the intricate dance of photons and electrons by which your television works. The cotton you wear was spun and woven by machines of a type whose original inventors are long-dead heroes of the industrial revolution. The bread you eat was first cross-bred by a Neolithic Mesopotamian and baked in a way that was first invented by a Mesolithic hunter-gatherer. Their knowledge is enduringly embodied in machines, recipes and programmes from which you benefit. Unlike Louis, you number among your servants John Logie Baird, Alexander Graham Bell, Sir Tim Berners-Lee, Thomas Crapper, Jonas Salk and myriad assorted other inventors. For you get the benefit of their labours, too, whether they are dead or alive.

The point of all this cooperation is to make (Adam Smith again) 'a smaller quantity of labour produce a greater quantity of work'. It is a curious fact that in return for this cornucopia of service, you produce only one thing. That is to say, having consumed the labour and embodied discoveries of thousands of people, you then produce and sell whatever it is you do at work – haircuts, ball bearings, insurance advice, nursing, dog walking. But each of those thousands of people who work 'for' you is equally monotonously employed. Each produces one thing. That is what the word 'job' means: it refers to the simplified, singular production to which you devote your working hours. Even those who have several paying jobs – say, freelance short-story writer/neuroscientist, or computer executive/photographer – have only two or three different occupations at most. But they each consume hundreds, thousands, of things. This is the diagnostic feature of modern life, the very definition of a high standard of living: diverse consumption, simplified production. Make one thing, use lots. The self-sufficient gardener, or his self-sufficient peasant or hunter-gatherer predecessor (who is, I shall argue, partly a myth in any case), is in contrast defined by his multiple production and simple consumption. He makes not just one thing, but many – his food, his shelter, his clothing, his entertainment. Because he only consumes what he produces, he cannot consume very much. Not for him the avocado, Tarantino or Manolo Blahnik. He is his own brand.

In the year 2005, if you were the average consumer you would have spent your after-tax income in roughly the following way:

- 20 per cent on a roof over your head
- 18 per cent on cars, planes, fuel and all other forms of transport
- 16 per cent on household stuff: chairs, refrigerators, telephones, electricity, water
- 14 per cent on food, drink, restaurants etc

- 6 per cent on health care
- 5 per cent on movies, music and all entertainment
- 4 per cent on clothing of all kinds
- 2 per cent on education
- 1 per cent on soap, lipstick, haircuts, and such like
- 11 per cent on life insurance and pensions (i.e., saved to secure future spending)
- and, alas from my point of view, only 0.3 per cent on reading

An English farm labourer in the 1790s spent his wages roughly as follows:

- 75 per cent on food
- 10 per cent on clothing and bedding
- 6 per cent on housing
- 5 per cent on heating
- 4 per cent on light and soap

A rural peasant woman in modern Malawi spends her time roughly as follows:

- 35 per cent farming food
- 33 per cent cooking, doing laundry and cleaning
- 17 per cent fetching water
- 5 per cent collecting firewood
- 9 per cent other kinds of work, including paid employment

Imagine next time you turn on the tap, what it must be like to walk a mile or more to the Shire River in Machinga province, hope you are not grabbed by a crocodile when filling your bucket (the UN estimates three crocodile deaths a month in the Machinga province, many of them of women fetching water), hope you have not picked up a cholera dose in your bucket, then walk back carrying the 20 litres that will have to last your family all day. I am not trying to make you feel guilty: I am trying to tease out what it is that makes you well off. It is having the hard work of living made easy by markets and machines and other people. There is probably nothing to stop you fetching free water from the nearest river in your home town, but you would rather pay something from your earnings to get it delivered clean and convenient from your tap.

So this is what poverty means. You are poor to the extent that you cannot afford to sell your time for sufficient price to buy the services you need, and rich to the extent that you can afford to buy not just the services you need but also those you crave. Prosperity, or growth, has been synonymous with moving from self-sufficiency to interdependence, transforming the family from a unit of laborious, slow and diverse production to a unit of easy, fast and diverse consumption paid for by a burst of specialised production.

## Self-sufficiency is poverty

It is fashionable these days to decry 'food miles'. The longer food has spent travelling to your plate, the more oil has been burnt and the more peace has been shattered along the way. But why single out food? Should we not protest against T-shirt miles, too, and laptop miles? After all, fruit and vegetables account for more than 20 per cent of all exports from poor countries, whereas most laptops come from rich countries, so singling out food imports for special discrimination means singling out poor countries for sanctions. Two economists recently concluded, after studying the issue, that the entire concept of food miles is 'a profoundly flawed sustainability indicator'. Getting food from the farmer to the shop causes just 4 per cent of all its lifetime emissions. Ten times as much carbon is emitted in refrigerating British food as in air-freighting it from abroad, and fifty times as much is emitted by the customer travelling to the shops. A New Zealand lamb, shipped to England, requires one-quarter as much carbon to get on to a London plate as a Welsh lamb; a Dutch rose, grown in a heated greenhouse and sold in London, has six times the carbon footprint of a Kenyan rose grown under the sun using water recycled through a fish farm, using geothermal electricity and providing employment to Kenyan women.

In truth, far from being unsustainable, the interdependence of the world through trade is the very thing that makes modern life as sustainable as it is. Suppose your local laptop manufacturer tells you that he already has three orders and then he is off on his holiday so he cannot make you one before the winter. You will have to wait. Or suppose your local wheat farmer tells you that last year's rains means he will have to cut his flour delivery in half this year. You will have to go hungry. Instead, you benefit from a global laptop and wheat market in which somebody somewhere has something to sell you so there are rarely shortages, only modest price fluctuations.

For example, the price of wheat approximately trebled in 2006–8, just as it did in Europe in 1315–18. At the earlier date, Europe was less densely populated, farming was entirely organic and food miles were short. Yet in 2008, nobody ate a baby or pulled a corpse from a gibbet for food. Right up until the railways came, it was cheaper for people to turn into refugees than to pay the exorbitant costs of importing food into a hungry district. Interdependence spreads risk.

The decline in agricultural employment caused consternation among early economists. François Quesnay and his fellow 'physiocrats' argued in eighteenth-century France that manufacturing produced no gain in wealth and that switching from agriculture to industry would decrease a country's wealth: only farming was true wealth creation. Two centuries later the decline in industrial employment in the late twentieth century caused a similar consternation among economists, who saw services as a frivolous distraction from the important business of manufacturing. They were just as wrong. There is no such thing as unproductive employment, so long as people are prepared to buy the service you are offering. Today, 1 per cent works in agriculture and 24 per cent in industry, leaving 75 per cent to offer movies, restaurant meals, insurance broking and aromatherapy.

## Arcadia redux

Yet, surely, long ago, before trade, technology and farming, human beings lived simple, organic lives in harmony with nature. That was not poverty: that was 'the original affluent society'. Take a snapshot of the life of hunter-gathering human beings in their heyday, say at 15,000 years ago well after the taming of the dog and the extermination of the woolly rhinoceros but just before the colonisation of the Americas. People had spear throwers, bows and arrows, boats, needles, adzes, nets. They painted exquisite art on rocks, decorated their bodies, traded foods, shells, raw materials and ideas. They sang songs, danced rituals, told stories, prepared herbal remedies for illnesses. They lived into old age far more frequently than their ancestors had done.

They had a way of life that was sufficiently adaptable to work in almost any habitat or climate. Where every other species needed its niche, the hunter-gatherer could make a niche out of anything: seaside or desert, arctic or tropical, forest or steppe.

A Rousseauesque idyll? The hunter-gatherers certainly looked like noble savages: tall, fit, healthy, and (having replaced stabbing spears with thrown ones) with fewer broken bones than Neanderthals. They ate plenty of protein, not much fat and ample vitamins. In Europe, with the help of increasing cold, they had largely wiped out the lions and hyenas that had both competed with and preyed upon their predecessors, so they had little to fear from wild animals. No wonder nostalgia for the Pleistocene runs through many of today's polemics against consumerism. Geoffrey Miller, for example, in his excellent book *Spent,* asks his readers to imagine a Cro-Magnon mother of 30,000 years ago living 'in a close-knit clan of family and friends ... gathering organic fruits and vegetables ... grooming, dancing, drumming and singing with people she knows, likes and trusts ... the sun rising over the six thousand acres of verdant French Riviera coast that her clan holds.'

Life was good. Or was it? There was a serpent in the hunter-gatherer Eden – a savage in the noble savage. Maybe it was not a lifelong camping holiday after all. For violence was a chronic and ever-present threat. It had to be, because – in the absence of serious carnivore predation upon human beings – war kept the population density below the levels that brought on starvation. '*Homo homini lupus*', said Plautus. 'Man is a wolf to man.' If hunter-gatherers appeared lithe and healthy it was because the fat and slow had all been shot in the back at dawn.

Here are the data. From the !Kung in the Kalahari to the Inuit in the Arctic, two-thirds of modern hunter-gatherers have proved to be in a state of almost constant tribal warfare, and 87 per cent to experience annual war. War is a big word for dawn raids, skirmishes and lots of posturing, but because these happen so often, death rates are high – usually around 30 per cent of adult males dying from homicide. The warfare death rate of 0.5 per cent of the population per year that was typical of many hunter-gatherer societies would equate to two billion people dying during the twentieth century (instead of 100 million). At a cemetery uncovered at Jebel Sahaba, in Egypt, dating from 14,000 years ago, twenty-four of the fifty-nine bodies had died from unhealed wounds caused by spears, darts and arrows. Forty of these bodies were women or children. Women and children generally do not take part in warfare – but they are frequently the object of the fighting. To be abducted as a sexual prize and see your children killed was almost certainly not a rare female fate in hunter-gatherer society. After Jebel Sahaba, forget the Garden of Eden; think Mad Max.

It was not just warfare that limited population growth. Hunter-gatherers are often vulnerable to famines. Even when food is abundant, it might take so much travelling and trouble to collect enough food that women would not maintain a sufficient surplus to keep themselves fully fertile for more than a few prime years. Infanticide was a common resort in bad times. Nor was disease ever far away: gangrene, tetanus and many kinds of parasite would have been big killers. Did I mention slavery? Common in the Pacific north-west. Wife beating? Routine in Tierra del Fuego. The lack of soap, hot water, bread, books, films, metal, paper, cloth? When you meet one of those people who go so far as to say they would rather have lived in some supposedly more delightful past age, just remind them of the toilet facilities of the Pleistocene, the transport options of Roman emperors or the lice of Versailles.

## The call of the new

None the less, you do not have to be starry-eyed about the Stone Age to find aspects of modern consumer society obscenely wasteful. Why, asks Geoffrey Miller, 'would the world's most intelligent primate buy a Hummer H1 Alpha sport-utility vehicle', which seats four, gets ten miles to the gallon, takes 13.5 seconds to reach 60 mph, and sells for $139,771? Because, he answers, human beings evolved to strive to signal social status and sexual worth. What this implies is that far from being merely materialist, human consumption is already driven by a sort of pseudo-spiritualism that seeks love, heroism and admiration. Yet this thirst for status then encourages people to devise recipes that rearrange the atoms, electrons or photons of the world in such a way as to make useful combinations for other people. Ambition is transmuted into opportunity. It was allegedly a young Chinese imperial concubine in 2600 BC who thought up the following recipe for rearranging beta pleated sheets of glycine-rich polypeptides into fine fabrics: take a moth caterpillar, feed it mulberry leaves for a month, let it spin a cocoon, heat it to kill it, put the cocoon in water to unstick the silk threads, carefully draw out the single kilometre-long thread from which the cocoon is made by reeling it on to a wheel, spin the thread and weave a fabric. Then dye, cut and sew, advertise and sell for cash. Rough guide on quantities: it takes about ten pounds of mulberry leaves to make 100 silkworm cocoons to make one necktie.

The cumulative accretion of knowledge by specialists that allows us each to consume more and more different things by each producing fewer and fewer is, I submit, the central story of humanity. Innovation changes the world but only because it aids the elaboration of the division of labour and encourages the division of time. Forget wars, religions, famines and poems for the moment. This is history's greatest theme: the metastasis of exchange, specialisation and the invention they have called forth, the 'creation' of time. The rational optimist invites you to stand back and look at your species differently, to see the grand enterprise of humanity that has progressed – with frequent set-backs – for 100,000 years. And then, when you have seen that, consider whether that enterprise is finished or if, as the optimist claims, it still has centuries and millennia to run. If, in fact, it might be about to accelerate to an unprecedented rate.

If prosperity is exchange and specialisation – more like the multiplication of labour than the division of labour – then when and how did that habit begin? Why is it such a peculiar attribute of the human species?

## CHAPTER 2

# The collective brain: exchange and specialisation after 200,000 years ago

He steps under the shower, a forceful cascade pumped down from the third floor. When this civilisation falls, when the Romans, whoever they are this time round, have finally left and the new dark ages begin, this will be one of the first luxuries to go. The old folk crouching by their peat fires will tell their disbelieving grandchildren of standing naked mid-winter under jet streams of hot clean water, of lozenges of scented soaps and of viscous amber and vermilion liquids they rubbed into their hair to make it glossy and more voluminous than it really was, and of thick white towels as big as togas, waiting on warming racks.

Ian McEwan

*Saturday*

## LIFE EXPECTANCY AT BIRTH (WORLD AVERAGE)

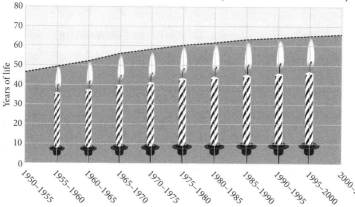

One day a little less than 500,000 years ago, near what is now the village of Boxgrove in southern England, six or seven two-legged creatures sat down around the carcass of a wild horse they had just killed, probably with wooden spears. Each took up a block of flint and began to fashion it into a hand axe, skilfully using hammers of stone, bone or antler to chip off flakes until all that remained was a symmetrical, sharp-edged, teardrop-shaped object in size and thickness somewhere between an i-phone and a computer mouse. The debris they left that day is still there, leaving eerie shadows of their own legs as they sat and worked. You can tell that they were right-handed. Notice: each person made his own tools.

The hand axes they made to butcher that horse are fine examples of 'Acheulean bifaces'. They are thin, symmetrical and razor-sharp along the edge, ideal for slicing through thick hide, severing the ligaments of joints and scraping meat from bones. The Acheulean biface is the stereotype of the Stone Age tool, the iconic flattened teardrop of the Palaeolithic. Because the species that made it has long been extinct we may never quite know how it was used. But one thing we do know. The creatures that made this thing were very content with it. By the time of the Boxgrove horse butchers, their ancestors had been making it to roughly the same design – hand-sized, sharp, double-sided, rounded – for about a million years. Their descendants would continue to make it for hundreds of thousands more years. That's the same technology for more than a thousand millennia, ten thousand centuries, thirty thousand generations – an almost unimaginable length of time.

Not only that; they made roughly the same tools in south and north Africa and everywhere in between. They took the design with them to the Near East and to the far north-west of Europe (though not to East Asia) and still it did not change. A million years across three continents making the same one tool. During those million years their brains grew in size by about one-third.

Here's the startling thing. The bodies and brains of the creatures that made Acheulean hand axes changed faster than their tools.

To us, this is an absurd state of affairs. How could people have been so unimaginative, so slavish, as to make the same technology for so long? How could there have been so little innovation, regional variation, progress, or even regress?

Actually, this is not quite true, but the detailed truth reinforces the problem rather than resolves it. There is a single twitch of progress in biface hand-axe history: around 600,000 years ago, the design suddenly becomes a little more symmetrical. This coincides with the appearance of a new species of hominid which replaces its ancestor throughout Eurasia and Africa. Called *Homo heidelbergensis*, this creature has a much bigger brain, possibly 25 per cent bigger than late *Homo erectus*. Its brain was almost as big as a modern person's. Yet not only did it go on making hand axes and very little else; the hand-axe design sank back into stagnation for another half a million years. We are used to thinking that technology and innovation go together, yet here is strong evidence that when human beings became tool makers, they did not experience anything remotely resembling cultural progress. They just did what they did very well. They did not change.

Bizarre as this may sound, in evolutionary terms it is quite normal. Most species do not change their habits during their few million years on earth or alter their lifestyle much in different parts of their range. Natural selection is a conservative force. It spends more of its time keeping species the same than changing them. Only towards the edge of its range, on an isolated island, or in a remote valley or on a lonely hill top, does natural selection occasionally cause part of a species to morph into something different. That different sport sometimes then spreads to conquer a broader ecological empire, perhaps even returning to replace the ancestral species – to topple the dynasty from which it sprang. There is constant ferment of change within the species' genes as it adapts to its parasites and they to it. But there is little progressive alteration of the organism. Evolutionary change happens largely by the replacement of species by daughter species, not by the changing of habits in species. What is surprising about the human story is not the mind-bogglingly tedious stasis of the Acheulean hand axe, but that the stasis came to an end.

The Boxgrove hominids of 500,000 years ago (who were members of *Homo heidelbergensis*) had their ecological niche. They had a way of getting food and shelter in their preferred habitat, of seducing mates and rearing babies. They walked on two feet, had huge brains, made spears and hand axes, taught each other traditions, perhaps spoke or signalled to each other grammatically, almost certainly lit fires and cooked their food, and undoubtedly killed big animals. If the sun shone, the herds of game were plentiful, the spears were sharp and diseases kept at bay, they may have sometimes thrived and populated new land. At other times, when food was scarce, the local population just died out. They could not change their ways much; it was not in their natures. Once they had spread all across Africa and Eurasia, their populations never really grew. On average death rates matched birth rates. Starvation, hyenas, exposure, fights and accidents claimed most of their lives before they were elderly enough to get chronically ill. Crucially, they did not expand or shift their niche. They remained trapped within it. Nobody woke up one day and said 'I'm going to make my living a different way.'

Think of it this way. You don't expect to get better and better at walking in each successive generation – or breathing, or laughing, or chewing. For Palaeolithic hominids, hand-axe making was like walking, something you grew good at through practice and never thought about again. It was almost a bodily function. It was no doubt passed on partly by imitation and learning, but unlike modern cultural traditions it showed little regional and local variation. It was part of what Richard Dawkins called 'the extended phenotype' of the erectus hominid species, the external expression of its genes. It was instinct, as inherent to the human behavioural repertoire as a certain design of nest is to a certain species of bird. A song thrush lines its nest with mud, a European robin lines its nest with hair and a chaffinch lines its nest with feathers – they always have and they always will. It's innate for them to do so. Making a teardrop-shaped sharp-edged stone tool takes no more skill than making a bird's nest and was probably just as instinctive: it was a natural expression of human development.

Indeed, the analogy with a bodily function is quite appropriate. There is now little doubt that hominids spent much of those million and a half years eating a lot of fresh meat. Some time after two million years ago, ape-men had become more carnivorous. With their feeble teeth and with finger nails where they should have had claws, they needed sharp tools to cut the skins of their kills. Because of their sharp tools they could tackle even the pachydermatous rhinos and elephants. Biface axes were like external canine teeth. The rich meat diet also enabled erectus hominids to grow a larger brain, an organ that burns energy at nine times the rate of the rest of the body. Meat enabled them to cut down on the huge gut that their ancestors had found necessary to digest raw vegetation and raw meat, and thus to grow a bigger brain instead. Fire and cooking in turn then released the brain to grow bigger still by making food more digestible with an even smaller gut – once cooked, starch gelatinises and protein denatures, releasing far more calories for less input of energy. As a result, whereas other primates have guts weighing four times their brains, the human brain weighs more than the human intestine. Cooking enabled hominids to trade gut size for brain size.

Erectus hominids, in other words, had almost everything we might call human: two legs, two hands, a big brain, opposable thumbs, fire, cooking, tools, technology, cooperation, long childhoods, kindly demeanour. And yet there was no sign of cultural take-off, little progress in technology, little expansion of range or niche.

## *Homo dynamicus*

Then there appeared upon the earth a new kind of hominid, which refused to play by the rules. Without any changes in its body, and without any succession of species, it just kept changing its habits. For the first time its technology changed faster than its anatomy. This was an evolutionary novelty, and you are it.

When this new animal appeared is hard to discern, and its entrance was low-key. Some anthropologists argue that in east Africa and Ethiopia the toolkit was showing signs of change as early as 285,000 years ago. Certainly, by at least 160,000 years ago a new, small-faced 'sapiens' skull was being worn on the top of the spine in Ethiopia. Around the same time at Pinnacle Point in South Africa, people – yes, I shall call them people for the first time – were cooking mussels and other shellfish in a cave close to the sea as well as making primitive 'bladelets', small flakes of sharp stone, probably for hafting on to spears. They were also using red ochre, perhaps for decoration, implying thoroughly modern symbolic minds.

This was during the ice age before last, when Africa was mostly a desert. And yet apparently nothing much came of this experiment. Consistent evidence of smart behaviour and a fancy toolkit peters out again. Genetic evidence suggests human beings were still rare even in Africa, eking out a precarious existence in pockets of savannah woodland when it was dry, or possibly on the margins of lakes and seas. In the Eemian interglacial period of 130,000–115,000 years ago, the climate grew warmer and much wetter and sea level rose. Some skulls from what is now Israel suggest that a few slender-headed Africans did begin to colonise the Middle East towards the end of the Eemian, before a combination of cold weather and Neanderthals drove them back again. It was during this mild spell that a fancy new toolkit first appeared in caves in what is now Morocco: flakes, toothed scrapers and retouched points. One of the most extraordinary clues comes in the form of a simple estuarine snail shell called *Nassarius*. This little winkle keeps popping up in archaeological sites, with unnatural holes in its shells. The oldest certain *Nassarius* find is at Grottes des Pigeons near Taforalt in Morocco, where forty-seven perforated shells, some smeared with red ochre, date from certainly more than 82,000 years ago and perhaps as much as 120,000 years ago. Similar shells, harder to date, have been found at Oued Djebanna in Algeria and Skhul in Israel, and perforated shells of the same genus but a different species are found at Blombos cave in South Africa from about 72,000 years ago along with the earliest bone awls. These shells were surely beads, probably worn on a string. Not only do they hint at a very modern attitude to personal ornament, symbolism or perhaps even money; they also speak eloquently of trade. Taforalt is 25 miles and Oued Djebanna 125 miles from the nearest coast. The beads probably travelled hand to hand by exchange. Likewise, there are hints from east Africa and Ethiopia that the volcanic glass known as obsidian may have begun to move over long distances around this time too, or even earlier, presumably by trade, but the dates and sources are still uncertain.

Just across the strait of Gibraltar from where these bead-wearing, flake-making people lived were the ancestors of Neanderthals, whose brains were just as big but who showed no signs of making beads or flake tools, let alone doing long-distance trade. There was clearly something different about the Africans. Over the next few tens of millennia there were sporadic improvements, but no great explosion. There may have been a collapse of human populations. The African continent was plagued by 'megadroughts' at this time, during which desiccating winds blew the dust of extensive deserts into Lake Malawi, whose level dropped 600 metres. Only well after 80,000 years ago, so genetic evidence attests, does something big start to happen again. This time the evidence comes from genomes, not artefacts. According to DNA scripture, it was then that one quite small group of people began to populate the entire African continent, starting either in East or South Africa and spreading north and rather more slowly west. Their genes, marked by the L3 mitochondrial type, suddenly expanded and displaced most others in Africa, except the ancestors of the Khoisan and pygmy people. Yet even now there was no hint of what was to come, no clue that this was anything but another evolutionary avatar of a precariously successful predatory ape. The new African form, with its fancy tools, ochre paint and shell-bead ornaments, might have displaced its neighbours, but it would now settle down to enjoy its million years in the sun before gracefully giving way to something new. This time, however, some of the L3 people promptly spilled out of Africa and exploded into global dominion. The rest, as they say, is history.

## Starting to barter

Anthropologists advance two theories to explain the appearance in Africa of these new technologies and people. The first is that it was driven by climate. The volatility of the African weather, sucking human beings into deserts in wet decades and pushing them out again in dry ones, would have placed a premium on adaptability, which in turn selected for new capabilities. The trouble with this theory is first that climate had been volatile for a very long time without producing a technologically adept ape, and second that it applies to lots of other African species too: if human beings, why not elephants and hyenas? There is no evidence from the whole of the rest of biology that desperate survival during unpredictable weather selects intelligence or cultural flexibility. Rather the reverse: living in large social groups on a plentiful diet both encourages and allows brain growth.

The second theory is that a fortuitous genetic mutation triggered a change in human behaviour by subtly altering the way human brains were built. This made people fully capable of imagination, planning, or some other higher function for the first time, which in turn gave them the capacity to make better tools and devise better ways of making a living. For a while, it even looked as if two candidate mutations of the right age had appeared – in the gene called FOXP2, which is essential to speech and language in both people and songbirds. Adding these two mutations to mice does indeed seem to change the flexibility of wiring in their brain in a way that may be necessary for the rapid flicker of tongue and lung that is called speech, and perhaps coincidentally the mutations even change the way mice pups squeak without changing almost anything else about them. But recent evidence confirms that Neanderthals share the very same two mutations, which suggests that the common ancestor of Neanderthals and modern people, living about 400,000 years ago, may have already been using pretty sophisticated language. If language is the key to cultural evolution, and Neanderthals had language, then why did the Neanderthal toolkit show so little cultural change?

Moreover, genes would undoubtedly have changed during the human revolution after 200,000 years ago, but more in response to new habits than as causes of them. At an earlier date, cooking selected mutations for smaller guts and mouths, rather than vice versa. At a later date, milk drinking selected for mutations for retaining lactose digestion into adulthood in people of western European and East African descent. The cultural horse comes before the genetic cart. The appeal to a genetic change driving evolution gets gene-culture co-evolution backwards: it is a top-down explanation for a bottom-up process.

Besides, there is a more fundamental objection. If a genetic change triggered novel human habits, why do its effects appear gradually and erratically in different places at different times but then accelerate once established? How could the new gene have a slower effect in Australia than in Europe? Whatever the explanation for the modernisation of human technology after 200,000 years ago, it must be something that gathers pace by feeding upon itself, something that is auto-catalytic.

As you can tell, I like neither theory. I am going to argue that the answer lies not in climate, nor genetics, nor in archaeology, nor even entirely in 'culture', but in economics. Human beings had started to do something to and with each other that in effect began to build a collective intelligence. They had started, for the very first time, to exchange things between unrelated, unmarried individuals; to share, swap, barter and trade. Hence the *Nassarius* shells moving inland from the Mediterranean. The effect of this was to cause specialisation, which in turn caused technological innovation, which in turn encouraged more specialisation, which led to more exchange – and 'progress' was born, by which I mean technology and habits changing faster than anatomy. They had stumbled on what Friedrich Hayek called the catallaxy: the ever-expanding possibility generated by a growing division of labour. This is something that amplifies itself once begun.

Exchange needed to be invented. It does not come naturally to most animals. There is strikingly little use of barter in any other animal species. There is sharing within families, and there is food-for-sex exchange in many animals including insects and apes, but there are no cases in which one animal gives an unrelated animal one thing in exchange for a different thing. 'No man ever saw a dog make fair and deliberate exchange of a bone with another dog,' said Adam Smith.

I need to digress here: bear with me. I am not talking about swapping favours – any old primate can do that. There is plenty of 'reciprocity' in monkeys and apes: you scratch my back and I scratch yours. Or, as Leda Cosmides and John Tooby put it, 'One party helps another at one point in time, in order to increase the probability that when their situations are reversed at some (usually) unspecified time in the future, the act will be reciprocated.' Such reciprocity is an important human social glue, a source of cooperation and a habit inherited from the animal past that undoubtedly prepared human beings for exchange. But it is not the same thing as exchange. Reciprocity means giving each other the same thing (usually) at different times. Exchange – call it barter or trade if you like – means giving each other different things (usually) at the same time: simultaneously swapping two different objects. In Adam Smith's words, 'Give me that which I want, and you shall have this which you want.'

Barter is a lot more portentous than reciprocity. After all, delousing aside, how many activities are there in life where it pays to do the same thing to each other in turn? 'If I sew you a hide tunic today, you can sew me one tomorrow' brings limited rewards and diminishing returns. 'If I make the clothes, you catch the food' brings increasing returns. Indeed, it has the beautiful property that *it does not even need to be fair*. For barter to work, two individuals do not need to offer things of equal value. Trade is often unequal, but still benefits both sides. This is a point that nearly everybody seems to miss. In the grasslands of Cameroon, for example, in past centuries the palm-oil producers, who lived on the periphery of the region on the poorest soils, worked hard to produce a low-value product that they exchanged for cereal, livestock and iron with their neighbours. On average it took them thirty days to afford the price of an iron hoe that had cost its makers just seven person-days of work. Yet palm oil was still the most profitable product they could make on their own land and with their own resources. The cheapest way for them to get an iron hoe was to make more palm oil. Or imagine a Trobriand island tribe on the coast that has ample fish and an inland tribe that has ample fruit: as long as two people are living in different habitats, they will value what each other has more than what they have themselves, and trade will pay them both. And the more they trade, the more it will pay them to specialise.

Evolutionary psychologists have assumed that it is rare for conditions to exist in which two people simultaneously have value to offer to each other. But this is just not true, because people can value highly what they do not have access to. And the more they rely on exchange, the more they specialise, which makes exchange still more attractive. Exchange is therefore a thing of explosive possibility, a thing that breeds, explodes, grows, auto-catalyses. It may have built upon an older animal instinct of reciprocity, and it may have been greatly and uniquely facilitated by language – I am not arguing that these were not vital ingredients of human nature that allowed the habit to get started. But I am saying that barter – the simultaneous exchange of different objects – was itself a human breakthrough, perhaps even the chief thing that led to the ecological dominance and burgeoning material prosperity of the species. Fundamentally, other animals do not do barter.

I still don't quite know why, but I have a lot of trouble getting this point across to both economists and biologists. Economists see barter as just one example of a bigger human habit of general reciprocity. Biologists talk about the role that reciprocity played in social evolution, meaning 'do unto others as they do unto you'. Neither seems to be interested in the distinction that I think is vital, so let me repeat it here once more: at some point, after millions of years of indulging in reciprocal back-scratching of gradually increasing intensity, one species, and one alone, stumbled upon an entirely different trick. Adam gave Oz an object in exchange for a different object. This is not the same as Adam scratching Oz's back now and Oz scratching Adam's back later, or Adam giving Oz some spare food now and Oz giving Adam some spare food tomorrow. The extraordinary promise of this event was that Adam potentially now had access to objects he did not know how to make or find; and so did Oz. And the more they did it, the more valuable it became. For whatever reason, no other animal species ever stumbled upon this trick – at least between unrelated individuals.

Do not take my word for it. The primatologist Sarah Brosnan tried to teach two different groups of chimpanzees about barter and found it very problematic. Her chimps preferred grapes to apples to cucumbers to carrots (which they liked least of all). They were prepared sometimes to give up carrots for grapes, but they almost never bartered apples for grapes (or vice versa), however advantageous the bargain. They could not see the point of giving up food they liked for food they liked even more. Chimpanzees and monkeys can be taught to exchange tokens for food, but this is a long way from spontaneously exchanging one thing for another: the tokens have no value to the chimpanzees, so they are happy to give them up. True barter requires that you give up something you value in exchange for something else you value slightly more.

This is reflected in the ecology of wild chimpanzees. Whereas in human beings, each sex eats 'not only from the food items they have collected themselves, but from their partners' finds,' says Richard Wrangham, 'not even a hint of this complementarity is found among nonhuman primates.' It is true that male chimps hunt monkeys more than females do and that having killed a monkey, a male sometimes allows others to share it if they beg to, especially a fertile female or a close partner to whom he owes a favour. But the one thing you do not see is trade of one food for another. There is never barter of meat for nuts. The contrast with human beings, who show an almost obsessive interest not just in sharing food with each other from an early age, but in swapping one item for another, is striking. Birute Galdikas reared a young orang-utan in her home alongside her daughter Binti, and was struck by the contrasting attitudes to food sharing of the two infants. 'Sharing food seemed to give Binti great pleasure,' she wrote. 'In contrast, Princess, like any orang-utan would beg, steal and gobble food at every opportunity'.

My argument is that this habit of exchanging, this appetite for barter, had somehow appeared in our African ancestors some time before 100,000 years ago. Why did human beings acquire a taste for barter as other animals did not? Perhaps it has something to do with cooking. Richard Wrangham makes a persuasive case that control of fire had a far-reaching effect on human evolution. Beyond making it safe to live on the ground, beyond liberating human ancestors to grow big brains on high-energy diets, cooking also predisposed human beings to swapping different kinds of food. And that maybe got them bartering.

## Hunting for gathering

As the economist Haim Ofek has argued, fire itself is hard to start, but easy to share; likewise cooked food is hard to make but easy to share. The time spent in cooking is subtracted from the time spent in chewing: wild chimpanzees spend six hours or more each day just masticating their food. Carnivores might not chew their meat (they are often in a hurry to eat before it is stolen), but they spend hours grinding it in muscular stomachs, which comes to much the same thing. So cooking adds value: the great advantage of cooked food is that though it takes longer to prepare than raw food, it takes just minutes to eat, and this means that somebody else can eat as well as the person who prepares it. A mother can feed her children for many years. Or a woman can feed a man.

In most hunter-gatherers, women spend long hours gathering, preparing and cooking staple foods while men are out hunting for delicacies. There is, incidentally, no hunter-gatherer society that dispenses with cooking. Cooking is the most female-biased of all activities, the only exceptions being when men prepare some ritual feasts or grill a few snacks while out on the hunt. (Does this ring any modern bells? Fancy chefs and barbecuing are the two most masculine forms of cooking today.) On average, across the world, each sex contributes similar quantities of calories, though the pattern varies from tribe to tribe: in Inuits, for example, most food is obtained by men, whereas in the Kalahari Khoisan people, most is gathered by women. But – and here is the crucial point – throughout the human race, males and females specialise and then share food.

In other words, cooking encourages specialisation by sex. The first and deepest division of labour is the sexual one. It is an iron rule documented in virtually all foraging people that 'men hunt, women and children gather'. The two sexes move 'through the same habitat, making strikingly different decisions about how to obtain resources within that habitat, and often returning to a central location with the results of their labour.' So, for example, while Hiwi women in Venezuela travel by foot to dig roots, pound palm starch, pick legumes and collect honey, their menfolk go hunting, fishing or collecting oranges by canoe; while Ache men in Paraguay hunt pigs, deer and armadillos for up to seven hours a day, the women follow them collecting fruit, digging for roots, gathering insects or pounding starch – and sometimes catching armadillos, too; while Hadza women in Tanzania collect tubers, fruit and nuts, men hunt antelope; while Greenland Inuit men hunt seals, women make stews, tools and clothing from the animals. And so on, through example after example. Even the apparent exceptions to the rule, where women do hunt, are instructive, because there is still a division of labour. Agta women in the Philippines hunt with dogs; men hunt with bows. Martu women in western Australia hunt goanna lizards; men hunt bustards and kangaroos. As one anthropologist put it after living with the Khoisan, 'Women demand meat as their social right, and they get it – otherwise they leave their husbands, marry elsewhere or make love to other men.'

What is true of extant hunter-gatherers was equally true of extinct ways of life, as far as can be ascertained. Cree Indian women hunted hares; men hunted moose. Chumash women in California gathered shellfish; men harpooned sea lions. Yahgan Indians (in Tierra del Fuego) hunted otters and sea lions; women fished. In the Mersey estuary near Liverpool are preserved dozens of 8,000-year-old footprints: the women and children appear to have been collecting razor clams and shrimps; the men's prints are moving fast and paralleling those of red and roe deer.

An evolutionary bargain seems to have been struck: in exchange for sexual exclusivity, the man brings meat and protects the fire from thieves and bullies; in exchange for help rearing the children, the woman brings veg and does much of the cooking. This may explain why human beings are the only great apes with long pair bonds.

Just to be clear, this argument has nothing to do with the notion that 'a woman's place is in the home' while men go out to work. Women work hard in hunter-gatherer societies, often harder than men. Neither gathering nor hunting is especially good evolutionary preparation for sitting at a desk answering the telephone. Anthropologists used to argue that the sexual division of labour came about because of the long, helpless childhood of human beings. Because women could not abandon their babies, they could not hunt game, so they stayed near the home and gathered and cooked food of the kind that was compatible with caring for children. With a baby strapped to your back and a toddler giggling at your feet, it is undoubtedly easier to gather fruit and dig roots than it is to ambush an antelope. The anthropologists have been revising the view that the division of labour by sex is all about childcare constraints, though. They have found that even when hunter-gatherer women do not face a hard choice between child care and hunting, they still seek out different kinds of food from their menfolk. In the Alyawarre aborigines of Australia, while young women care for children, older women go out looking for goanna lizards, not for the kangaroos and emus that their menfolk hunt. A sexual division of labour would exist even without childcare constraints.

When did this specialisation begin? There is a neat economic explanation for the sexual division of labour in hunter-gatherers. In terms of nutrition, women generally collect dependable, staple carbohydrates whereas men fetch precious protein. Combine the two – predictable calories from women and occasional protein from men – and you get the best of both worlds. At the cost of some extra work, women get to eat some good protein without having to chase it; men get to know where the next meal is coming from if they fail to kill a deer. That very fact makes it easier for them to spend more time chasing deer and so makes it more likely they will catch one. Everybody gains – gains from trade. It is as if the species now has two brains and two stores of knowledge instead of one – a brain that learns about hunting and a brain that learns about gathering.

Neat, as I say. There are untidy complications to the story, including that men seem to strive to catch big game to feed the whole band – in exchange for both status and the occasional seduction – while women feed the family. This can lead to men being economically less productive than they might be. Hadza men spend weeks trying to catch a huge eland antelope when they could be snaring a spring-hare each day instead; men on the island of Mer in the Torres Strait stand with spears at the fringe of the reef hoping to harpoon giant trevally while their women gather twice as much food by collecting shellfish. Yet even allowing for such conspicuous generosity or social parasitism – depends on how you view it – the economic benefits of food sharing and specialised sex roles are real. They are also unique to human beings. There are a few birds in which the sexes have slightly different feeding habits – in the extinct Huia of New Zealand male and female even had different beak shapes – but collecting different foods and sharing them is something no other species does. It is a habit that put an end to self-sufficiency long ago and that got our ancestors into the habit of exchange.

When was the sexual division of labour invented? The cooking theory points to half a million years ago or much more, but two archaeologists argue otherwise. Steven Kuhn and Mary Stiner think that modern, African-origin *Homo sapiens* had a sexual division of labour and Neanderthals did not, and that this was the former's crucial ecological advantage over the latter when they came head-to-head in Eurasia 40,000 years ago. In advancing this notion they are contradicting a long-held tenet of their science, first advocated by Glyn Isaac in 1978 – that different sex roles started with food sharing millions of years ago. They point out that there is just no sign of the kind of food normally brought by gatherer women in Neanderthal debris, nor of the elaborate clothing and shelters that Inuit women make while their men are hunting. There are occasional shellfish, tortoises, eggshells and the like – foods easily picked up while hunting – but no grindstones and no sign of nuts and roots. This is not to deny that Neanderthals cooperated, and cooked. But it is to challenge the notion that the sexes had different foraging strategies and swapped the results. Either the Neanderthal women sat around doing nothing, or, since they were as butch as most modern men, they went out hunting with the men. That seems more likely.

This is a startling shift of view. Instead of talking about

'hunter-gathering' as the natural state of humanity effectively since forever, as they are apt to do, scientists must begin to consider the possibility that it is a comparatively recent phase, an innovation of the last 200,000 years or so. Is the sexual division of labour a possible explanation of what made a small race of Africans so much better at surviving in a time of megadroughts and volatile climate change than all other hominids on the planet?

Perhaps. Remember how few are the remains from Neanderthal sites. But at least the burden of proof has shifted a bit. Even if the habit is more ancient, it may have been the predisposing factor that then conditioned the African race to the whole notion of specialisation and exchange. Having trained themselves to specialise and exchange between the sexes, having got into the habit of exchanging labour with others, the thoroughly modern Africans had then begun to extend the idea a little bit further and tentatively try a new and still more portentous trick, of specialising within the band and then between bands. This latter step was very hard to take, because of the homicidal relationships between tribes. Famously, no other species of ape can encounter strangers without trying to kill them, and the instinct still lurks in the human breast. But by 82,000 years ago, human beings had overcome this problem sufficiently to be able to pass *Nassarius* shells hand to hand 125 miles inland. Barter had begun.

## Beachcombing east

Barter was the trick that changed the world. To paraphrase H.G. Wells, 'We had struck our camp forever, and were out upon the roads.' Having conquered much of Africa by about 80,000 years ago, the modern people did not stop there. Genes tell an almost incredible story. The pattern of variation in the DNA of both mitochondrial and Y chromosomes in all people of non-African origin attests that some time around 65,000 years ago, or not much later, a group of people, numbering just a few hundred in all, left Africa. They probably crossed the narrow southern end of the Red Sea, a channel much narrower then than it is now. They then spread along the south coast of Arabia, hopping over a largely dry Persian Gulf, skirting round India and a then-connected Sri Lanka, moving gradually down through Burma, Malaya and along the coast of a landmass called Sunda in which most of the Indonesian islands were then embedded, until they came to a strait somewhere near Bali. But they did not stop there either. They paddled across at least eight straits, the largest at least forty miles wide, presumably on canoes or rafts, working their way through an archipelago to land, probably around 45,000 years ago, on the continent of Sahul, in which Australia and New Guinea were conjoined.

This great movement from Africa to Australia was not a migration, but an expansion. As bands of people feasted on the coconuts, clams, turtles, fish and birds on one part of the coast and grew fat and numerous, so they would send out pioneers (or exile troublemakers?) to the east in search of new camp sites. Sometimes these emigrants would have to leapfrog others already in possession of the coast by trekking inland or taking to canoes.

Along the way they left tribes of hunter-gatherer descendants, a few of whom survive to this day genetically unmixed with other races. On the Malay Peninsula, forest hunter-gatherers called the Orang Asli ('original people') look 'negrito' in appearance and prove to have mitochondrial genes that branched off from the African tree about 60,000 years ago. In New Guinea and Australia, too, the genetics tell an unambiguous story of almost complete isolation since the first migration. Most remarkable of all, the native people of the Andaman islands, black-skinned, curly-haired and speaking a language unrelated to any other, have Y-chromosome and mitochondrial genes that diverged from the common ancestor with the rest of humankind 65,000 years ago. At least this is true of the Jarawa tribe on Great Andaman. The North Sentinelese, on the nearby island of North Sentinel, have not volunteered to give blood – at least not their own. As the only hunter-gatherers who still resist 'contact', these fine-looking people – strong, slim, fit and stark naked except for a small plant-fibre belt round the waist – usually greet visitors with showers of arrows. Good luck to them.

To reach the Andaman islands (then closer to the Burmese coast, but still out of sight) and Sahul, however, the migrants of 65,000 years ago must have been proficient canoeists. It was in the early 1990s that the African-born zoologist Jonathan Kingdon first suggested that the black skin of many Africans, Australians, Melanesians and 'negrito' Asians hinted at a maritime past. For a hunter-gatherer on the African savannah, a very black skin is not needed, as the relatively pale Khoisan and pygmies prove. But out on an exposed reef or beach, or in a fishing canoe, maximum sunscreen is called for. Kingdon believed that the 'Banda strandlopers', as he called them, had returned to conquer Africa from Asia, rather than the other way round, but he was ahead of the genetic evidence in coining the idea of an essentially maritime Palaeolithic race.

This remarkable expansion of the human race along the shore of Asia, now known as the 'beachcomber express', has left few archaeological traces, but that is because the then coastline is now 200 feet under water. It was a cool, dry time with vast ice sheets in high latitudes and big glaciers on mountain ranges. The interior of many of the continents was inhospitably dry, windy and cold. But the low-lying coasts were dotted with oases of freshwater springs. The low sea level not only exposed more springs, but increased the relative pressure on underground aquifers to discharge near the coast. All along the coast of Asia, the beachcombers would have found fresh water bubbling up and flowing into streams that meandered down to the ocean.

The coast is also rich in food if you have the ingenuity to find it, even on desert shores. It made sense to stick to the beach.

The evidence of DNA attests that some of these beach-combers, on reaching India and apparently not before, must have eventually moved inland, because by 40,000 years ago 'modern' people were pressing west into Europe and east into what is now China. Abandoning the crowded coast, they resumed their old African ways of hunting game and gathering fruits and roots, becoming gradually more dependent on hunting once more as they inched north into the steppes grazed by herds of mammoths, horses and rhinoceroses. Soon they came across their distant cousins, the descendants of *Homo erectus*, with whom they last shared an ancestor half a million years before. They got close enough to acquire the latter's lice to add to their own, so louse genes suggest, and conceivably even close enough to acquire a smattering of their cousins' genes by interbreeding. But inexorably they rolled back the territory of these Eurasian erectus hominids till the last survivor, of the European cold-adapted sort known as Neanderthal, died with his back to the Strait of Gibraltar about 28,000 years ago. Another 15,000 years saw some of them spilling into the Americas from north-east Asia.

They were very good at wiping out not only their distant cousins, but also much of their prey, something previous hominid species had not managed. The earliest of the great cave painters, working at Chauvet in southern France 32,000 years ago, was almost obsessed with rhinoceroses. A more recent artist, working at Lascaux 15,000 years later, depicted mostly bisons, bulls and horses – rhinoceroses were rare or extinct in Europe by then. At first, modern human beings around the Mediterranean relied mostly on large mammals for meat. They ate small game only if it was slow-moving – tortoises and limpets were popular. Then, gradually and inexorably, starting in the Middle East, they switched their attention to smaller animals, and especially to fast-breeding species, such as rabbits, hares, partridges and smaller gazelles. They gradually stopped eating tortoises. The archaeological record tells this same story at sites in Israel, Turkey and Italy.

The reason for this shift, say Mary Stiner and Steven Kuhn, was that human population densities were growing too high for the slower-reproducing prey such as tortoises, horses and elephants. Only the fast-breeding rabbits, hares and partridges, and for a while gazelles and deer, could cope with such hunting pressure. This trend accelerated about 15,000 years ago as large game and tortoises disappeared from the Mediterranean diet altogether – driven to the brink of extinction by human predation. (A modern parallel: in the Mojave Desert of California, ravens occasionally kill tortoises for food. But only when landfills provided the ravens with ample alternative food and boosted – subsidised – their numbers did the tortoise numbers start to collapse from raven predation. So modern people, subsidised by hare meat, could extinguish mammoths.)

It is rare for a predator to wipe out its prey altogether. In times of prey scarcity, erectus hominids, like other predators, had simply suffered local depopulation; that in turn would have saved the prey from extinction and the hominid numbers could recover in time. But these new people could innovate their way out of trouble; they could shift their niche, so they continued to thrive even as they extinguished their old prey. The last mammoth to be eaten on the Asian plain was probably thought a rare delicacy, a nice change from hare and gazelle stew. As they adjusted their tactics to catch smaller and faster prey, so the moderns developed better weapons, which in turn enabled them to survive at high densities, though at the expense of extinguishing more of the larger and slower-breeding prey. This pattern of shifting from big prey to small as the former were wiped out was characteristic of the new ex-Africans wherever they went. In Australia, almost all larger animal species, from diprotodons to giant kangaroos, became extinct soon after human beings arrived. In the Americas, human arrival coincided with a sudden extinction of the largest, slowest-breeding beasts. Much later in Madagascar and New Zealand mass extinctions of large animals also followed with human colonisation. (Incidentally, given the obsession of 'show-off' male hunters with catching the largest beasts with which to buy prestige in the tribe, it is worth reflecting that these mass extinctions owe something to sexual selection.)

## Shall we trade?

Meanwhile, the stream of new technologies gathered pace. From around 45,000 years ago, the people of western Eurasia had progressively revolutionised their toolkit. They struck slim, sharp blades from cylindrical rock 'cores' – a trick that produces ten times as much cutting edge as the old way of working, but is far harder to pull off. By 34,000 years ago they were making bone points for spears, and by 26,000 they were making needles. Bone spear throwers, or atlatls – which greatly increase the velocity of javelins – appear by 18,000 years ago. Bows and arrows came soon afterwards. 'Microburin' borers were used for drilling the holes in needles and beads. Of course, stone tools would have been only a tiny tip of a technological iceberg, dominated by wood, which has long since rotted away. Antler, ivory and bone were just as important. String, made from plant fibres or leather, was almost certainly in use by then to catch fish and rabbits in nets or snares, and to make bags for carrying things in.

Nor was this virtuosity confined to practicalities. As well as bone and ivory, shells, fossil coral, steatite, jet, lignite, hematite, and pyrite were used to make ornaments and objects. A flute made from the bone of a vulture dates from 35,000 years ago at Hohle Fels and a tiny horse, carved from mammoth ivory and worn smooth by being used as a pendant, dates from 32,000 years ago at Vogelherd – both in Germany. By the time of

Sungir, an open-air settlement from 28,000 years ago at a spot near the city of Vladimir, north-east of Moscow, people were being buried in clothes decorated with thousands of laboriously carved ivory beads, and even little wheel-shaped bone ornaments had appeared. At Mezherich, in what is now Ukraine, 18,000 years ago, jewellery made of shells from the Black Sea and amber from the Baltic implied trade over hundreds of miles.

This is in striking contrast to the Neanderthals, whose stone tools were virtually always made from raw material available within an hour's walk of where the tool was used. To me this is a vital clue to why the Neanderthals were still making hand axes, while their African-origin competitors were making ever more types of tool. Without trade, innovation just does not happen. Exchange is to technology as sex is to evolution. It stimulates novelty. The remarkable thing about the moderns of west Asia is not so much the diversity of artefacts as the continual innovation. There is more invention between 80,000 and 20,000 years ago than there had been in the previous million. By today's standards, it was very slow, but by the standards of *Homo erectus* it was lightning-fast. And the next ten millennia would see still more innovations: fish hooks, all sorts of implements, domesticated wolves, wheat, figs, sheep, money.

If you are not self-sufficient, but are working for other people, too, then it pays you to spend some time and effort to improve your technology and it pays you to specialise. Suppose, for example, that Adam lives in a grassy steppe where there are herds of reindeer in winter, but some days' walk away is a coast, where there are fish in summer. He could spend winter hunting, then migrate to the coast to go fishing. But that way he would not only waste time travelling, and probably run a huge risk crossing the territory of another tribe. He would also have to get good at two quite different things.

If, instead, Adam sticks to hunting and then gives some dried meat and reindeer antlers – ideal for fashioning hooks from – to

Oz, a coastal fisherman, in exchange for fish, he has achieved the goal of varying his diet in a less tiring or dangerous way. He has also bought an insurance policy. And Oz would be better off, because he could now catch (and spare) more fish. Next Adam realises that instead of giving Oz raw antlers, he can give him pieces of antler already fashioned into hooks. These are easier to transport and fetch a better price in fish. He got the idea when he once went to the trading point and noticed others selling antlers that had already been cut up into easy segments. One day, Oz asks him to make barbed hooks. And Adam suggests that Oz dries or smokes his fish so it lasts longer. Soon Oz brings shells, too, which Adam buys to make jewellery for a young woman he fancies. After a while, depressed by the low price fetched by hooks of even high quality, Adam hits on the idea of tanning some extra hides and bringing those to the trading point, too. Now he finds he is better at making hides than hooks, so he specialises in hides, giving his antlers to somebody from his own tribe in exchange for his hides. And so on, and on and on.

Fanciful, maybe. And no doubt wrong in all sorts of details. But the point is how easy it is to envisage both opportunities for trade among hunter-gatherers – meat for plants, fish for leather, wood for stone, antler for shells – and how easy it is for Stone Age people to discover mutual gains from trade and then to enhance that effect by further specialising and further dividing labour. The extraordinary thing about exchange is that it breeds: the more of it you do, the more of it you can do. And it calls forth innovation.

Which only raises another question: why did economic progress not accelerate towards an industrial revolution there and then? Why was progress so agonisingly slow for so many millennia? The answer, I suspect, lies in the fissile nature of human culture. Human beings have a deep capacity for isolationism, for fragmenting into groups that diverge from each other. In New Guinea, for instance, there are more than 800 languages, some spoken in areas just a few miles across yet as unintelligible to those on either side as French and English. There are still 7,000 languages spoken on earth and the people who speak each one are remarkably resistant to borrowing words, traditions, rituals or tastes from their neighbours. 'Whereas vertical transmission of cultural traits goes largely unnoticed, horizontal transmission is far more likely to be regarded with suspicion or even indignation,' say the evolutionary biologists Mark Pagel and Ruth Mace. 'Cultures, it seems, like to shoot messengers.' People do their utmost to cut themselves off from the free flow of ideas, technologies and habits, limiting the impact of specialisation and exchange.

## Ricardo's magic trick

Divisions of labour beyond the pair bond had probably been invented in the Upper Palaeolithic. Commenting on the ten thousand mammoth-ivory beads with which the clothing of two 28,000-year-old child corpses at Sungir in Russia were decorated, the anthropologist Ian Tattersall remarks: 'It's hardly probable that these young people had made their richly adorned vestments themselves. It's much more likely that the sheer diversity of material production in their society was the result of the specialisation of individuals in different activities.' The carvers of mammoth beads at Sungir, the painter of rhinoceroses at Chauvet, the striker of blades from rock cores, the maker of rabbit nets – perhaps these were all specialists, exchanging their labour for that of others. Perhaps there had been different roles within each band of human beings ever since the first emergence of modern people over 100,000 years ago.

It is such a human thing to do, and so obvious an explanation of the thing that needs explaining: the capacity for innovation. Specialisation would lead to expertise, and expertise would lead to improvement. Specialisation would also give the specialist an excuse for investing time in developing a laborious new technique. If you have a single fishing harpoon to make, there's no sense in building a clever tool for making harpoons first, but if you have to make harpoons for five fishermen, then maybe there is sense and time-saving in first making the harpoon-making tool.

Specialisation would therefore create and increase the opportunities for gains from trade. The more Oz goes fishing, the better he gets at it, so the less time it takes him to catch each fish. The more hooks Adam the reindeer hunter makes, the better he gets at it, so the less time he takes to make each one. So it pays Oz to spend his day fishing and buy his hooks off Adam by giving him some fish. And it pays Adam to spend his day making hooks and get his fish delivered by Oz.

And, wonderfully, this is true even if Oz is better at hook-making than Adam. Suppose Adam is a clumsy fool, who breaks half his hooks, but he is an even clumsier fisherman who cannot throw a line to save his life. Oz, meanwhile, is one of those irritating paragons who can whittle a bone hook with little trouble and always catches lots of fish. Yet it still pays Oz to get his hooks made for him by clumsy Adam. Why? Because with practise Adam has at least become better at making hooks than he is at fishing. It takes him three hours to make a hook, but four hours to catch a fish. Oz takes only an hour to catch a fish, but good as he is he still needs two hours to make a hook. So if each is self-sufficient, then Oz works for three hours (two to make the hook and one to catch the fish), while Adam works for seven hours (three to make the hook and four to catch a fish). If Oz catches two fish and swaps one for a hook from Adam, he only has to work two hours. If Adam makes two hooks and uses one to buy a fish from Oz, he only works for six hours. Both are better off than when they were self-sufficient. Both have gained an hour of leisure time.

I have done nothing here but retell, in Stone Age terms, the notion of comparative advantage as defined by the stockbroker David Ricardo in 1817. He used the example of England trading cloth for Portuguese wine, but the argument is the same:

> England may be so circumstanced, that to produce the cloth may require the labour of 100 men for one year; and if she attempted to make the wine, it might require the labour of 120 men for the same time. England would therefore find it in her interest to import wine, and to purchase it by the exportation of cloth. To produce the wine in Portugal, might require only the labour of 80 men for one year, and to produce the cloth in the same country, might require the labour of 90 men for the same time. It would therefore be advantageous for her to export wine in exchange for cloth. This exchange might even take place, notwithstanding that the commodity imported by Portugal could be produced there with less labour than in England.

Ricardo's law has been called the only proposition in the whole of the social sciences that is both true and surprising. It is such an elegant idea that it is hard to believe that Palaeolithic people took so long to stumble upon it (or economists to define it); hard to understand why other species do not make use of it, too. It is rather baffling that we appear to be the only species that routinely exploits it. Of course, that is not quite right. Evolution has discovered Ricardo's law and applied it to symbioses, such as the collaboration between alga and fungus that is a lichen plant or the collaboration between a cow and a bacterium in a rumen. Within species, too, there are clear gains from trade between cells of a body, polyps of a coral colony, ants of an ant colony, or mole-rats of a mole-rat colony. The great success of ants and termites – between them they may comprise one-third of all the animal biomass of land animals – is undoubtedly down to their division of labour. Insect social life is built not on increases in the complexity of individual behaviour, 'but instead on specialization among individuals'. In the leafcutter ants of the Amazon rainforest, colonies may number millions, and workers grow into one of four distinct castes: minors, medias, majors and supermajor. In one species a supermajor (or soldier) may weigh the same as 500 minors.

But the big difference is that in every other species than human beings, the colonies consist of close relatives – even a city of a million ants is really just a huge family. Yet reproduction is the one task that people never delegate to a specialist, let alone a queen. What gave people the chance to exploit gains from trade, without waiting for Mother Nature's tedious evolutionary crawl, was technology. Equipped with the right tool, a human being can become a soldier or a worker (maybe not a queen), and he can switch between the roles. The more you do something, the better you get at it. A band of hunter-gatherers in west Eurasia, 15,000 years ago, dividing labour not just by gender but by individual as well, would have been formidably more efficient than an undifferentiated band. Imagine, say, 100 people in the band. Some of them make tools, others make clothes, others hunt, others gather. One tiresome bloke insists on prancing around in a deer skull chanting spells and prayers, adding little to the general well-being, but then maybe he is in charge of the lunar calendar so he can tell people when the tides will be lowest for limpet-picking expeditions.

True, there is not much specialisation in modern hunter-gatherers. In the Kalahari or the Australian desert, apart from the gathering women, the hunting men and maybe the shaman, there are not too many distinct occupations in each band. But these are the simple societies left in the harsh habitats. In the relatively fertile lands of west Eurasia after 40,000 years ago, when bands of people were larger and lines of work were diverse, specialisation had probably grown up within each band. The Chauvet rhino painter was so good at his job (and yes, archaeologists think it was mainly one artist) that he must surely have had plenty of time off hunting duties to practise. The Sungir bead maker must have been working for a wage of some kind, because he cannot surely have had time to hunt for himself. Even Charles Darwin reckoned that 'primeval man practised a division of labour; each man did not manufacture his own tools or rude pottery, but certain individuals appear to have devoted themselves to such work, no doubt receiving in exchange the produce of the chase.'

## Innovation networks

According to the anthropologist Joe Henrich, human beings learn skills from each other by copying prestigious individuals, and they innovate by making mistakes that are very occasionally improvements – that is how culture evolves. The bigger the connected population, the more skilled the teacher, and the bigger the probability of a productive mistake. Conversely, the smaller the connected population, the greater the steady deterioration of the skill as it was passed on. Because they depended on wild resources, hunter-gatherers could rarely live in bands larger than a few hundred and could never achieve modern population densities. This had an important consequence. It meant that there was a limit to what they could invent. A band of a hundred people cannot sustain more than a certain number of tools, for the simple reason that both the production and the consumption of tools require a minimum size of market. People will only learn a limited set of skills and if there are not enough experts to learn one rare skill from, they will lose that skill. A good idea, manifest in bone, stone or string, needs to be kept alive by numbers. Progress can easily falter and turn into regress.

Where modern hunter-gatherers have been deprived of access to a large population of trading partners – in sparsely populated Australia, especially Tasmania, and on the Andaman islands, for example – their technological virtuosity was stunted and barely progressed beyond those of Neanderthals. There was nothing special about the brains of the moderns; it was their trade networks that made the difference – their collective brains.

The most striking case of technological regress is Tasmania. Isolated on an island at the end of the world, a population of less than 5,000 hunter-gatherers divided into nine tribes did not just stagnate, or fail to progress. They fell steadily and gradually back into a simpler toolkit and lifestyle, purely because they lacked the numbers to sustain their existing technology. Human beings reached Tasmania at least 35,000 years ago while it was still connected to Australia. It remained connected – on and off – until about 10,000 years ago, when the rising seas filled the Bass Strait. Thereafter the Tasmanians were isolated. By the time Europeans first encountered Tasmanian natives, they found them not only to lack many of the skills and tools of their mainland cousins, but to lack many technologies that their own ancestors had once possessed. They had no bone tools of any kind, such as needles and awls, no cold-weather clothing, no fish hooks, no hafted tools, no barbed spears, no fish traps, no spear throwers, no boomerangs. A few of these had been invented on the mainland after the Tasmanians had been isolated from it – the boomerang, for instance – but most had been made and used by the very first Tasmanians. Steadily and inexorably, so the archaeological history tells, these tools and tricks were abandoned. Bone tools, for example, grew simpler and simpler until they were dropped altogether about 3,800 years ago. Without bone tools it became impossible to sew skins into clothes, so even in the bitter winter, the Tasmanians went nearly naked but for seal-fat grease smeared on their skin and wallaby pelts over their shoulders. The first Tasmanians caught and ate plenty of fish, but by the time of Western contact they not only ate no fish and had eaten none for 3,000 years, but they were disgusted to be offered it (though they happily ate shellfish).

The story is not quite that simple, because the Tasmanians did invent a few new things during their isolation. Around 4,000 years ago they came up with a horribly unreliable form of canoe-raft, made of bundles of rushes and either paddled by men or pushed by swimming women (!), which enabled them to reach offshore islets to harvest birds and seals. The raft would become waterlogged and disintegrate or sink after a few hours, so it was no good for re-establishing contact with the mainland. As far as innovation goes, it was so unsatisfactory that it almost counts as an exception to prove the rule. The women also learnt to dive up to twelve feet below the water to prise clams off the rocks with wooden wedges and to grab lobsters. This was dangerous and exhausting work, which they were very skilled at: the men did not take part. So it was not that there was no innovation; it was that regress overwhelmed progress.

The archaeologist who first described the Tasmanian regress, Rhys Jones, called it a case of the 'slow strangulation of the mind', which perhaps understandably enraged some of his academic colleagues. There was nothing wrong with individual Tasmanian brains; there was something wrong with their collective brains. Isolation – self-sufficiency – caused the shrivelling of their technology. Earlier I wrote that division of labour was made possible by technology. But it is more interesting than that. Technology was made possible by division of labour: market exchange calls forth innovation.

Now, at last, it becomes clear why the erectus hominids saw such slow technological progress. They, and their descendants the Neanderthals, lived without trade (recall how Neanderthal stone tools were sourced within an hour's walk of their use). So in effect each erectus hominid tribe occupied a virtual Tasmania, cut off from the collective brain of the wider population. Tasmania is about the size of the Irish Republic. By the time

Abel Tasman pitched up in 1642 it held probably about 4,000 hunter-gatherers divided into nine tribes, and they lived mainly off seals, seabirds and wallabies, which they killed with wooden clubs and spears. That means that there were only a few hundred young adults on the entire island who were learning new skills at any one time. If, as seems to be the case everywhere, culture works by faithful imitation with a bias towards imitating prestigious individuals (in other words, copy the expert, not the parent or the person closest to hand), then all it would take for certain skills to be lost would be a handful of unlucky accidents in which the most prestigious individual had forgotten or mislearned a crucial step or even gone to his grave without teaching an apprentice. Suppose, for example, that an abundance of seabirds led one group to eschew fishing for a number of years until the last maker of fishing tackle had died. Or that the best barbed-spear maker on the island fell off a cliff one day leaving no apprentice. His barbs went on being used for some years, but once they had all broken, suddenly there was nobody who could make them. Acquiring a skill costs a lot of time and effort; nobody could afford to learn barb-making from scratch. People concentrated on learning the skills that they could watch first-hand.

Bit by bit, Tasmanian technology simplified. The most difficult tools and complex skills were lost first, because they were the hardest to master without a master to learn from. Tools are in effect a measure of the extent of the division of labour and, as Adam Smith argued, the division of labour is limited by the extent of the market. The Tasmanian market was too small to sustain many specialised skills. Imagine if 4,000 people from your home town were plonked on an island and left in total isolation for ten millennia. How many skills and tools do you think they could preserve? Wireless telephony? Double-entry book-keeping? Suppose one of the people in your town was an accountant. He could teach double-entry book-keeping to a youth, but would the youth or the youth's youth pass it on – for ever?

On other Australian islands much the same thing happened as on Tasmania. On Kangaroo Island and Flinders Island, human occupation petered out, probably by extinction, a few thousand years after isolation. Flinders is a fertile island that should be a paradise. But the hundred or so people it could support were far too small a human population to sustain the technology of hunter-gathering. The Tiwi people, isolated on two islands north of Darwin for 5,500 years, also reversed the ratchet of accumulating skills and slipped back to a simpler tool set. The Torres islanders lost the art of canoe making, causing the anthropologist W.H.R. Rivers to puzzle over the 'disappearance of the useful arts'. It seems the hunter-gathering lifestyle was doomed if too isolated. The Australian mainland, by contrast, experienced steady technological progress. Where Tasmanian spears merely had fire-hardened wood points, on the mainland spears acquired detachable tips, stone barbs and 'woomera' spear throwers. It is no coincidence that the mainland had long-range trade, so that inventions and luxuries could be sourced from distant parts of the land. Shell beads had been moving long distances across Australia since at least 30,000 years ago. Pearl and baler shell pendants from the north coast moved through at least eight tribal areas to reach the far south more than a thousand miles from where they had been harvested, growing in sacredness as they went. 'Pitchera' – a tobacco-like plant – moved west from Queensland. The best stone axes travelled up to 500 miles from where they were mined.

In contrast to Tasmania, Tierra del Fuego – an island not much bigger than Tasmania, home to not many more people and generally rather colder and less hospitable – possessed a race of people who, when Charles Darwin met them in 1834, set bait for fish, nets for seals and snares for birds, used hooks and harpoons, bows and arrows, canoes and clothing – all made with specialised tools and skills. The difference is that the Fuegians were in fairly frequent contact with other people across the Strait of Magellan so that they could relearn lost skills or import new tools from time to time. All it took was an occasional incomer from the mainland to keep technology from regressing.

## Networking in the near-east

The lesson is stark. Self-sufficiency was dead tens of thousand years ago. Even the relatively simple lifestyle of a hunter-gatherer cannot exist without a large population exchanging ideas and skills. The importance of this notion cannot be emphasised too strongly. The success of human beings depends crucially, but precariously, on numbers and connections. A few hundred people cannot sustain a sophisticated technology: trade is a vital part of the story.

Vast though it is, Australia itself may have suffered from this isolation effect. Recall that it was colonised 45,000 years ago by pioneering beachcombers spreading east from Africa along the shore of Asia. The vanguard of such a migration must have been small in number and must have travelled comparatively light. The chances are they had only a sample of the technology available to their relatives back at the Red Sea crossing. This may explain why Australian aboriginal technology, although it developed and elaborated steadily over the ensuing millennia, was lacking in so many features of the Old World – elastic weapons, for example, such as bows and catapults, were unknown, as were ovens. It was not that they were 'primitive' or that they had mentally regressed: it was that they had arrived with only a subset of technologies and did not have a dense enough population and therefore a large enough collective brain to develop them much further.

The 'Tasmanian effect' may also explain why technological

progress had been so slow and erratic in Africa after 160,000 years ago. It explains the periodic bursts of modern tools found at South African sites like Pinnacle Point, Blombos Cave and Klasies River. Despite the invention of exchange, the continent was like a patchwork of virtual Tasmanias. As Steve Shennan and his colleagues have calculated, whenever the right combination of (say) seafood, freshwater and fertile savannahs produced local population explosions, technology would have grown sophisticated in proportion to the number of people networked by exchange to sustain and develop it – in proportion to the scale of the collective intelligence. But when a river dried up or deserts advanced and human populations collapsed or shrank, technology would simplify again. Human cultural progress is a collective enterprise and it needs a dense collective brain.

Thus the extraordinary change in technology and cultural tradition that seems to have flourished more than 30,000 years ago in western Asia and the Near East – the so-called Upper Palaeolithic Revolution – may be explained by a dense population. Fed by an increasingly intensive and vegetarian hunter-gathering lifestyle, and with close contact between tribes, the people of south-west Asia were in a position to accumulate more and more skills and technologies than any previous human populations. A large, interconnected population meant faster cumulative invention – a surprising truth even to this day, as Hong Kong and Manhattan islands demonstrate. As the economist Julian Simon put it, 'population growth leading to diminishing returns is fiction; the induced increase in productivity is scientific fact'. And one of those inventions was farming, which is the subject of Chapter 4.

It is right to end the hunter-gatherer chapter, though, by remembering what happened to the Tasmanians. In the early 1800s, white sealers began to arrive along the island's coasts and it was not long before the Tasmanians were eagerly meeting the sealers to trade with them, proving that 10,000 years of limited exchange had done nothing to dampen their innate enthusiasm for barter. The sealers' dogs were especially sought after, being deerhounds that could easily run down kangaroos. In exchange, sad to relate, Tasmanians sold women to the sealers as concubines. Once white farmers arrived, relations between the two peoples deteriorated and eventually the whites sent bounty hunters to kill the natives, then rounded up the survivors and exiled them to Flinders Island, where they eked out their last days in misery.

## CHAPTER 3

# The manufacture of virtue: barter, trust and rules after 50,000 years ago

Money is not metal. It is trust inscribed.
NIALL FERGUSON
*The Ascent of Money*

**HOMICIDE RATE IN EUROPE**

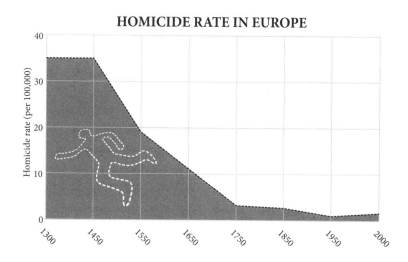

There is a scene in the film *The Maltese Falcon* in which Humphrey Bogart is about to be given $1,000 by Sydney Greenstreet and will have to share some of it with Mary Astor. Greenstreet whispers to Bogart that he'd like to give him a word of advice: that he assumes that Bogart is going to give her some of the money, but that if he does not give her as much as she thinks she ought to have, he should be careful. The scene prefigures a game, invented by Werner Guth in the late 1970s and much loved by economists, called the Ultimatum Game, which opens a little window into the human spirit. The first player is given some money and told to divide it with the second player. The second player is told he can accept or refuse the offer, but not change it. If he accepts, he receives the money; if he refuses, neither he nor the first player gets a penny. The question is, how much money should the first player offer the second player? Rationally, he should tender almost nothing, and the second player should accept it, because however small the sum, refusal will only make the second player worse off than acceptance. But in practice, people usually offer close to half the money. Generosity seems to come naturally, or rather, ungenerous behaviour is irrationally foolish, because the second player will – and does – consider a derisory offer worth rejecting, if only to punish the selfishness of the first player.

The lesson of the ultimatum game and hundreds like it is that again and again people emerge from such experiments as nicer than you think. But the even more surprising lesson is that the more people are immersed in the collective brain of the modern commercial world, the more generous they are. As the economist Herb Gintis puts it, 'societies that use markets extensively develop a culture of co-operation, fairness and respect for the individual'. His evidence comes from a fascinating study in which people in fifteen mostly small-scale tribal societies were enticed to play the Ultimatum Game. Those societies with the least experience of dealing with outsiders were the most hard-hearted, ungenerous and narrowly 'rational'. Machiguenga slash-and-burn farmers from the Amazon most often offered just 15 per cent of the sum to their co-subjects, and in all but one cases, the second player accepted. Likewise, a Hadza hunter-gatherer from Tanzania usually makes a very small offer and experiences few rejections. On the other hand, players from those societies that are most integrated into modern markets, such as the Orma nomads of Kenya or the Achuar subsistence gardeners of Ecuador, will usually offer half the money just as a Western undergraduate would. The whale-hunting Lamalera of the island of Lembata in Indonesia, who need to coordinate large teams of strangers on hunts, offer on average 58 per cent – as if investing the windfall in acquiring new obligations. Much the same happens in two New Guinea tribes, the Au and Gnau, whose members often make 'hyper-fair' offers and yet see them rejected: in such cultures, gifts can be a burden to the receiver because they carry an obligation to reciprocate.

The lesson of this study is that, on the whole, having to deal with strangers teaches you to be polite to them, and that in order for such generosity to emerge, costly punishment of selfishness may be necessary. Rejecting the offer is costly for the second player, but he reckons it is worth it to teach the first player a lesson. The argument is not that exchange teaches people to be kind; it is that exchange teaches people to recognise their en-lightened self-interest lies in seeking cooperation. Here, then, lies a clue to the unique human attribute of being able to deal with strangers, to extend the division of labour to include even your enemies.

Cooperation, exchange and specialisation within a family group are routine throughout the animal kingdom: among chimpanzees and dolphins, among wolves and lions, among individuals of almost any social species. A meerkat or a scrub jay trusts its relative on sentry duty to sound the alarm if an eagle appears and shares the duty. A worker ant divides labour with its queen, with soldiers and with its sisters in other castes of worker. All these societies are just large families. Collaboration between unrelated strangers seems to be a uniquely human achievement. In no other species can two individuals that have never before met exchange goods or services to the benefit of each other, as happens routinely each time you visit a shop or a restaurant or a website. Indeed, in other group-living species, such as ants or chimpanzees, the interactions between members of different groups are almost always violent. Yet human beings can treat strangers as honorary friends.

Taking the first step to proffer the hand of cooperation to a homicidal enemy must have been momentous and almost impossibly difficult, which is perhaps why it is such a rare trick in the animal kingdom. It took primatologists such as Sarah Hrdy and Frans de Waal to notice just how peculiar this is: how inconceivable it would be for an orderly queue of stranger chimpanzees to board an aeroplane, or sit down in a restaurant, without turning violently on each other. And generally speaking the more cooperative a species is within groups, the more hostility there is between groups. As a highly 'groupish' species ourselves, still given to mutual aid within groups and mutual violence between groups, it is an extraordinary thing that people can overcome their instincts enough to have social commerce with strangers.

I think the first overtures may have been ventured first by human females. After all, homicidal raids against neighbouring groups are – in human beings and in most other primates – conducted always by males. So encounters between strange females are not necessarily going to turn violent. Moreover, in all apes females are the sex that leaves the group into which they were born when they mate; in monkeys, curiously, it is males that leave. Assuming human beings follow the ape pattern – as they do to this day in most human societies – then women would have had close relations in other groups in the shape of their mothers, fathers and brothers with whom to build relationships. There is even a curious, much later echo of such a female-centred pattern in the trading patterns of south-east Asia before the arrival of Westerners. The traders of Malaysia, Indonesia and the Philippines were often women, who were taught to calculate and to account from an early age.

Again and again throughout history, trust has to start with relatives before it can be extended to strangers; sending relatives abroad as agents has a long history. The trading ports of Asia each had their own communities of Gujaratis, Fujianese, Persians, Armenians, Jews and Arabs, just as the ports of Europe had their separate communities of Genoese, Florentine, Dutch, English and Hanseatic merchants, keeping the trust within the family as their diasporas spread. The financing of Wellington's armies in Spain in 1809–12 was made possible because the British government trusted a Jewish lender named Nathan Rothschild to trust his brothers on the continent to buy bullion with British paper.

## Finding a trade buddy

In 2004, a series of volunteer undergraduates sat down at computer screens at George Mason University in Virginia to play games for money. In the game each person found himself in a virtual village with his own house and field in which he could produce and consume red and blue virtual 'units' during brief sessions of the game. In each case, he knew that the more he acquired and the closer he got to a certain ratio of blue and red units (e.g., 3:1) the more real money he went home with. But unknown to him, he was either an 'odd' player, who was pro-grammed to be faster at making red units, or an 'even' player, faster at making blue units. On his screen each player could see what other players (two, four or eight in total) were up to and he could chat with them on-screen during each run and in the 100-

second gaps between runs. On one run of the game, in session six, two players had the following exchange:

> 'wonder if u can give me objects'
> 'oh yeah.'
> 'heyyy, i make blues faster, what color do u make faster?'
> 'red'
> 'lol ok'
> 'LOL'
> 'so ill make all blues and u make all reds'
> 'then drop them to each other's houses?'
> 'yea do it'
> 'ok 100% red'
> '100% blue'

The purpose of the experiment, run by Bart Wilson, Vernon Smith and their colleagues, was of course to see if people discovered exchange and specialisation for themselves with no rules or instructions. In the game, specialising is risky because the pay-off for ending up with units of only one colour is zero, but specialisation with exchange allows three times the pay-off of self-sufficiency. Yet there were no clues that trading was even possible. Though some players remained stuck in low-yielding self-sufficiency, most eventually discovered gains from trade. 'Prior to exchange,' comment the experimenters, 'near-autarky prevails, and once the "power of exchanging" is discovered, specialisation gradually evolves.' Intriguingly, the players began by trading bilaterally and personally – that is, each player developed a trading relationship with another player and only later extended the invitation to others.

That trade began as a bilateral and personal affair seems plausible. In the nineteenth century among the Yir Yoront aborigines, in northern Australia, each man's family camp had at least one highly valued stone axe. The axes all came from a quarry jealously guarded and systematically worked by the Kalkadoon tribe at Mount Isa, 400 miles to the south, far beyond the Yir Yoront lands, and they passed through the hands of many trading partners to reach the tribe. Each older man had a trading partner to the south whom he met once a year in the dry season at a ceremonial gathering. In exchange for a dozen sting-ray barbs, to be used as spear tips, he received an axe. In turn he had obtained some of the barbs from his other trading partner to the north – to whom he gave an axe in return. Another 150 miles to the south, the exchange rate was different: one axe for one barb. There were arbitrage profits all along the chain.

So perhaps the first steps to trade with strangers began as individual friendships. A woman could trust her daughter who had married into an allied band within the same tribal grouping. Then perhaps the woman's husband could learn to trust his son-in-law. The alliance between the bands in the face of a common enemy allowed the barrier of suspicion to be breached long enough for one to discover that the other had a surplus of stone for making axes, or of sting-ray barbs for making spear tips. Gradually, step by step, the habit of trade began to grow alongside the habit of xenophobia, complicating the ambitions of men and women.

Most people assume that long-distance trade among strangers and the very concept of the market was a comparatively late development in human history, coming long after agriculture. But, as the Australian aborigines suggest, this is bunk. There is no known human tribe that does not trade. Western explorers, from Christopher Columbus to Captain Cook, ran into many confusions and misunderstandings when they made first contact with isolated peoples. But the principle of trading was not one of them, because the people they met in every case already had a notion of swapping things. Within hours or days of meeting a new tribe, every explorer is bartering. In 1834 in Tierra del Fuego a young naturalist named Charles Darwin came face to face with some hunter-gatherers: 'Some of the Fuegians plainly showed that they had a fair notion of barter. I gave one man a large nail (a most valuable present) without making any signs for a return; but he immediately picked out two fish, and handed them up on the point of his spear.' Darwin and his new friend needed no common language to understand the bargain they were agreeing. Likewise, New Guinea highlanders, when first contacted by Michael Leahy and his fellow prospectors in 1933, gave them bananas in exchange for cowrie shells. Pre-contact, the New Guineans had been trading stone axes over large distances for a very long time. In Australia, baler shells and stone axes had been crossing the entire continent by trade for untold generations. The people of the Pacific coast of North America were sending seashells hundreds of miles inland, and importing obsidian from even farther afield. In Europe and Asia in the Old Stone Age, amber, obsidian, flint and seashells were travelling farther than individual people could possibly have carried them. In Africa, obsidian, shells and ochre were being traded long distances by 100,000 years ago. Trade is prehistoric and ubiquitous.

Moreover, some ancient hunter-gatherer societies reached such a pitch of trade and prosperity as to live in dense, sophisticated hierarchical societies with much specialisation. Where the sea produced a rich bounty, it was possible to achieve a density of the kind that normally requires agriculture to support it – complete with chiefs, priests, merchants and conspicuous consumption. The Kwakiutl Americans, living off the salmon runs of the Pacific North West, had family property rights to streams and fishing spots, had enormous buildings richly decorated with sculptures and textiles, and engaged in bizarre rituals of conspicuous consumption such as the giving of rich copper gifts to each other, or the burning of candlefish oil, just for the prestige of being seen to be philanthropic. They also employed slaves. Yet they were strictly speaking hunter-gatherers. The Chumash of the Californian channel islands, well fed on sea food and seal meat, included specialist craftsmen who fashioned beads from abalone shells to use as currency in a sophisticated and long-range canoe trade. Trade with strangers, and the trust that underpins it, was a very early habit of modern human beings.

## The trust juice

But is trade made possible by the milk of human kindness, or the acid of human self-interest? There was once a German philosophical conundrum known as *Das Adam Smith Problem*, which professed to find a contradiction between Adam Smith's two books. In one he said that people were endowed with instinctive sympathy and goodness; in the other, that people were driven largely by self-interest. 'How selfish soever man may be supposed, there are evidently some principles in his nature, which interest him in the fortunes of others, and render their happiness necessary to him, though he derives nothing from it, except the pleasure of seeing it,' he wrote in *Theory of Moral Sentiments*. 'Man has almost constant occasion for the help of his brethren, and it is in vain for him to expect it from their benevolence only. He will be more likely to prevail if he can interest their self-love in his favour,' he wrote in *The Wealth of Nations*.

Smith's resolution of the conundrum is that benevolence and friendship are necessary but not sufficient for society to function, because man 'stands at all times in need of the co-operation and assistance of great multitudes, while his whole life is scarce sufficient to gain the friendship of a few persons'. In other words, people go beyond friendship and achieve common interest with strangers: they turn strangers into honorary friends, to use Paul Seabright's term. Smith brilliantly confused the distinction between altruism and selfishness: if sympathy allows you to please yourself by pleasing others, are you being selfish or altruistic? As the philosopher Robert Solomon put it, 'What I want for myself is your approval, and to get it I will most likely do what you think I should do.'

This ability to transact with strangers as if they were friends is made possible by an intrinsic, instinctive human capacity for trust. Often the very first thing you do when you meet a stranger and begin to transact with him or her, say a waiter in a restaurant, is to smile – a small, instinctive gesture of trust. The human smile, the glowing embodiment of Smith's innate sentiment of sympathy, can reach right into the brain of another person and influence her thoughts. In the extreme case, a baby smiling causes particular circuits in its mother's brain to fire and make her feel good. No other animal smiles in this way. But even among adults, a touch, a massage, or, as experiments have shown, a simple act of financial generosity, can cause the release of the hormone oxytocin in the brain of the recipient, and oxytocin is the chemical that evolution uses to make mammals feel good about each other – whether parents about their babies, lovers about their mates or friends about their friends. It works the other way, too: squirting oxytocin up the noses of students will cause them to trust strangers with their money more readily than those who receive a placebo squirted up their noses. 'Oxytocin is a physiologic signature of empathy,' says the neuroeconomist Paul Zak, who conducts these experiments, 'and appears to induce a temporary attachment to others.'

In 2004 Zak, together with Ernst Fehr and other colleagues, conducted one of the most revealing experiments in the history of economics, which showed just how specific the trusting effect of oxytocin is. They recruited 194 male students from Zurich (the experiment must not be done with females, because if one happens to be pregnant without knowing it, oxytocin might trigger labour) and made them play one of two games. In the first game, the trust game, a player called the investor is given twelve monetary units and told that if he hands some of it over to another player, the trustee, that amount will be quadrupled by the experimenter. Thus if he hands over all twelve units, the trustee will receive forty-eight. The trustee may pay some of it back to the investor, but has absolutely no obligation to do so. So the investor risks losing all his money, but if he can trust the trustee to be generous, he might stand to make a good profit. The question is: how much will the investor hand over?

The results were remarkable. Investors who receive a squirt of oxytocin up their noses before the experiment begins hand over 17 per cent more money than those who receive a squirt of inert saline solution up their noses, and the median transfer is ten units rather than eight. The oxytocin investors are more than twice as likely to hand over the full twelve units as the controls. Yet oxytocin has no such effect on the back transfers offered by the trustees, who are just as generous without oxytocin as with. So – as animal experiments have suggested – oxytocin does not affect reciprocity, just the tendency to take a social risk, to go out on a limb. Moreover, a second game, identical to the first except that the generosity of the trustees is randomly decided, shows no effect of oxytocin on the investors. So oxytocin specifically increases trusting, rather than general risk-taking. As with lovers and mothers, the hormone enables animals to take the risk of approaching other members of the species – it 'links the overcoming of social avoidance with the activation of brain circuits implicated in reward'. It does this partly by suppressing the activity of the amygdala, the organ that expresses fear. If human economic progress has included a crucial moment when human beings learned to treat strangers as trading partners, rather than enemies, then oxytocin undoubtedly played a vital role.

People are surprisingly good at guessing who to trust. Robert Frank and his colleagues set up an experiment in which the volunteer subjects had conversations in groups of three for half an hour. After that, they were sent to separate rooms to play, with their conversation partners, the prisoner's dilemma game (in which each player must decide whether to cooperate in the hope of a mutual gain or defect in the hope of a selfish gain if the other player cooperates). First, though, each player filled in a form not only saying how she would play with each partner, but also predicting what strategy each partner would adopt. As so often in this game, three-quarters of subjects said they would cooperate, reinforcing Smith's point that people are innately nice (economics students, who have been taught the self-interested nature of human beings, are twice as likely to defect!). Remarkably, the subjects were very good at predicting who would cooperate and who would defect: people who were pre-dicted to cooperate did so 81 per cent of the time, compared with 74 per cent for the group as a whole. People who were predicted to defect did so 57 per cent of the time, compared with 26 per cent for the group as a whole. Most people, says the economist Robert Frank, can think of an unrelated friend who they would trust to return to them a wallet that had been lost in a crowded concert. Conversely, people acutely remember the faces of those who cheat them.

Thus, the entire edifice of human cooperation and exchange, upon which prosperity and progress are built, depends on a fortunate biological fact. Human beings are capable of empathy, and are discerning trusters. Is that it, then? That human beings can build complicated societies and experience prosperity is down to the fact that they have a biological instinct that encourages cooperation? If only it were that simple. If only the arguments of Hobbes and Locke, of Rousseau and Voltaire, of Hume and Smith, of Kant and Rawls, could be brought to such a neat and reductionist conclusion. However, the biology is only the start. It is something that makes prosperity possible, but it is not the whole explanation.

Besides, there is still no evidence that any of this biology is uniquely developed in human beings. Capuchin monkeys and chimpanzees are just as resentful of unfair treatment as human beings are and just as capable of helpful acts towards kin or group members. The more you look at altruism and co-operation, the less uniquely human it appears. Oxytocin is common to all mammals, and is used for mother-love in sheep and lover-love in voles, so the chances are that it is available to underpin trust in almost any social mammal. It is necessary, but not sufficient to explain the human propensity to exchange. On the other hand, it is highly likely that during the past 100,000 years human beings have developed peculiarly sensitive oxy-tocin systems, much more ready to fire with sympathy, as a result of natural selection in a trading species. That is to say, just as the genes for digesting milk as an adult have changed in response to the invention of dairying, so the genes for flushing your brain with oxytocin have probably changed in response to population growth, urbanisation and trading – people have become oxytocin-junkies far more than many other animals.

Moreover, finding the underlying physiology of trust does little to explain why some human societies are much better at generating trust than others. As a broad generalisation, the more people trust each other in a society, the more prosperous that society is, and trust growth seems to precede income growth. This can be measured by a combination of questionnaires and experiments – leaving a wallet on the street and seeing if it is returned, for instance. Or asking people, in their native tongue, 'generally speaking, would you say that most people can be trusted, or that you cannot be too careful in dealing with people?' By these measures, Norway is heaving with trust (65 per cent trust each other) and wealthy, while Peru is wallowing in mistrust (5 per cent trust each other) and poor. 'A 15% increase in the proportion of people in a country who think others are trustworthy,' says Paul Zak, 'raises income per person by 1% per year for every year thereafter.' This is most unlikely to be because Norwegians have more oxytocin receptors in their brains than Peruvians, but it does suggest that Norwegian society is better designed to elicit the trust systems than Peruvian.

It is not at all clear what comes first: the trust instinct or trade. It is most unlikely that the oxytocin system fortuitously mutated into a sensitive form, which then enabled human beings to develop trading. Much more plausibly, human beings began tentatively to trade, capturing the benefits of comparative advantage and collective brains, which in turn encouraged natural selection to favour mutant forms of the human mind that were especially capable of trust and empathy – and even then to do so cautiously and suspiciously. I shall be amazed if the genetics of the oxytocin system do not show evidence of having changed rapidly and recently in response to the invention of trade, by gene-culture co-evolution.

## The shadow of the future

A trillion generations of unbroken parental generosity stand behind a bargain with your mother. A hundred good experiences stand behind your reliance on a friend. The long shadow of the future hangs over any transaction with your local shopkeeper. He surely knows that in making a quick buck now by ripping you off he risks losing all future purchases you might make. What is miraculous is that in modern society you can trust and be trusted by a shopkeeper you do not know. Almost invisible, the guarantors of trust lurk beneath every modern market transaction: the sealed packaging, the warranty, the customer feedback form, the consumer legislation, the brand itself, the credit card, the 'promise to pay the bearer' on the money. When I go into a well-known supermarket and pick up a tube of toothpaste of a well-known brand, I do not need to open the package and squirt a little toothpaste on to my finger to test that the tube is not filled with water; I do not even need to know that the shop is subject to laws that would prosecute it for selling false goods. I just need to know that this big retailing company, and the big company that made the toothpaste, are both keen to keep me coming back year after year, that the shadow of reputational risk hangs over this simple transaction, ensuring that I can trust this toothpaste seller without a moment's thought.

There is a vast history behind the trustworthiness of a tube of toothpaste, a long path of building trust inch by inch. Once that path is trodden, though, trust can be borrowed for new products and new media with surprising ease. The remarkable thing about the early days of the internet was not how hard it proved to enable people to trust each other in the anonymous reaches of the ether, but how easy. All it took was for eBay to solicit feedback from customers after each transaction and post the comments of buyers about the sellers. Suddenly every deal lay under the shadow of the future; suddenly, every eBay user felt the hot breath of reputation on his neck as surely as a Stone Age reindeer hide salesman returning to a trading place after selling a rotten hide the year before. When Pierre Omidyar founded eBay, few believed as he did that trust between anonymous strangers would prove easy to create in the new medium. But by 2001, fewer than 0.01 per cent of all transactions on the site were fraud attempts. John Clippinger draws an optimistic conclusion: 'The success of trust-based peer organizations such as eBay, Wikipedia, and the open-source movement, indicates that trust is a highly expandable network property.' Perhaps the internet has returned us to a world a bit like the Stone Age in which there is no place for a fraudster to hide.

That response would be naïve. There is plenty of innovative and destructive cyber-crime to come. None the less, the internet is a place where the problem of trust between strangers is solved daily. Viruses can be avoided, spam filters can work, Nigerian emails that con people into divulging their bank account details can be marginalised, and as for the question of trust between buyer and seller, companies like eBay have enabled their customers to police each other's reputations by the simple practice of feedback. The internet, in other words, may be the best forum for crime, but it is also the best forum for free and fair exchange the world has ever seen.

My point is simply this: with frequent setbacks, trust has gradually and progressively grown, spread and deepened during human history, because of exchange. Exchange breeds trust as much as vice versa. You may think you are living in a suspicious and dishonest world, but you are actually the beneficiary of immense draughts of trust. Without that trust the swapping of fractions of labour that goes to make people richer could not happen. Trust matters, said J.P. Morgan to a congressional hearing in 1912, 'before money or anything else. Money cannot buy it ... because a man I do not trust could not get money from me on all the bonds in Christendom.' Google's code of conduct echoes Morgan: 'Trust is the foundation upon which our success and prosperity rest, and it must be re-earned every day, in every way, by every one of us.' (And, yes, one day people will probably look back on Google's founders as robber barons, too.) If people trust each other well, then mutual service can evolve with low transactional friction; if they do not, then prosperity will seep away. That is, of course, a large part of the story of the banking crisis of 2008. Banks found themselves holding bits of paper that told lies – that said they were worth far more than they were. Transactions collapsed.

## If trust makes markets work, can markets generate trust?

A successful transaction between two people – a sale and purchase – should benefit both. If it benefits one and not the other, it is exploitation, and it does nothing to raise the standard of living. The history of human prosperity, as Robert Wright has argued, lies in the repeated discovery of non-zero-sum bargains that benefit both sides. Like Portia's mercy in *The Merchant of Venice*, exchange is 'twice blest: it blesseth him that gives and him that takes.' That's the Indian rope trick by which the world gets rich. Yet it takes only a few sidelong glances at your fellow human beings to realise that remarkably few people think this way. Zero-sum thinking dominates the popular discourse, whether in debates about trade or in complaints about service providers. You just don't hear people coming out of shops saying, 'I got a great bargain, but don't worry, I paid enough to be sure that the shopkeeper feeds his family, too.' Michael Shermer thinks that is because most of the Stone Age transactions rarely benefited both sides: 'during our evolutionary tenure, we lived in a zero-sum (win-lose world), in which one person's gain meant another person's loss'.

This is a shame, because the zero-sum mistake was what made so many -isms of past centuries so wrong. Mercantilism said that exports made you rich and imports made you poor, a fallacy mocked by Adam Smith when he pointed out that Britain selling durable hardware to France in exchange for perishable wine was a missed opportunity to achieve the 'incredible augmentation of the pots and pans of the country'. Marxism said that capitalists got rich because workers got poor, another fallacy. In the film *Wall Street*, the fictional Gordon Gekko not only says that greed is good; he also adds that it's a zero-sum game where somebody wins and somebody loses. He is not necessarily wrong about some speculative markets in capital and in assets, but he is about markets in goods and services. The notion of synergy, of both sides benefiting, just does not seem to come naturally to people. If sympathy is instinctive, synergy is not.

For most people, therefore, the market does not feel like a virtuous place. It feels like an arena in which the consumer does battle with the producer to see who can win. Long before the credit crunch of 2008 most people saw capitalism (and therefore the market) as necessary evils, rather than inherent goods. It is almost an axiom of modern debate that free exchange encourages and demands selfishness, whereas people were kinder and gentler before their lives were commercialised, that putting a price on everything has fragmented society and cheapened souls. Perhaps this lies behind the extraordinarily widespread view that commerce is immoral, lucre filthy and that modern people are good despite being enmeshed in markets rather than because of it – a view that can be heard from almost any Anglican pulpit at any time. 'Marx long ago observed the way in which unbridled capitalism became a kind of mythology, ascribing reality, power and agency to things that had no life in themselves,' said the Archbishop of Canterbury in 2008.

Like biological evolution, the market is a bottom-up world with nobody in charge. As the Australian economist Peter Saunders argues, 'Nobody planned the global capitalist system, nobody runs it, and nobody really comprehends it. This particularly offends intellectuals, for capitalism renders them redundant. It gets on perfectly well without them.' There is nothing new about this. The intelligentsia has disdained commerce throughout Western history. Homer and Isaiah despised traders. St Paul, St Thomas Aquinas and Martin Luther all considered usury a sin. Shakespeare could not bring himself to make the persecuted Shylock a hero. Of 1900, Brink Lindsey writes: 'Many of the brightest minds of the age mistook the engine of eventual mass deliverance – the competitive market system – for the chief bulwark of domination and oppression.' Economists like Thorstein Veblen longed to replace the profit motive with a combination of public-spiritedness and centralised government decision-taking. In the 1880s Arnold Toynbee, lecturing working men on the English industrial revolution which had so enriched them, castigated free enterprise capitalism as a 'world of gold-seeking animals, stripped of every human affection' and 'less real than the island of Lilliput'. In 2009 Adam Phillips and Barbara Taylor argued that 'capitalism is no system for the kind-hearted. Even its devotees acknowledge this while insisting that, however tawdry capitalist motives may be, the results are socially beneficial.' As the British politician Lord Taverne puts it, speaking of himself: 'a classical education teaches you to despise the wealth it prevents you from earning.'

But both the premise and the conclusion are wrong. The notion that the market is a necessary evil, which allows people to be wealthy enough to offset its corrosive drawbacks, is wide of the mark. In market societies, if you get a reputation for unfairness, people will not deal with you. In places where traditional, honour-based feudal societies gave way to commercial, prudence-based economies – say, Italy in 1400, Scotland in 1700, Japan in 1945 – the effect is civilising, not coarsening. When John Padgett at the University of Chicago compiled data on the commercial revolution in fourteenth-century Florence, he found that far from self-interest increasing, it withered, as a system of 'reciprocal credit' emerged in which business partners gradually extended more and more trust and support to each other. There was a 'trust explosion'. 'Wherever the ways of man are gentle, there is commerce, and wherever there is commerce, the ways of men are gentle,' observed Charles, Baron de Montesquieu. Voltaire pointed out that people who would otherwise have tried to kill each other for worshipping the wrong god were civil when they met on the floor of the Exchange in London. David Hume thought commerce 'rather favourable to liberty, and has a natural tendency to preserve, if not produce a free government' and that 'nothing is more favourable to the rise of politeness and learning, than a number of neighbouring and independent states, connected together by commerce and policy'. It dawned on Victorians such as John Stuart Mill that a rule of Rothschilds and Barings was proving rather more pleasant than one of Bonapartes and Habsburgs, that prudence might be a less bloody virtue than courage or honour or faith. (Courage, honour and faith will always make better fiction.) True, there was always a Rousseau or a Marx to carp, and a Ruskin or a Goethe to scoff, but it was possible to wonder, with Voltaire and Hume, if commercial behaviour might make people more moral.

## Coercion is the opposite of freedom

Perhaps Adam Smith was right, that in turning strangers into honorary friends, exchange can transmute base self-interest into general benevolence. The rapid commercialisation of lives since 1800 has coincided with an extraordinary improvement in human sensibility compared with previous centuries, and the process began in the most commercial nations, Holland and England. Unimaginable cruelty was commonplace in the pre-commercial world: execution was a spectator sport, mutilation a routine punishment, human sacrifice a futile tragedy and animal torture a popular entertainment. The nineteenth century, when industrial capitalism drew so many people into dependence on the market, was a time when slavery, child labour and pastimes like fox tossing and cock fighting became unacceptable. The late twentieth century, when life became still more commercialised, was a time when racism, sexism and child molesting became unacceptable. In between, when capitalism gave way to various forms of state-directed totalitarianism and their pale imitators, such virtues were noticeable by their retreat – while faith and courage revived. The twenty-first century, when commercialisation has so far continued to spread, is already a time when battery farming and unilaterally declaring war have just about become unacceptable. Random violence makes the news precisely because it is so rare; routine kindness does not make the news precisely because it is so commonplace.

Charitable giving has been growing faster than the economy as a whole in recent decades. The internet reverberates with people sharing tips for free.

Of course, these trends could be nothing more than coincidence: we happen to be becoming nicer as we become more irretrievably dependent on markets and free enterprise. But I do not think so. It was the 'nation of shopkeepers' that first worried about abolishing slave trading, emancipating Catholics and feeding the poor. Just as it was the nouveau riche merchants, with names like Wedgwood and Wilberforce, who financed and led the anti-slavery movement before and after 1800, while the old county money looked on with indifference, so today it is the new money of entrepreneurs and actors that funds compassion for people, pets and planets. There is a direct link between commerce and virtue. 'Far from being a vice,' says Eamonn Butler, 'the market system makes self interest into something thoroughly virtuous.' This is the extraordinary feature of markets: just as they can turn many individually irrational individuals into a collectively rational outcome, so they can turn many individually selfish motives into a collectively kind result.

For instance, as evolutionary psychologists confirm, sometimes the motivation behind conspicuous displays of virtue by the very rich are far from pure. When shown a photograph of an attractive man and asked to write a story about an ideal date with him, a woman will say she is prepared to spend time on conspicuous pro-social volunteering. By contrast, a woman shown a photograph of a street scene and asked to write about ideal weather for being there, shows no such sudden urge to philanthropy. (A man in the same 'mating-primed' condition will want to spend more on conspicuous luxuries, or on heroic acts.) That Charles Darwin's wealthy spinster aunt Sarah Wedgwood's funding of the anti-slavery movement (she was the movement's biggest donor) may have a hint of unconscious sexual motives, is a charming surprise. But it does not detract from the good she did, or from the fact that commerce paid for that good.

This applies among the poor as well as the rich. The working poor give a much higher proportion of their income to good causes than the rich do, and crucially they give three times as much as people on welfare do. As Michael Shermer comments, 'Poverty is not a barrier to charity, but welfare is.' Those of libertarian bent often prove more generous than those of a socialist persuasion: where the socialist feels that it is government's job to look after the poor using taxes, libertarians think it is their duty. I am not saying that the market is the only source of charity. Clearly not: religion and community provide much motivation to philanthropy too. But the idea that the market destroys charity by teaching selfishness is plainly wide of the mark. When the market economy booms so does philanthropy. Ask Warren Buffett and Bill Gates.

It is not just cruelty and indifference to the disadvantaged that have retreated with the spread of the collective brain. So has illiteracy and ill health. So has crime: your chances of being murdered have fallen steadily since the seventeenth century in every European country, but once again beginning with the trade-mad Holland and England. Murder was ten times as common before the industrial revolution in Europe, per head of population, as it is today. The fall in crime rates turned into a plummet at the turn of the twenty-first century – and use of illegal drugs fell too. So has pollution, which was far worse under communist regimes than in the free-market, democratic West. There is now a pretty well established rule of thumb (known as the environmental Kuznets curve) that when per capita income reaches about $4,000, people demand a clean-up of their local streams and air. Universal access to education came about during a time when Western societies were unusually devoted to free enterprise. Flexible working hours, occupational pensions, safety at work – all of these improved in the postwar West because people were enriching themselves and demanding higher standards, as much as because higher standards were imposed on recalcitrant firms by saintly politicians; the decline in workplace accidents was just as steep before the occupational safety and health act as after it. Again, some of these trends might have happened anyway, without the commercialisation of life, but don't bet on it. The taxes that paid for sewers were generated by commerce.

Commerce is good for minorities, too. If you don't like the outcome of an election you have to lump it; if you don't like your hairdresser, you can find another. Political decisions are by definition monopolistic, disenfranchising and despotically majoritarian; markets are good at supplying minority needs. The other day I bought a device for attaching a fly-fishing rod to my car. How long would I have had to wait in 1970s Leningrad before some central planner had the bright idea of supplying such a trivial need? The market found it. Moreover, thanks to the internet, the economy is getting better and better at meeting the desires of minorities. Because the very few people in the world who need fishing rod attachments or books on fourteenth-century suicide can now find suppliers on the web, niches are thriving. The 'long tail' of the distribution – the very many products that are each wanted by very few, rather than vice versa – can be serviced more and more easily.

Freedom itself owes much to commerce. The great drive to universal suffrage, religious tolerance and female emancipation began with pragmatic enthusiasts for free enterprise, like Ben Franklin, and was pressed forward by the urban bourgeoisie as a response to economic growth. Right into the twentieth century tsars and general secretaries found it an awful lot easier to dictate a tyranny of peasants than a demos of bourgeois consumers. Parliamentary reform began in Britain in the 1830s because of the grotesque under-representation of the growing manufac-turing towns. Even Marx was subsidised by Engels's father's textile mill. It was the now-unfashionable philosopher Herbert Spencer who insisted that freedom would increase along with commerce. 'My aim,' he wrote in 1842 (anticipating John Stuart Mill by nine years), 'is the liberty of each limited alone by the like liberty of all.' Yet he foresaw that the battle to persuade leaders not to believe in coercion was far from over: 'Though we no longer coerce men for their spiritual good, we still think ourselves called upon to coerce them for their material good: not seeing that the one is as unwarrantable as the other.' The inherent illiberalism of the bureaucracy, not to mention its tendency to corruption and extravagance, was a threat Spencer warned against in vain.

A century later, the gradual dismantling of apartheid and segregation was helped by commercialisation, too. The American civil rights movement drew its strength partly from a great economic migration. More African-Americans left the South between 1940 and 1970 than Poles, Jews, Italians or Irish had arrived in America as immigrants during their great migrations. Lured by better jobs or displaced by mechanical cotton pickers, black share-croppers came to the cities of the industrial North and began to discover their economic and political voice. They then began to challenge the system of prejudice and discrimination they had left behind. The first victory along that road was an exercise in consumer power – the Montgomery bus boycott of 1955–6.

The sexual and political liberation of women in the 1960s followed directly their domestic liberation from the kitchen by labour-saving electrical machinery. Lower-class women had always worked for wages – tilling in fields, sewing in sweatshops, serving in parlours. Among the upper-middle classes, though, it was a badge of rank, handed down from the feudal past, to be or to have a non-working (or at least housekeeping) wife. In the 1950s many suburban men, returning from war, found they too could afford such an accessory, and many women were pressured into giving their battleship-welding jobs back to men. In the absence of economic change, that is probably how it would have stayed, but soon the opportunities to work outside the home grew as the time spent on increasingly mechanised housework dwindled, and it was this, as much as any political awakening, that enabled the feminist movement to gain traction in the 1960s.

The lesson of the last two centuries is that liberty and welfare march hand in hand with prosperity and trade. Countries that lose their liberty to tyrants today, through military coups, are generally experiencing falling per capita income at an average rate of 1.4 per cent at the time – just as it was falling per capita income that helped turn Russia, Germany and Japan into dictatorships between the two world wars. One of the great puzzles of history is why this did not happen in America in the 1930s, where on the whole pluralism and tolerance not only survived the severe economic shocks of the 1930s, but thrived. Perhaps it nearly did happen: Father Coughlin tried, and had Roosevelt been more ambitious or the constitution weaker, who knows where the New Deal might have led? Perhaps some democracies were just strong enough for their values to survive. Today there is much argument about whether democracy is necessary for growth, China seeming to prove that it is not. But there can be little doubt that China would – indeed may yet – see either more revolution or more repression if its growth rate were to fall to nothing.

I am happy to cheer, with Deirdre McCloskey: 'Hurrah for late twentieth-century enrichment and democratisation. Hurrah for birth control and the civil rights movement. Arise ye wretched of the earth'. Interdependence through the market made these things possible. Politically, as Brink Lindsey has diagnosed, the coincidence of wealth with toleration has led to the bizarre paradox of a conservative movement that embraces economic change but hates its social consequences and a liberal movement that loves the social consequences but hates the economic source from which they come. 'One side denounced capitalism but gobbled up its fruits; the other cursed the fruits while defending the system that bore them.'

Contrary to the cartoon, it was commerce that freed people from narrow materialism, that gave them the chance to be different. Much as the intelligentsia continued to despise the suburbs, it was there that tolerance and community and voluntary organisation and peace between the classes flourished; it was there that the refugees from cramped tenements and tedious farms became rights-conscious consumers – and parents of hippies. For it was in the suburbs that the young, seizing their economic independence, did something other than meekly follow father and mother's advice. By the late 1950s, teenagers were earning as much as whole families had in the early 1940s. It was this prosperity that made Presley, Ginsberg, Kerouac, Brando and Dean resonate. It was the mass affluence of the 1960s (and the trust funds it generated) that made possible the dream of free-love communes. Just as material progress subverts the economic order, so it also subverts the social order – ask Osama bin Laden, the ultimate spoilt rich kid.

## The corporate monster

Yet for all the liberating effects of commerce, most modern commentators see a far greater threat to human freedom from the power of corporations that free markets inevitably throw up. The fashionable cultural critic sees himself or herself as David slinging stones at vast, corrupt and dehumanising Goliath-like corporations that punish, pollute and profiteer with impunity. To my knowledge, no large company has yet featured in a Hollywood movie without its boss embarking on a sinister plot to kill people (in the latest one I watched, Tilda Swinton somewhat predictably tried to kill George Clooney for exposing her company's poisoning of people with pesticides). I hold no brief for large corporations, whose inefficiencies, complacencies and anti-competitive tendencies often drive me as crazy as the next man. Like Milton Friedman, I notice that 'business corporations in general are not defenders of free enterprise. On the contrary, they are one of the chief sources of danger.' They are addicted to corporate welfare, they love regulations that erect barriers to entry to their small competitors, they yearn for monopoly and they grow flabby and inefficient with age.

But I detect that the criticism is increasingly out of date, and that large corporations are ever more vulnerable to their nimbler competitors in the modern world – or would be if they were not granted special privileges by the state. Most big firms are actually becoming frail, fragile and frightened – of the press, of pressure groups, of government, of their customers. So they should be. Given how frequently they vanish – by take-over or bankruptcy – this is hardly surprising. Coca-Cola may wish its customers were 'serfs under feudal brandlords', in the words of one critic, but look what happened to New Coke. Shell may have tried to dump an oil-storage device in the deep sea in 1995, but a whiff of consumer boycott and it changed its mind. Exxon may have famously stood out from the consensus by funding scepticism of climate change (while Enron funded climate alarmism) – but by 2008 it had been bullied into recanting.

Companies have a far shorter half-life than government agencies. Half of the biggest American companies of 1980 have now disappeared by take-over or bankruptcy; half of today's biggest companies did not even exist in 1980. The same is not true of government monopolies: the Internal Revenue Service and the National Health Service will not die, however much incompetence they might display. Yet most anti-corporate activists have faith in the good will of the leviathans that can force you to do business with them, but are suspicious of the behemoths that have to beg for your business. I find that odd.

Moreover, for all their eventual sins, entrepreneurial corporations can do enormous good while they are young and growing. Consider the case of discount retailing. The burst of increasing productivity that countries like America and Britain rather unexpectedly experienced in the 1990s at first puzzled many economists. They wanted to credit computers, but as the economist Robert Solow had quipped in 1987, 'you can see the computer everywhere but in the productivity statistics', and those of us who experienced how easy it was to waste time using a computer in those days agreed. A study by McKinsey concluded that the 1990s surge in the United States was caused by (drum roll of excitement) logistical changes in business (groan of disappointment), especially in the retail business and especially in just one firm – Wal-Mart. Efficient ordering, ruthless negotiating, hyper-punctual time keeping (suppliers must sometimes hit a thirty-second window for deliveries), merciless cost control and ingenious responses to customers' preferences had given Wal-Mart a 40 per cent efficiency advantage over its competitors by the early 1990s. Wal-Mart's competitors rapidly followed suit, raising their own productivity by 28 per cent in the later 1990s, but Wal-Mart had not stood still, gaining another 22 per cent in the same time, even as it opened an average of seven new three-acre supercentres a month for a decade. According to Eric Beinhocker of McKinsey, these 'social-technology' innovations in the retail sector alone accounted for fully a quarter of all United States productivity growth. Tesco probably had a similar effect in Britain.

Sam Walton's determination in 1950s Arkansas to sell everyday items for less than his competitors was hardly a new idea. It is difficult to describe it as an innovation, although things like 'cross-docking' where goods go from suppliers' trucks to distributor's trucks without spending time in warehouses in between were indeed new. Yet the way in which he pursued and resolutely stuck to that simple idea ended up delivering a huge boost to American living standards. Like corrugated iron and container shipping, discount merchandising is among the most unsophisticated yet enriching innovations of the twentieth century. A single, routine, minuscule Wal-Mart decision in the 1990s – not to sell deodorant in cardboard boxes – saved America $50 million a year, half of which was passed on to customers. Charles Fishman writes: 'Whole forests have not fallen in part because of a decision made in the Wal-Mart home office ... to eliminate the [deodorant] box.'

On average, when it lands in a town, Wal-Mart causes a 13 per cent drop in its competitors' prices and saves its customers nationally $200 billion a year. Yet critics of corporate giants, who normally complain about profiteering, still disapprove of Wal-Mart, saying the low prices are a bad thing because smaller businesses can't compete or that Wal-Mart is 'the world's largest sweatshop' for paying low wages even though Wal-Mart pays twice the minimum wage (and as I was writing this announced $2 billion in bonuses to staff, despite the recession, because of record sales). It is true that the growth of Wal-Mart in the 1990s, just like the opening of a new Wal-Mart in a certain town, created turmoil. Competitors went bust or were forced into humiliating mergers. Suppliers found themselves driven to new practices. Unions lost their leverage over retailing workforces. Cardboard box makers went to the wall. Consumers changed their habits. Innovation, whether in the form of new technology or new ways of organising the world, can destroy as well as create. A Wal-Mart store drives small general retailers out of business as surely as the computer drove the typewriter out of business. But against this must be balanced the enormous benefits that (especially the poorest) customers reap in terms of cheaper, more varied and better goods.

It was Joseph Schumpeter who pointed out that the competition which keeps a businessman awake at night is not that from his rivals cutting prices, but that of entrepreneurs making his product obsolete. As Kodak and Fuji slugged it out for dominance in the 35mm film industry in the 1990s, digital photography began to extinguish the entire market for analogue film – as analogue records and analogue video cassettes had gone before. Creative destruction, Schumpeter called it. His point was that there is just as much creation going on as destruction – that the growth of digital photography would create as many jobs in the long run as were lost in analogue, or that the savings pocketed by a Wal-Mart customer are soon spent on other things, leading to the opening of new stores to service those new demands. In America, roughly 15 per cent of jobs are destroyed every year; and roughly 15 per cent created.

## Commerce and creativity

This turnover in itself ensures a steady improvement in working conditions. From Josiah Wedgwood, proud of conditions in his Etruria pottery factory, via Henry Ford, doubling the wages of his employees in 1914 to reduce staff turnover, to Larry Page, idealistically designing the Googleplex, each generation of entrepreneurs often tries to make work a better experience for their employees. In the early days of the internet, eBay was just one of many online auction companies. It succeeded where its competitors failed because it realised that a sense of shared community, not a competitive auction process, was key. 'This isn't about auctions,' said Meg Whitman, the chief executive of eBay, 'in fact it's not about economic warfare. It's the opposite.' It was survival of the nicest.

The turnover of firms is accelerating so much that most criticism of corporations is out of date already. Large companies not only fall more often these days – the disappearance in a month in 2008 of many banking names is merely an accelerated case in a particular industry – but increasingly they fragment and decentralise, too. As islands of top-down planning in a bottom-up sea, big companies have less and less of a future (the smaller the scale, the better planning works). AIG and General Motors may have been kept alive by taxpayers, but they are in corporate comas. The stars of the modern market economy are as different from the giants of industrial capitalism, eBay from Exxon, as capitalism is from socialism. Nike, born in 1972, grew into a huge company merely by contracting between factories in Asia and shops in America from a relatively small head office. Wikipedia has a paid staff of fewer than thirty and makes no profit. Whereas the typical firm was once a team of workers, hierarchically arranged and housed on a single site, increasingly it is a nebulous and ephemeral coming together of creative and marketing talent to transmit the efforts of contracting individuals towards the satisfying of consumer preferences.

In that sense 'capitalism' is dying, and fast. The size of the average American company is down from twenty-five employees to ten in just twenty-five years. The market economy is evolving a new form in which even to speak about the power of corporations is to miss the point. Tomorrow's largely self-employed workers, clocking on to work online in bursts for different clients when and where it suits them, will surely look back on the days of bosses and foremen, of meetings and appraisals, of time sheets and trade unions, with amusement. I repeat: firms are temporary aggregations of people to help them do their producing in such a way as to help others do their consuming.

Nor can there be any doubt that the collective brain enriches culture and stimulates the spirit. The intelligentsia generally looks down on commerce as irredeemably philistine, conventional and lowering in its taste. But for anybody who thinks great art and great philosophy have nothing to do with commerce, let him visit Athens and Baghdad to ask how Aristotle and al-Khwarizmi had the leisure time to philosophise. Let him visit Florence, Pisa and Venice and inquire into how Michelangelo,

Galileo and Vivaldi were paid. Let him go to Amsterdam and London and ask what funded Spinoza, Rembrandt, Newton and Darwin. Where commerce thrives, creativity and compassion both flourish.

## Rules and tools

Even if the world is indeed becoming a more trusting and less violent place as it becomes more commercial, that does not mean that commerce is in itself either the only way to make the world trusting, or enough on its own to create trust. As well as new tools, there had to be new rules. The innovations that made the world nicer, it may be argued, are institutions, not technologies: things like the golden rule, the rule of law, respect for private property, democratic government, impartial courts, credit, consumer regulation, the welfare state, a free press, religious teaching of morality, copyright, the custom that you do not spit at the table and the convention that you always drive on the right (or left if in Japan, Britain, India, Australia and much of Africa). These rules made trustful, safe commerce possible, at least as much as vice versa.

The aborigines of Australia or the Khoisan of southern Africa lacked not only steel and steam when they first met Westerners; they also lacked courts and Christmas. Certainly, the imposition of a new rule has often enabled a society to capture the benefits of exchange and specialisation ahead of its rivals, and to better the lives of its citizens in moral as well as material ways. Looking around the world, there are plainly societies which manage their citizens' lives well with good rules and societies which manage their citizens' lives badly with bad rules. Good rules reward exchange and specialisation; bad rules reward confiscation and politicking. South and North Korea spring to mind. One is generally a fair and free place, where people are mostly becoming more rich and happy; the other an arbitrary, hungry and cruel place whence people are fleeing as desperate refugees whenever they can. The difference – which results in fifteen times as much prosperity per head for the South – is plainly in the way they are ruled, in their institutions. Later in this book I will argue that the wrong kind of government can be a disastrous long-term impoverishing factor – the Ming empire is my prime example. Zimbabwe today needs better rules before it can have better markets. But note here that a country's economic freedom predicts its prosperity better than its mineral wealth, education system or infrastructure do. In a sample of 127 countries, the sixty-three with the higher economic freedom had more than four times the income per capita and nearly twice the growth rate of the countries that did not.

A few years ago the World Bank published a study of 'intangible wealth' – trying to measure the value of education, the rule of law and other such nebulous things. It simply added up the natural capital (resources, land) and produced capital (tools, property) and measured what was left over to explain each country's per capita income. It concluded that Americans can draw upon more than ten times as much intangible capital as Mexicans, which explains why a Mexican who crosses the border can quadruple his productivity almost immediately. He has access to smoother institutions, clearer rules, better-educated customers, simpler forms – that sort of thing. 'Rich countries,' concluded the Bank, 'are largely rich because of the skills of their populations and the quality of the institutions supporting economic activity.' In some countries, intangible capital may be minute or even negative. Nigeria, for example, scores so low on the rule of law, education and the probity of its public institutions that even its immense oil reserves have failed to enrich it.

So perhaps I am wrong to seek the flywheel of human progress in the gradual development of exchange and specialisation. Perhaps they are symptoms, not causes, and it was the invention of institutions and rules that then made exchange possible. The rule against revenge killing, for example, must have greatly helped society to settle down. It must have been quite a break-through to say that 'do unto others' applies only to charity, not to homicide, and that handing the matter of revenge over to the state to pursue on your behalf through due process would be of general benefit to all. Both *Orestes* and *Romeo and Juliet* (and *The Godfather* and *Dirty Harry*, for that matter) capture societies in the act of wrestling with the issue: all can agree that the rule of law is better than the rule of reciprocal revenge, though it makes less good theatre, but not all can overcome their instincts and customs to achieve it.

True enough, but I see these rules and institutions as evolutionary phenomena, too, emerging bottom-up in society rather than being imposed top-down by fortuitously Solomonic rulers. They come through the filter of cultural selection just as surely as do technologies. And if you look at the history of, for instance, merchant law, you find exactly this: merchants make it up as they go along, turning their innovations into customs, ostracising those who break the informal rules and only later do monarchs subsume the rules within the laws of the land. That is the story of the *lex mercatoria* of the medieval period: the great law-giving kings of England, such as Henry II and John, were mostly codifying what their trading subjects had already agreed among themselves when trading with strangers in Bruges, Brabant and Visby. Indeed, it is the whole point of common law. When Michael Shermer and three friends started a bicycle race across America in the 1980s, they began with virtually no rules. Only with experience did they have to bring in rules about how to deal with being arrested for causing a traffic jam on a hill in Arizona and other such unexpected complications.

So while it is true that institutional innovators in the public sphere are just as vital as technological innovators in the private, I suspect that specialisation is the key to both. Just as becoming a specialist axe maker for the whole tribe gives you the time, the capital and the market to develop a new and better form of axe, so becoming the specialist bicycle racer enables you to make up rules about bicycle racing. Human history is driven by a co-evolution of rules and tools. The increasing specialisation of the human species, and the enlarging habit of exchange, are the root cause of innovation in both.

# CHAPTER 4

# The feeding of the nine billion: farming after 10,000 years ago

Whoever could make two ears of corn, or two blades of grass, to grow upon a spot of ground where only one grew before, would deserve better of mankind, and do more essential service to his country, than the whole race of politicians put together.

JONATHAN SWIFT
*Gulliver's Travels*

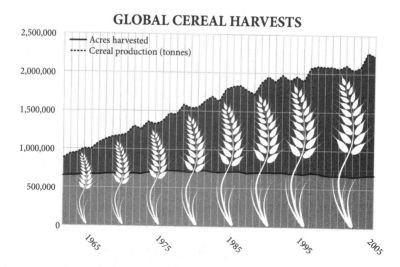

**GLOBAL CEREAL HARVESTS**

Oetzi, the mummified 'iceman' found high in the Alps in 1991, was carrying as much equipment on him as the hikers who found him. He had tools made of copper, flint, bone and six kinds of wood: ash, viburnum, lime, dogwood, yew and birch. He wore clothes made of woven grass, tree bark, sinew and four kinds of leather: bearskin, deer hide, goat hide and calf skin. He carried two species of fungus, one as medicine, another as part of a tinder kit that included a dozen plants and pyrite for making sparks. He was a walking encyclopedia of accumulated knowledge – knowledge of how to fashion tools and clothes and from what materials to make them. He carried the inventions of scores, perhaps thousands, of people upon him, their insights manifest in his kit. If he had had to invent from scratch all his equipment he would have had to be a genius. But even knowing what to make and how to make it, if Oetzi had spent his days collecting all the raw materials he needed for just his food and his clothing (let alone his shelter or his tools), he would have been stretched to breaking point, let alone if he then had to smelt, tan, weave, sew, shape and sharpen everything. He was undoubtedly consuming the labour of many other people, and giving his own in exchange.

He was also consuming the specialised labour of other species. Oetzi lived about 5,300 years ago in an Alpine valley. This was 2,000 years after agriculture reached southern Europe. Compared with his hunter-gatherer ancestors, Oetzi had cattle and goats that spent all day working for him gathering grass and turning it into leather and meat; wheat plants that gathered sunlight and turned it into grain. Under human genetic tutelage these species had grown specialised in doing so at the expense of their other biological imperatives. That is the point of agriculture: it diverts the labour of other species to providing services for human beings. The biologist Lee Silver was once watching chickens coming 'home to roost' in a village in southeast Asia and it struck him that they were like the farmer's tools:

they had been gathering food for him in the forest all day. Farming is the extension of specialisation and exchange to include other species.

Oetzi was also the beneficiary of capital investment. He lived right at the beginning of the metal age, when copper was first being smelted. His pristine copper axe, 99.7 per cent pure, had been smelted in a furnace that had consumed a lot of somebody's capital to build. The chaff in his clothing came from a grain crop grown with invested capital in the form of stored seeds and stored labour. For Adam Smith capital is 'as it were, a certain quantity of labour stocked and stored up to be employed, if necessary, upon some other occasion'.

If you can store the labour of others for future use, then you can spare yourself the time and the energy of working for your own immediate needs, which means you can invest in something new that will bring even greater reward. Once capital had arrived on the scene, innovation could accelerate, because time and property could be invested in projects that initially generated no benefit. Few hunter-gatherers, for example, could ever afford the time off 'work' to build a furnace and slowly and laboriously smelt enough metal to make a copper axe: they would starve in the meantime – even if they could find a market for the axes.

In the conventional account it was agriculture that made capital possible by generating stored surpluses and stored surpluses could be used in trade. Before farming, nobody could hoard a surplus. There is some truth in this, but to some degree it gets the story the wrong way round. Agriculture was possible because of trade. Trade provided the incentive to specialise in farmed goods and to generate surplus food.

Agriculture started to appear independently in the Near East, the Andes, Mexico, China, the highlands of New Guinea, the Brazilian rainforest and the African Sahel – all within a few thousand years. Something made it inevitable, almost compulsory around this time: however much it eventually resulted in misery, disease and despotism in the long run, it clearly gave its first practitioners competitive advantage. Yet farming was not an overnight transition. It was the culmination of a long, slow intensification of human diet that took tens of thousands of years. In search of extra calories people gradually 'moved down the trophic pyramid' – i.e., became more vegetarian. By 23,000 years ago the people of what is now Israel and Syria had become dependent on acorns, pulses and even grass seeds, as well as fish and birds, garnished with the occasional gazelle – perhaps supplied by other hunting tribes through trade. At one remarkable site, Ohalo II, now submerged except in dry years by Lake Kinneret (the Sea of Galilee), direct evidence has emerged of the eating of wild grains long before farming. In the remains of one of six brushwood huts, there is a flat stone apparently used for grinding seeds, and on it, preserved for 23,000 years by lake sediments, are microscopic starch grains from wild barley seeds. Nearby is what appears to be a stone oven for baking. By grinding grain to flour and baking it, the users would have nearly doubled the energy they could get from it.

So bread is far older than farming. It would be an astonishing 12,000 years after Ohalo II before anybody started planting and reaping cereals such as rye, wheat and barley, and 4,000 years after that before modern, genetically hexaploid wheat, with its heavy, free-threshing seeds, was invented – and began its long career as humankind's biggest and most widespread source of calories. The inescapable conclusion is that the people of the Near East were no fools. They captured the benefits of cereals – milled and baked starch – long before they took on the hard graft of farming them. Why spend months tending your own field of corn, when you can spend hours harvesting a wild one? One study notes an 'extreme reluctance to shift to domestic foods'.

By 13,000 years ago the people of the Near East, known now

as the Natufian culture, were using stone sickle blades to harvest the heads of the grasses, rather than beating the seeds into baskets. They lived in settlements that were sufficiently stable to be plagued by house mice. They were as close to farming as you can get without genetic domestication of crops. Yet, at this moment, on the brink of making history, they regressed. They abandoned their settlements, returned to nomadism and broadened their diet again. The same happened in Egypt about the same time – a retreat from grinding grain to hunting and fishing (except in Egypt's case it was much longer before the proto-farming experiment resumed). The probable cause of this hiatus was a cold snap, over a thousand years long, known as the 'Younger Dryas'. The probable cause of the cold snap was the North Atlantic suddenly cooling either from the bursting of a series of vast ice dams on the North American continent, or from the sudden outflow of water from the Arctic ocean. Once the cold snap had begun, not only was it colder and drier, but the weather fluctuated wildly from year to year, with changes of up to seven degrees in a single decade. Unable to rely on local rainfall, or local summer ripening, the people could not sustain their intensive cereal-feeding lifestyle. They must have starved in great numbers, and the survivors took to nomadic hunter-gathering again.

Then, around 11,500 years ago the temperature of the Greenland ice cap shot up by ten degrees (centigrade) in half a century; throughout the world conditions became dramatically warmer, wetter and more predictable. In the Levant intensi-fication of cereal use could resume, the Natufians could return to settled homes and soon something prompted somebody to start deliberately saving seed to plant. Chickpeas may have been the first crop, then rye and einkorn wheat, though figs had probably been cultivated and dogs domesticated some millennia before. Can there be any doubt that it was woman, the diligent gatherer, rather than man, the dilettante hunter, who first had the idea of sowing grain? A well planted crop, sown into riverbank mud or some other bare land, then carefully weeded and guarded from birds, would have meant new and harder work, but would have brought rewards in yield to the family of the woman who tried it. It would have brought a surplus of flour that could be exchanged with hunters for meat, so it would have kept not only the field's owner and her children alive, but perhaps a couple of other hunting families too. The exchange of grain for meat effectively subsidised hunting, or raised the 'price' of meat, putting more pressure on the hares and gazelles and so gradually making the entire settlement more dependent on the farm – and bringing a new incentive to the first man who thought of raising an orphaned goat kid rather than eating it. Farming would have become a necessity for all the people living there, and the hunter-gatherer way of life would have gradually atrophied. It was undoubtedly a long and slow process: farmers supplemented their diet with hunted 'bushmeat' for many millennia after they first started cultivating the land. In most of North America, the natives combined crops with seasonal hunts. In parts of Africa, many still do.

The Fertile Crescent was probably the place where agriculture first took hold, and from there the habit gradually spread south to Egypt, west into Asia Minor and east to India, but farming was quickly invented in at least six other places in a short time, driven by the same ratchet of trade, population growth, stable climate and increasingly vegetarian intensification. Squashes and then peanuts were cultivated in Peru by 9,200 years ago, millet and rice in China by 8,400 years ago, maize in Mexico by 7,300 years ago, taro and bananas in New Guinea by 6,900 years ago, sunflowers in North America by 6,000 years ago, and sorghum in Africa by around the same time. This phenomenal coincidence, as bizarre as finding that an aborigine, an Inuit, a Polynesian and a Scotsman all invented steam engines in the same decade of the eighteenth century without contact of any kind, is explained by the stabilising climate after the ice age ended. In the words of a recent paper, 'agriculture was impossible during the last glacial, but compulsory in the Holocene'. It is no accident that modern Australia, with its unpredictable years of drought followed by years of wet, still looks a bit like that volatile glacial world. Australians were probably quite capable of farming: they knew how to grind grass seeds, burn the bush to improve kangaroo grazing and encourage favoured plants; and they certainly knew how to alter the flow of rivers to encourage and harvest eels. But they also knew, or found out the hard way, that farming does not work in a highly volatile climate.

## No farming without trade

One of the intriguing things about the first farming settlements is that they also seem to be trading towns. From 14,000 years ago, much-valued obsidian (volcanic glass) from the Cappadocian volcanoes in Anatolia was being transported south along the upper Euphrates, through the Damascus basin and down the Jordan Valley. Seashells from the Red Sea were going the other way. This is precisely where the first farming settlements are – at Catalhoyuk, Abu Hureyra and Jericho. Such settlements were sited in oases where springs of fresh water from the mountains spilled out on to the western edge of the desert: places where soil nutrients, moisture and sunshine came together nicely – and also places where people mixed with their neighbours because of trade. This is surprising only because it is easy to think of early farmers as sedentary, self-sufficient folk. But they were exchanging harder in this region than anywhere else, and it is a reasonable guess that one of the pressures to invent agriculture was to feed and profit from wealthy traders – to generate a surplus that could be exchanged for obsidian, shells or other more perishable goods. Trade came first.

In the 1960s, Jane Jacobs suggested in her book *The Economy of Cities* that agriculture was invented to feed the first cities, rather than cities being made possible by the invention of agriculture. This goes too far, and archaeologists have discredited the idea of urban centres preceding the first farms. The largest permanent settlements of hunter-gatherers cannot be described as urban even among the fishermen of the Pacific coast of North America. None the less, there was a germ of truth in her idea: the first farmers were already enthusiastic traders breaking free of subsistence through exchange, and farming was just another expression of trade.

In Greece, farmers arrived suddenly and dramatically around 9,000 years ago. Stone tools suggest that they were colonists from Anatolia or the Levant who probably came by boat deliberately seeking to colonise new land. Moreover, these very earliest Greek farmers were also apparently enthusiastic traders with each other and were very far from being self-sufficient: they relied upon specialist craftsmen to produce obsidian tools from raw material imported from elsewhere. This is once again not what conventional wisdom envisages. Trade comes first, not last. Farming works precisely because it is embedded in trading networks.

Some time later, at 7,600 years ago, farmers who were happily cultivating the fertile plains around the 'Euxine lake' suffered a rude shock, when rising sea levels burst over the Hellespont and flooded into the lake's basin, filling it at a rate of six inches a day till it became the modern Black Sea. Baffled refugees presumably fled up the Danube into the heart of Europe. Within just a few hundred years, they had reached the Atlantic coast, peopling all of the southern half of Europe with farmers, sometimes by infecting their neighbours with enthusiasm for the new trick of farming, but more often (so the genetic evidence suggests) displacing and violently overwhelming hunter-gatherers as they went. It took another thousand years to reach the Baltic, chiefly because fishermen inhabited that coast at high densities and had no need to start farming. The crops the farmers took with them changed little, despite the new conditions they encountered. Some crops, like lentils and figs, had to be left behind on the Mediterranean. Others, like emmer and einkorn wheat, adapted readily to the wetter and cooler lands of Northern Europe. By 5,000 years ago farmers had reached Ireland, Spain, Ethiopia and India.

Other descendants of the Black Sea refugees took to the plains of what is now Ukraine where they domesticated the horse and developed a new language, Indo-European, that would come to dominate the western half of the Eurasian continent, and of which Sanskrit and Gaelic are both descendants. It was also somewhere near the Baltic or the Black Sea between 6,000 and 10,000 years ago that a genetic mutation, substituting G for A in a control sequence upstream of a pigment gene called OCA2, gave adults blue eyes for the first time. It was a mutation that would eventually be inherited by nearly 40 per cent of Europeans. Because it went with unusually pale skin, it probably helped those people who were trying to live on vitamin-D-deficient grain in sunless northern climates: sunlight enables the body to synthesise vitamin D. The gene's frequency speaks of the fecundity of farmers.

One of the reasons that farming spreads so rapidly once it starts is that the first few crops are both more productive and more easily grown than later crops, so farmers are always happy to move on to virgin land. If you burn down a forest, you are left with a soft, friable soil seasoned with fertilising ash. All you need do is poke a digging stick into the ground and plant a seed and sit back and wait for it to grow. After a few years, however, the soil is compacted and needs breaking up with a hoe, and weeds have proliferated. If you now leave the ground fallow to allow the fertility to build up again, the tough roots of grasses need to be broken up and buried to make a good seedbed – and for that you need a plough and an ox to pull it. But the ox needs feeding, so you need pasture as well as arable land. No wonder that shifting agriculture – slash and burn – remains so much more popular with many tribal people in forests to this day. In Neolithic Europe, the smoke of fires must have hung heavy in the air as the expanding front of farming spread west. The carbon dioxide released by the fires may even have helped to warm the climate to its 6,000-years-ago balmy maximum, when the Arctic ice retreated from Greenland's northern coast in summer. This is because early farming used probably nine times as much land per head of population as farming does today, so the small populations of the day generated lots of carbon dioxide per head.

## Capital and metal

Wherever they went, the farmers also brought their habits: not just sowing, reaping and threshing, but baking, fermenting, hoarding and owning. Hunter-gatherers have to travel light; even if they are not seasonally nomadic, they must be ready to move at any time. Farmers, by contrast, have to store grain or protect herds or guard fields before they are harvested. The first person to plant a wheat field must have faced the dilemma of how to say 'This is mine; only I may harvest it.' The first signs of private property are the stamp seals of the Halaf people, 8,000 years ago on the borders of Syria and Turkey: similar seals were later used for denoting ownership. This land rush presumably left the remaining hunters baffled spectators as their game lands were carved up, possibly by 'poorer', more desperate people. Perhaps Cain was a farmer; Abel a hunter.

Meanwhile, as farming replaced gathering, so herding replaced hunting. The Neolithic settlements of the Middle East probably grew up as markets where shepherds from the hills could meet cereal farmers from the plains and exchange their surpluses. The hunter-gatherer market now became the herder-farmer market. Haim Ofek writes: 'On the human level, nothing could be more handy at the onset of agriculture than a well-established propensity to exchange, for nothing could better reconcile the need for specialisation in food production with the need for diversification in food consumption.'

Copper smelting was a practice that makes no sense for an individual trying to meet his own needs, or even for a self-sufficient tribe. It requires a stupendous effort to mine the ore and then by virtue of elaborate bellows to smelt it in a charcoal fire at more than 1,083 °C, just to produce a few ingots of a metal that is strong and malleable, but not very hard. Imagine: you have to cut wood and make charcoal from it, make ceramic crucibles for the smelting, dig and crush the ore, then mould and hammer the copper. Only by consuming the stored labour of others – by living off capital – could you even finish the job. Then, even if you can sell copper axes to other hunter-gatherers, the market is likely to be too small to make it worth your while setting up a smelting operation. But once agriculture has provided the capital, increased the density of people, and given them a good reason for chopping down trees, then there might be a market large enough to support a community of full-time copper smelters, so long as they can sell the copper to neighbouring tribes. Or, in the words of two theorists: 'The denser societies made possible by agriculture can realize considerable returns to better exploitation of the potential of co-operation, co-ordination and the division of labour.'

Hence, the invention of metal smelting was an almost inevitable consequence of the invention of agriculture (though some very early mining of pure copper-metal deposits around Lake Superior was apparently done by hunter-gatherers, perhaps supplying the almost agricultural salmon ranchers of the Pacific coast). Copper was produced throughout the Alps, where some of the best ores are to be found, but it was exported to the rest of Europe for several thousand years after Oetzi's death, only later being displaced by copper mined in Cyprus. A little more than a thousand years after Oetzi died, and a short distance to the west in the Mitterberg region of what is now Austria, there were settlements inhabited by people who apparently did little else but mine and smelt copper from lodes in the nearby mountains. Living in a cold mountain valley, they found it more profitable to make copper and exchange it for, say, meat and grain from the Danube plains, rather than to raise their own cattle. It seems not to have made them very rich – nor would Cornish tin, Peruvian silver, or for that matter Welsh coal enrich their miners in the millennia to come. Compared with the farmers on the Danube Plain, the Mitterberg copper miners left behind few ornaments or luxuries. But they were better off than they could be trying to live self-sufficiently in the mountains raising their own food. They were not supplying a need; they were making a living, responding to economic incentives as clearly as any modern person. *Homo economicus* was not an eighteenth-century Scottish invention. Their copper, turned into ingots and sickles, standardised for weight, then broken up and circulated far and wide, would soon become a primitive form of money widely used throughout Europe to lubricate exchange.

Conventional wisdom has probably underestimated the extent of specialisation and trade in the Neolithic age. There is a tendency to think that everybody was a farmer. But in Oetzi's world, there were farmers who grew einkorn and maybe farmers who grew grass for weaving into cloaks; coppersmiths who made axes and maybe bear hunters who made hats and shoes. And yet there were things that Oetzi no doubt made for himself: his bow was unfinished and so were some of his arrows. At a rough estimation, typical modern non-industrial people, living in traditional societies, directly consume between one-third and two-thirds of what they produce, and exchange the rest for other goods. Up to about 300 kilograms of food per head per year, people eat what they grow; after that they start to exchange surplus food for clothing, shelter, medicine or education. Almost by definition, the more wealthy somebody is, the more things he acquires from specialists. The characteristic signature of prosperity is increasing specialisation. The characteristic signature of poverty is a return to self-sufficiency. Go to a poor village in Malawi or Mozambique today and you will find few specialists and people consuming a high proportion of what they produce. They are 'not in the market', as an economist might say. And quite possibly they are less 'in the market' than ancient agrarian folk like Oetzi were.

Indulge me in a little sermon. The tradition among many anthropologists and archaeologists has been to treat the past as a very different place from the present, a place with its own mysterious rituals. To cram the Stone Age or the tribal South Seas into modern economic terminology is therefore an anachronistic error showing capitalist indoctrination. This view was promulgated especially by the anthropologist Marshall Sahlins, who distinguished pre-industrial economies based on 'reciprocity' from modern economies based on markets. Stephen Shennan satirises the attitude thus: 'We engage in exchanges to make some sort of profit; they do so in order to cement social relationships; we trade commodities; they give gifts.' Like Shennan, I think this is patronising bunk. I think people respond to incentives and always have done. People weigh costs and benefits and do what profits them. Sure, they take into account non-economic factors, such as the need to remain on good terms with trading partners and to placate malevolent deities. Sure, they give better deals to families, friends and patrons than they do to strangers. But they do that today as well. Even the most market-embedded modern financial trader is enmeshed in a web of ritual, etiquette, convention and obligation, not excluding social debt for a good lunch or an invitation to a football match. Just as modern economists often exaggerate the cold-hearted rationality of consumers, so anthropologists exaggerate the cuddly irrationality of pre-industrial people.

The 'kula' system of the south Pacific is a favourite case history of those who like to argue that markets were unknown to pre-industrial people. According to Bronislaw Malinowski, the people of fourteen different island groups exchanged armshells for necklaces in such a way that the armshells travelled in an anti-clockwise circle around the entire island group, while the necklaces went clockwise. After two years or more, an item might have returned to its original owner. To describe such a system as a market is plainly absurd: the exchange itself, not profit, must be the point. But look closer and kula becomes less peculiar. It was only one of many kinds of exchange practised in these islands; the fact that Westerners give each other cards and socks at Christmas speaks to the importance in their lives of the social meaning of exchange, but does not mean they do not also seek profits in markets. An anthropologist from the South Pacific might study Western Christmas and conclude that an utterly pointless and profitless but frantic midwinter commercial activity, inspired by religion, dominates the lives of Westerners. Pacific islanders were and are acutely aware of the importance of getting a good bargain when trading with a stranger. In any case, further research since the days of Malinowski has demonstrated that he had rather exaggerated the circular nature of the system, which is a mere side effect of the fact that traders who are exchanging useful items also like to give each other useless but pretty gifts that then sometimes end back where they started.

## Ignoble savage?

In the first half of the twentieth century, the Neolithic Revolution was interpreted by Gordon Childe and his followers as a bettering of the human condition, which brought obvious benefits: stored food with which to survive famines; new forms of nutrition close at hand, such as milk and eggs; less need for exhausting, dangerous and often fruitless treks through the wilderness; work that the unfit and injured could still do; perhaps more spare time in which to invent civilisation.

In the last third of the twentieth century, a prosperous yet nostalgic time, farming came to be reinterpreted as an invention born of desperation rather than inspiration, and perhaps even 'the worst mistake in the history of the human race'. The pessimists, led by Mark Cohen and Marshall Sahlins, argued that farming was a back-breaking treadmill that brought a monotonous diet deficient in nutrients to a people plagued by pollution, squalor, infectious diseases and early death. More people could now live upon the land, but with unchecked fertility, they would have to work harder. More babies were born, but more people died young. Whereas extant hunter-gatherers such as the Dobe !Kung seemed to have ample leisure and to live in 'the original affluent society' (Sahlins's phrase), limiting their reproduction and so preventing overpopulation, skeletons of the first farmers seemed to show wear and tear, chronic deformity, toothache and short stature. Meanwhile, they would catch measles from cattle, smallpox from camels, tuberculosis from milk, influenza from pigs, plague from rats, not to mention worms from using their own excrement as fertiliser and malaria from mosquitoes in their ditches and water butts.

They also got a bad attack of inequality for the first time. Extant hunter-gatherers are remarkably egalitarian, a state of affairs dictated by their dependence on sharing each other's hunting and gathering luck. (They sometimes need to enforce this equality with savage reprisals against people who get ideas above their stations.) A successful farmer, however, can soon afford to store some provisions with which to buy the labour of other less successful neighbours, and that makes him more successful still, until eventually – especially in an irrigated river valley, where he controls the water – he can become an emperor using servants and soldiers to impose his despotic whim upon subjects.

Worse still, as Friedrich Engels was the first to argue, agriculture may have worsened sexual inequality. It is certainly painfully obvious that in many peasant farming communities, men make women do much of the hard work. In hunter-gathering, men have many tiresome sexist habits, but they do at least contribute. When the plough was invented around 6,000 years ago, men took over the work of driving the oxen that cultivated fields, because it required greater strength, but this only exacerbated inequality. Now women were treated increasingly as the chattels of men, loaded with bracelets and ankle rings to indicate their husband's wealth. Now art became dominated by the symbols of male power and competition – arrows, axes and daggers. Now polygamy probably increased and the wealthiest men acquired harems and patriarchal status: at Branc in Slovakia, more women than men were buried with elaborate grave goods, indicating not that they were wealthy, so much as that their polygamous husbands were wealthy while other men languished in celibate poverty. In this way, polygamy enables poor women to share in prosperity more than poor men. It was an age of patriarchy.

Yet there is no evidence that early farmers behaved any worse than hunter-gatherers. Those few hunter-gatherer societies that became fat and prosperous on a dependable and rich local resource – most notably, the salmon-fishing tribes of the American north-west – soon indulged in patriarchy and inequality, too. The 'original affluence' of the modern hunter-gatherer !Kung was only possible because of modern tools, trade with farmers and even the odd helping hand from anthropologists. Their low fertility owed more to sexually transmitted infections than birth control. As for the deformities of early farmers, skeletons may not be representative and may tell you more about the injuries and diseases that were survived, rather than proved fatal. Even the gender equality of hunter-gatherers may prove wishful thinking. After all, Fuegian men, who could not swim, left their wives to anchor canoes in kelp beds and swim ashore in snow storms. The truth is that both hunter-gathering and farming could produce affluence or misery depending on the abundance of food and the relative density of people. One commentator writes: 'All pre-industrial economies, no matter how simple or complex, are capable of generating misery and will do so given enough time.'

The chronic and perpetual violence of the hunter-gatherer world had not ended with the invention of farming. Oetzi died a violent death, shot from behind by an arrow that pierced an artery in his shoulder, after – so DNA suggests – killing two men with one of his own arrows and carrying a wounded comrade on his back. The blood of a fourth man was on his knife. In the process he sustained a deep cut to his thumb and a fatal blow to the head. This was no small skirmish. His position in death suggests that his killer turned him over to retrieve the arrow, but the stone arrow head broke off inside his body. The archaeologist Steven LeBlanc says that the evidence of constant violence in the ancient past has been systematically overlooked by Rousseauesque wishful thinking among academics. He cites his own discoveries of innumerable sling shots and doughnut-shaped stones in Turkish sites from around 8,000 years ago. In the 1970s when he worked there he thought these were used by shepherds to chase away wolves and by farmers to weigh down their hoes. Now he realises that they were weapons of violence: the stones were mace heads and the sling shots were stockpiled for defence.

Wherever archaeologists look, they find evidence that early farmers fought each other incessantly and with deadly effect. The early inhabitants of Jericho dug a defensive ditch thirty feet deep and ten feet wide into solid rock without metal tools. In the Merzbach valley in Germany, the arrival of agriculture brought five centuries of peaceful population growth followed by the building of defensive earthworks, the dumping of corpses in pits and the abandonment of the whole valley. At Talheim around 4900 BC, an entire community of thirty-four people was massacred by blows to the head and arrows in the back, apart from the adult women who are missing – presumably abducted as sexual prizes. The killers were doing no more than Moses later ordered his followers in the Bible. After a successful battle against the Midianites and a massacre of the adult males, he told them to finish the job by raping the virgins: 'Now therefore kill every male among the little ones, and kill every woman that hath known man by lying with him. But all the women children, that have not known a man by lying with him, keep alive for yourselves.' (Numbers 31)

Likewise, wherever anthropologists look, from New Guinea to the Amazon and Easter Island, they find chronic warfare among today's subsistence farmers. Pre-emptively raiding your neighbours lest they raid you is routine human behaviour. As Paul Seabright has written: 'Where there are no institutional restraints on such behaviour, systematic killing of unrelated individuals is so common among human beings that, awful though it is, it cannot be described as exceptional, pathological or disturbed.'

Nor can it be denied that such violence was habitually accompanied by cruelty to a degree that turns the modern stomach. When Samuel Champlain accompanied (and assisted with his arquebus) a successful Huron raid upon the Mohawks in 1609, he had to watch as his allies sacrificed a captive by branding his torso with glowing sticks from the fire, periodically reviving him with buckets of water if he passed out, from dusk till dawn. Only when the sun rose were they permitted by their tradition to disembowel and then eat the unfortunate victim, during which procedure he gradually died.

## The fertiliser revolution

The Neolithic Revolution provided posterity with almost limitless calories. There would be famines aplenty in the millennia to come, but they would never again reduce human population density to the hunter-gatherer level. Inch by inch, trick by trick and crop by crop, people would find a way to coax food from even the poorest soils, and calories from even the poorest foods and would crystallise insights of almost miraculous perspicacity as to how to do so. Fast-forward from the Neolithic a few thousand years to the industrial revolution, when population began to explode rather than expand and stand amazed that you and your ancestors came through that explosion better fed, not starving. In 1798 Robert Malthus famously predicted in his *Essay on Population* that food supply could not keep pace with population growth because of the finite productivity of land. He was wrong, but it was no easy matter; in the nineteenth century it was at times touch-and-go. Even though steamships, railways, the Erie Canal, refrigeration and the binder-and-reaper enabled the Americas to send vast amounts of wheat back east to feed the industrial masses of Europe, directly and in the form of beef and pork, famine was never far away.

It would have been worse but for a strange windfall discovered in about 1830. On dry bird islands off the South American and South African coasts, where no rain leached away the cormorant, penguin and booby droppings, immense deposits of nitrogen and phosphorus had accumulated over centuries. Guano mining became a very profitable, and very grim, business. The tiny island of Ichaboe yielded 800,000 tonnes of guano in a few short years. Between 1840 and 1880, guano nitrogen made a colossal difference to European agriculture. But soon the best deposits were exhausted. The miners turned to rich mineral saltpetre deposits in the Andes (which proved to be ancient guano islands lifted up by South America's westward drift), but these could barely keep pace with demand. By the turn of the twentieth century the fertiliser crisis was desperate. In 1898, the centenary of Malthus's pessimistic prognostication, the eminent British chemist Sir William Crookes gave a similar jeremiad in his presidential address to the British Association entitled 'The Wheat Problem'. He argued that, given the growing population and the lack of suitable new acres to plough in the Americas, 'all civilisations stand in deadly peril of not having enough to eat,' and unless nitrogen could be chemically 'fixed' from the air by some scientific process, 'the great Caucasian race will cease to be foremost in the world, and will be squeezed out of existence by races to whom wheaten bread is not the staff of life.'

Within fifteen years his challenge had been met. Fritz Haber and Carl Bosch invented a way of making large quantities of inorganic nitrogen fertiliser from steam, methane and air. Today nearly half the nitrogen atoms in your body passed through such an ammonia factory. But an even bigger factor in averting Crookes's disaster was the internal combustion engine. The first tractors had few advantages over the best horses, but they did have one enormous benefit as far as the world was concerned: they did not need land to grow their fuel. America's horse population peaked at twenty-one million animals in 1915; at the time about one-third of all agricultural land was devoted to feeding them. So the replacement of draught animals by machines released an enormous acreage of land to grow food for human consumption. At the same time motorised transport was bringing land within reach of railheads. As late as 1920, over three million acres of good agricultural land in the American Midwest lay uncultivated because it was more than eighty miles from a railway, which meant a five-day trip by horse wagon costing up to 30 per cent more than the value of the grain.

In 1920 plant breeders developed a vigorous and hardy new variety of wheat, 'Marquis', by crossing a Himalayan and an American plant, which could survive further north in Canada. So thanks to tractors, fertilisers and new varieties, by 1931, the year in which Crookes had chosen to place his potential future famine, the supply of wheat had so far exceeded the demand that the price of wheat had plummeted and wheat land was being turned over to pasture all over Europe.

## Borlaug's genes

The twentieth century would continue to confound the Malthusian pessimists, most spectacularly in the 1960s in Asia. For two years in the mid-1960s, India seemed to be on the brink of mass famine. Crops were failing in a drought, and people were starving in growing numbers. Hunger had never been absent from the subcontinent for long, and memories of the great Bengal famine of 1943 were raw. With over 400 million people, the country was in the midst of an unprecedented population explosion. The government had put agriculture at the top of its agenda, but the state monopolies charged with finding new varieties of wheat and rice had nothing to offer. There was little new land to bring into cultivation. Five million tonnes of food aid a year from America were all that stood between India and a terrible fate, and those shipments could surely not continue for ever.

Yet even amid such defeatism, India's wheat production was taking off, because of a sequence of events that had begun more than twenty years before. On General Douglas MacArthur's team in Japan at the end of the Second World War was an agricultural scientist named Cecil Salmon. Salmon collected sixteen varieties of wheat including one called 'Norin 10'. It grew just two feet tall, instead of the usual four – thanks, it is now known, to a single mutation in a gene called Rht1, which makes the plant less responsive to a natural growth hormone. Salmon collected some seeds and sent them back to the United States, where they reached a scientist named Orville Vogel in Oregon in 1949. At the time it was proving impossible to boost the yield of tall wheat by adding artificial fertiliser. The fertiliser caused the crop to grow tall and thick, whereupon it fell over, or 'lodged'. Vogel began crossing Norin 10 with other wheats to make new short-strawed varieties. In 1952 Vogel was visited by a scientist working in Mexico called Norman Borlaug, who took some Norin and Norin-Brevor hybrid seeds back to Mexico and began to grow new crosses. Within a few short years Borlaug had produced wheat that yielded three times as much as before. By 1963, 95 per cent of Mexico's wheat was Borlaug's variety, and the country's wheat harvest was six times what it had been when Borlaug set foot in Mexico. Borlaug started training agricultural scientists from other countries, including Egypt and Pakistan.

Between 1963 and 1966 Borlaug and his Mexican dwarf wheats faced innumerable hurdles to acceptance in Pakistan and India. Jealous local researchers deliberately underfertilised the experimental plots. Customs officials in Mexico and America – not to mention race riots in Los Angeles – delayed shipments of seed so they arrived late for the planting season. Over-enthusiastic fumigation at customs killed half the seeds. The Indian state grain monopolies lobbied against the seeds, spreading rumours that they were susceptible to disease. The Indian government refused to allow increased fertiliser imports, because it wanted to build up an indigenous fertiliser industry, until Borlaug shouted at the deputy prime minister. To cap it all, war broke out between the two countries.

But gradually, thanks to Borlaug's persistence, the dwarf wheats prevailed. The Pakistani agriculture minister took to the radio extolling the new varieties. The Indian agriculture minister ploughed and planted his cricket pitch. In 1968, after huge shipments of Mexican seed, the wheat harvest was extraordinary in both countries. There were not enough people, bullock carts, trucks or storage facilities to cope with the crop. In some towns grain was stored in schools.

In March of that year India issued a postage stamp celebrating the wheat revolution. That was the very same year the environmentalist Paul Ehrlich's book *The Population Bomb* was published declaring it a fantasy that India would ever feed itself. His prediction was wrong before the ink was dry. By 1974, India was a net exporter of wheat. Wheat production had tripled. Borlaug's wheat – and dwarf rice varieties that followed – ushered in the Green Revolution, the extraordinary transformation of Asian agriculture in the 1970s that banished famine from almost the entire continent even as population was rapidly expanding. In 1970 Norman Borlaug was awarded the Nobel Peace Prize.

In effect, Borlaug and his allies had unleashed the power of fertiliser, made with fossil fuels. Since 1900 the world has increased its population by 400 per cent; its cropland area by 30 per cent; its average yields by 400 per cent and its total crop harvest by 600 per cent. So per capita food production has risen by 50 per cent. Great news – thanks to fossil fuels.

## Intensive farming saves nature

Taking all cereal crops together worldwide, in 2005 twice as much grain was produced from the same acreage as in 1968. That intensification has spared land on a vast scale. Consider this extraordinary statistic, calculated by the economist Indur Goklany. If the average yields of 1961 had still prevailed in 1998, then to feed six billion people would have required the plough-ing of 7.9 billion acres, instead of the 3.7 billion acres actually ploughed in 1998: an extra area the size of South America minus Chile. And that's optimistically assuming that yields would have remained at the same level in the newly cultivated land, taken from the rainforests, the swamps and the semi-deserts. If yields had not increased, therefore, rainforests would have been burnt, deserts irrigated, wetlands drained, tidal flats reclaimed, pastures ploughed – to a far greater extent than actually happened. To put it another way, today people farm (i.e., plough, crop or graze) just 38 per cent of the land area of the earth, whereas with 1961 yields they would have to farm 82 per cent to feed today's population. Intensification has saved 44 per cent of this planet for wilderness. Intensification is the best thing that ever happened – from the environmental perspective. There are now over two billion acres of 'secondary' tropical rainforest, regrowing after farmers left for the cities, and it is already almost as rich in biodiversity as primary forest. That is because of intensive farming and urbanisation.

Some argue that the human race already appropriates for itself an unsustainable fraction of the planet's primary production and that if it uses any more, the ecosystem of the entire globe will collapse. Human beings comprise about 0.5 per cent by weight of the animals on the planet. Yet they beg, borrow and steal for themselves roughly 23 per cent of the entire primary production of land plants (the number is much lower if the oceans are included). This number is known to ecologists as the HANPP – the 'human appropriation of net primary productivity'. That is to say, of the 650 billion tonnes of carbon potentially absorbed from the air by land plants each year, eighty are harvested, ten are burnt and sixty are prevented from grow-ing by ploughs, streets and goats, leaving 500 to support all the other species.

That may seem to leave some room for growth yet, but is it

really practical to expect a planet to go on supporting such a dominant monoculture of one ape? To answer this question, break the numbers down by region. In Siberia and the Amazon perhaps 99 per cent of plant growth supports wildlife rather than people. In much of Africa and central Asia, people reduce the productivity of land even as they appropriate a fifth of the production – an overgrazed scrubland supports fewer goats than it would support antelopes if it were wilderness. In western Europe and eastern Asia, however, people eat nearly half the plant production yet barely reduce the amount left over for other species at all – because they dramatically raise the productivity of the land with fertiliser: the grass meadow near my house, sprinkled with nitrate twice a year, supports a large herd of milking cows, but it is also teeming with worms, leatherjackets, dung flies – and the blackbirds, jackdaws and swallows that eat them. This actually gives great cause for optimism, because it implies that intensifying agriculture throughout Africa and central Asia could feed more people and still support more other species, too. Or, in academic-ese: 'These findings suggest that, on a global scale, there may be a considerable potential to raise agricultural output without necessarily increasing HANPP.'

Other trends too have made modern farming better for the planet. Now that weeds can be controlled by herbicides rather than ploughing (the main function of a plough is to bury weeds), more and more crops are sown directly into the ground without tilling. This reduces soil erosion, silt run-off and the massacre of innocent small animals of the soil that inevitably attends the ploughing of a field – as flocks of worm-eating seagulls attest. Food processing with preservatives, much despised by green-chic folk, has greatly reduced the amount of food that goes to waste. Even the confinement of chickens, pigs and cattle to indoor barns and batteries, though it troubles the consciences (mine included) of those who care for animal welfare, undoubtedly results in more meat produced from less feed with less pollution and less disease. When bird flu threatened, it was free-range flocks of chickens, not battery farms, that were at greatest risk. Some intensive farming of animals is unacceptably cruel; but some is no worse than some kinds of free-range farming, and its environmental impact is undoubtedly smaller.

Borlaug's genes, sexually recombined with Haber's ammonium and Rudolf Diesel's internal combustion engine, have rearranged sufficient atoms not only to ensure that Malthus was wrong for at least another half-century, but that tigers and toucans can still exist in the wild. So I am going to make an outrageous proposal: that the world could reasonably set a goal of feeding itself to a higher and higher standard throughout the twenty-first century without bringing any new land under the plough, indeed with a gradual reduction in farmland area. Could it be done? In the early 1960s the economist Colin Clark calculated that human beings could in theory sustain themselves on just twenty-seven square metres of land each. His reasoning went like this: an average person needs about 2,500 calories of food per day, equivalent to about 685 grams of grain. Double it for growing a bit of fuel, fibre and some animal protein: 1,370 grams. The maximum rate of photosynthesis on well-watered, rich soils is about 350 grams per square metre per day, but you can knock that down to about fifty for the best that farming is in practice able to achieve over a wide area. So it takes twenty-seven square metres to grow the 1,370 grams a person needs. On this basis and using the yields of the day, Clark calculated in the 1960s that the world could feed thirty-five billion mouths.

Well, let me assume that despite Clark's conservatism about photosynthesis, this is still wildly optimistic. Let me quadruple his number and assume that earth cannot feed an average human from less than 100 square metres. How close are we to that point? In 2004, the world grew about two billion tonnes of rice, wheat and maize on about half a billion hectares of land: an average yield of four tonnes to the hectare. Those three crops provided about two-thirds of the world's food, both directly and via beef, chicken and pork – equivalent to feeding four billion people. So a hectare fed about eight people, or about 1,250 square metres each, down from about 4,000 square metres in the 1950s. That is a long way above 100 square metres. In addition, the world cultivated another billion hectares growing other cereals, soybeans, vegetables, cotton and the like (pasture land is not part of this calculation) – that is about 5,000 square metres each. Even if you increase the number of people to nine billion, there is still an enormous amount of room for improvement before we start hitting the limit of agricultural productivity. You could double or quadruple yields and still be nowhere near the maximum practical yields of land, let alone the photosynthetic limit. If we all turned vegetarian, the amount of land we would need would be still less, but if we turned organic, it would be more: we would need extra acres to grow the cows whose manure would fertilise our fields: more precisely, to replace all the industrial nitrogen fertiliser now applied would mean an extra seven billion cattle grazing an extra thirty billion acres of pasture. (You will often hear organic champions extol the virtues of both manure and vegetarianism: notice the contradiction.) But these calculations show that even without vegetarianism, there will be a growing surplus of farmland.

So let's do it: let's continue to cut down the area of farmland per person to the point where we can begin to turn the rest over to wilderness.

Running out of land to capture sunlight is not going to be a problem for food production – not since Haber broke the fertiliser bottleneck. Running out of water could well be. Lester Brown points out that India depends heavily on a rapidly depleting aquifer and a slowly drying Ganges to irrigate crops, that salination caused by evaporation of irrigation water is an increasing problem all across the world and that fully 70 per cent of all the world's water usage is for crop irrigation. But he goes on to admit that the inefficiency of irrigation systems (i.e., the loss to evaporation) is falling fast, especially in China, and that there is already a well-used technique – drip irrigation – that could almost eliminate the problem. Countries like Cyprus, Israel and Jordan are already heavy users of drip irrigation. In other words, the wastefulness of irrigation is a product of the low price of water. Once it is properly priced by markets, water is not only used more frugally, but its very abundance increases through incentives to capture and store it.

This is what it would take to feed nine billion people in 2050: at least a doubling of agricultural production driven by a huge increase in fertiliser use in Africa, the adoption of drip irrigation in Asia and America, the spread of double cropping to many tropical countries, the use of GM crops all across the world to improve yields and reduce pollution, a further shift from feeding cattle with grain to feeding them with soybeans, a continuing relative expansion of fish, chicken and pig farming at the expense of beef and sheep (chickens and fish convert grain into meat three times as efficiently as cattle; pigs are in between) – and a great deal of trade, not just because the mouths and the plants will not be in the same place, but also because trade encourages specialisation in the best-yielding crops for any particular district. If price signals drive the world's farmers to take these measures it is quite conceivable that in 2050 there will be nine billion people feeding more comfortably than today off a smaller acreage of cropland, releasing large tracts of land for nature reserves. Imagine that: an immense expansion of wilderness throughout the world by 2050. It's a wonderful goal and one that can only be brought about by further intensi-fication and change, not by retreat and organic subsistence. Indeed, come to think of it, let's make farming a multi-storey business, with hydroponic drip-irrigation and electric lighting producing food year-round on derelict urban sites linked by conveyor belt directly to supermarkets. Let's pay for the buildings and the electricity by granting the developer tax breaks for retiring farmland elsewhere into forest, swamp or savannah. It is an uplifting and thrilling ideal.

Should the world decide, as a professor and a chef have both suggested on my radio recently, that countries should largely grow and eat their own food (why countries? Why not continents, or villages, or planets?), then of course a very much higher acreage will be needed. My country happens to be as useless at growing bananas and cotton as Jamaica is at growing wheat and wool. If the world decides, as it crazily started to do in the early 2000s, that it wants to grow its motor fuel in fields rather than extract it from oil wells, then again the acreage under the plough will have to balloon. And good night rainforests. But as long as some sanity prevails, then yes, my grandchildren can both eat well and visit larger and wilder nature reserves than I can. It is a vision I am happy to strive for. Intensive yields are the way to get there.

When human beings were all still hunter-gatherers, each needed about a thousand hectares of land to support him or her. Now – thanks to farming, genetics, oil, machinery and trade – each needs little more than a thousand square metres, a tenth of a hectare. (Whether the oil will last long enough is a different subject and one I tackle later in the book: briefly my answer is that substitutes will be adopted if the price rises high enough.) That is possible only because each square metre is encouraged to grow whatever it is good at growing and global trade distributes the result to ensure that everybody gets a bit of everything. Once again, the theme of specialised production/ diversified consumption turns out to be the key to prosperity.

# Organic's wrong call

Politicians can make my prediction fail. Should the world decide to go organic – that is, should farming get its nitrogen from plants and fish rather than direct from the air using factories and fossil fuels – then many of the nine billion will starve and all rainforests will be cut down. Yes, I wrote 'all'. Organic farming is low-yield, whether you like it or not. The reason for this is simple chemistry. Since organic farming eschews all synthetic fertiliser, it exhausts the mineral nutrients in the soil – especially phosphorus and potassium, but eventually also sulphur, calcium and manganese. It gets round this problem by adding crushed rock or squashed fish to the soil. These have to be mined or netted. Its main problem, though, is nitrogen deficiency, which it can reverse by growing legumes (clover, alfalfa or beans), which fix nitrogen from the air, and either ploughing them into the soil or feeding them to cattle whose manure is then ploughed into the soil. With such help a particular organic plot can match non-organic yields, but only by using extra land elsewhere to grow the legumes and feed the cattle, effectively doubling the area under the plough. Conventional farming, by contrast, gets its nitrogen from what are in effect point sources – factories, which fix it from the air.

Organic farmers also aspire to rely less on fossil fuels, but unless organic food is to be expensive, scarce, dirty and decaying, then it has to be intensively produced, and that means using fuel – in practice, a pound of organic lettuce, grown without synthetic fertilisers or pesticides in California, and containing eighty calories, requires 4,600 fossil-fuel calories to get it to a customer's plate in a city restaurant: planting, weeding, harvesting, refrigerating, washing, processing and transporting all use fossil fuel. A conventional lettuce requires about 4,800 calories. The difference is trivial.

Yet when a technology came along that promised to make organic farming both competitive and efficient, the organic movement promptly rejected it. That technology was genetic modification, which was first invented in the mid-1980s as a kinder, gentler alternative to 'mutation breeding' using gamma rays and carcinogenic chemicals. Did you know that this was the way many crops were produced over the last half-century? That much pasta comes from an irradiated variety of durum wheat? That most Asian pears are grown on irradiated grafts? Or that Golden Promise, a variety of barley especially popular with organic brewers, was first created in an atomic reactor in Britain in the 1950s by massive mutation of its genes followed by selection? By the 1980s, scientists had reached the point where, instead of this random scrambling of the genes of a target plant with unknown result and lots of collateral genetic damage, they could take a known gene, with known function, and inject it into the genome of a plant, where it would do its known job. That gene might come from a different species, so achieving the horizontal transfer of traits between species that happens relatively rarely among plants in nature (though it is common-place among microbes).

For example, many organic farmers happily adopted an insect-killing bacterium called *Bacillus thuringiensis* or bt, first commercialised in France as Sporeine in the 1930s, which they sprayed on crops to control pests. As a 'biological' not a chemical spray, it met their tests. By the 1980s lots of different variants of bt had been developed for different insects. All were regarded as organic. But then genetic engineers took the bt toxin and incorporated it into the cotton plant to produce bt-cotton, one of the first genetically modified crops. This had two huge advantages: it killed bollworms living inside the plant where sprays could not reach them easily; and it did not kill innocent insects that were not eating the cotton plant. Yet, though this was an officially organic product, biologically integrated into the plant, and obviously better for the environment, organic high priests rejected the technology. Bt cotton went on to transform the cotton industry and has now replaced more than a third of the entire cotton crop. Indian farmers, denied the technology by their government, rioted to demand it after seeing bootlegged crops growing in their neighbours' fields. Now most Indian cotton is bt, and the result has been a near-doubling of yield and a halving of insecticide use – win/win. In every study of bt cotton crops across the world from China to Arizona, the use of insecticides is down by as much as 80 per cent and the bees, butterflies and birds are back in abundance. Economically and ecologically, good news all round. Yet merely to board a passing bandwagon of protest publicity, the leaders of the organic movement locked themselves out of a new technology that has delivered huge reductions in the use of synthetic pesticides. One estimate puts the amount of pesticide *not* used because of genetic modification at over 200 million kilograms of active ingredients and climbing.

This is just one example of how the organic movement's insistence on freezing agricultural technology at a mid-twentieth-century moment means it misses out on environmental benefits brought by later inventions. 'I'm so tired of people who wouldn't visit a doctor who used a stethoscope instead of an MRI demanding that farmers like me use 1930s technology to raise food,' writes the Missouri farmer Blake Hurst. Organic farmers are happy to spray copper sulphate or nicotine sulphate, but forbid themselves the use of synthetic pyrethroids, which swiftly kill insects but have very low toxicity for mammals and do not persist in the environment causing collateral damage to non-pests. They forbid themselves herbicides, which means they have to weed by hand, using poorly paid labour, or by tilling and flame-throwing, which can devastate soil fauna, accelerate soil erosion and release greenhouse gases. They forbid themselves fertiliser made from air, but allow themselves fertiliser made from trawled fish.

In her classic book *Silent Spring*, Rachel Carson called upon scientists to turn their backs on chemical pesticides and seek 'biological solutions' to pest control instead. They have done so, and the organic movement has rejected them.

# The many ways of modifying genes

Of course, almost by definition, all crop plants are 'genetically modified'. They are monstrous mutants capable of yielding unnaturally large, free-threshing seeds or heavy, sweet fruit and dependent on human intervention to survive. Carrots are orange thanks only to the selection of a mutant first discovered perhaps as late as the sixteenth century in Holland. Bananas are sterile and incapable of setting seed. Wheat has three whole diploid (double) genomes in each of its cells, descended from three different wild grasses, and simply cannot survive as a wild plant – you never encounter wheat weeds. Rice, maize and wheat all share genetic mutations that alter the development of the plant to enlarge seeds, prevent shattering, and allow free threshing from chaff. These mutations were selected, albeit inadvertently, by the first farmers sowing and reaping them.

But modern genetic modification, using single genes, was a technology that came worryingly close to being stifled at birth by irrational fears fanned by pressure groups. First they said the food might be unsafe. A trillion GM meals later, with not a single case of human illness caused by GM food, that argument has gone. Then they argued that it was unnatural for genes to cross the species barrier. Yet wheat, the biggest crop of all, is an unnatural 'polyploid' merger of three wild plant species and horizontal gene transfer is showing up in lots of plants, such as Amborella, a primitive flowering plant, which proves to have DNA sequences borrowed from mosses and algae. (DNA has even been caught jumping naturally from snakes to gerbils with the help of a virus.) Then they said GM crops were produced and sold for profit, not to help farmers. So are tractors. Then they tried the bizarre argument that herbicide-resistant crops might cross-breed with wild plants and result in a 'super' weed that was impossible to kill – with that herbicide. This from people who were against herbicides anyway, so what could be more attractive to them than rendering the herbicide useless?

By 2008, less than twenty-five years after they were first invented, fully 10 per cent of all arable land, thirty million acres, was growing genetically modified crops: one of the most rapid and successful adoptions of a new technology in the history of farming. Only in parts of Europe and Africa were these crops denied to farmers and consumers by the pressure of militant environmentalists, with what Stewart Brand calls their 'customary indifference to starvation'. African governments, after intense lobbying by Western campaigners, have been persuaded to tie genetically modified food in red tape, which prevents them being grown commercially in all but three countries (South Africa, Burkina Faso and Egypt). In one notorious case Zambia in 2002 even turned down food aid in the middle of a famine after being persuaded by a campaign by groups, including Greenpeace International and Friends of the Earth, that because it was genetically modified it could be dangerous. A pressure group even told a Zambian delegation that GM crops might cause retroviral infections. Robert Paarlberg writes that, 'Europeans are imposing the richest of tastes on the poorest of people.' Ingo Potrykus, developer of golden rice, thinks that 'blanket opposition to all GM foods is a luxury that only pampered Westerners can afford.' Or as the Kenyan scientist Florence Wambugu puts it, 'You people in the developed world are certainly free to debate the merits of genetically modified foods, but can we eat first?'

Yet it is Africa that could stand to benefit the most from GM crops precisely because so many of its farmers are smallholders with little access to chemical pesticides. In Uganda, where a fungal disease called Black Sigatoka threatens the staple banana crop, and resistant strains with rice genes are still years from market because of regulations, the experimental GM plants have to be guarded by padlocked fences, not to protect them from trampling titled protesters, but to protect them from eager users. Per capita food production in Africa has fallen 20 per cent in thirty-five years; some 15 per cent of the African maize crop is lost to stem-borer moth larvae and at least as much again is lost in storage to beetles: bt maize is resistant to both pests. Nor is corporate ownership a problem: Western companies and foundations are keen to give such crops royalty-free to African farmers, through organisations like the African Agricultural Technology Foundation. There are glimmers of hope. Field trials begin in Kenya in 2010 of drought-resistant and insect-resistant maize, though years of safety testing will follow.

Ironically, the main result globally of the campaign against GM crops was to delay the retreat of chemical pesticides and ensure that only commodity crops could afford to find their way through the regulatory thicket to market, which effectively meant that the crops were denied to small farmers and charities. Genetic engineering remained for longer than it would have done the preserve of large corporations able to afford the regulations imposed by the pressure of the environmentalists. Yet the environmental benefits of GM crops are already immense – pesticide use is falling fast wherever GM cotton is grown and no-till cultivation is enriching soil wherever herbicide-tolerant soybeans are grown. But the benefits will not stop there. Plants that are resistant to drought, salt and toxic aluminium are on the way. Lysine-enriched soybeans may soon be feeding salmon in fish farms, so that wild stocks of other fish do not have to be plundered to make feed. By the time you read this, plants may already be on the market that absorb nitrogen more efficiently, so that higher yields can be achieved with less than half as much fertiliser, saving aquatic habitats from eutrophic runoff, saving the atmosphere from a greenhouse gas (nitrous oxide) that is 300 times as potent as carbon dioxide and cutting the amount of fossil fuel used to make fertiliser – not to mention saving farmers' costs. Some of this would be possible without gene transfer, but it is a lot quicker and safer with it. Greenpeace and Friends of the Earth still oppose it all.

There is one respect in which the environmental critique of modern agriculture has force. In the pursuit of quantity, science may have sacrificed nutritional quality of food. Indeed, the twentieth-century drive to provide a growing population with an ever faster-growing supply of calories has succeeded so magnificently that the diseases caused by too much bland food are rampant: obesity, heart disease, diabetes, and perhaps depression. For example, modern plant oils and plentiful red meat make for a diet low in omega-3 fatty acids, which may contribute to heart disease; modern wheat flour is rich in amylopectin starch, which may contribute to insulin resistance and hence diabetes; and maize is especially low in the amino acid tryptophan, a precursor of serotonin, the 'feel-good' neurotransmitter. Consumers will rightly be looking to the next generation of plant varieties to redress these deficiencies. They could do so by eating more fish, fruit and vegetables. But not only would this be a land-hungry option, it would suit the wealthy more than the poor, so it would exacerbate health inequalities. Arguing against vitamin-enriched rice, the Indian activist Vandana Shiva, echoing Marie-Antoinette, recommended that Indians should eat more meat, spinach and mangoes rather than relying on golden rice.

Instead, genetic modification provides an obvious solution: to insert healthy nutritional traits into high-yielding varieties: tryptophan into maize to fight depression, calcium transporter genes into carrots to help fight osteoporosis in people who cannot drink milk, or vitamins and minerals into sorghum and cassava for those who depend on them as staples. By the time this book is published soybeans with omega-3 fatty acids developed in South Dakota should be on the way to supermarkets in America. They promise to lower the risk of heart attacks and perhaps help the mental health of those who cook with their oil – and at the same time they can reduce the pressure on wild fish stocks from which fish oils are derived.

CHAPTER 5

# The triumph of cities: trade after 5,000 years ago

Imports are Christmas morning; exports are
January's MasterCard bill.

P.J. O'ROURKE
*On The Wealth of Nations*

## U.S. DEATH RATES FROM WATER-RELATED DISEASE

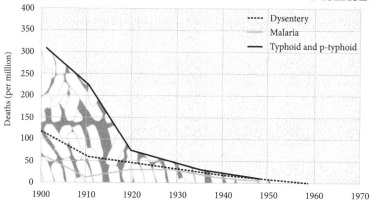

A modern combine harvester, driven by a single man, can reap enough wheat in a single day to make half a million loaves. Little wonder that as I write these words (around the end of 2008), for the very first time the majority of the world's population lives in cities – up from just 15 per cent in 1900. The mechanisation of agriculture has enabled, and been enabled by, a flood of people leaving the land to seek their fortune in the city, all free to make for each other things other than food.

Though some came to town with hope and ambition, and some with desperation and fear, almost all were drawn by the same aim: to take part in trade. Cities exist for trade. They are places where people come to divide their labour, to specialise and exchange. They grow when trade expands – Hong Kong's population grew by thirty times in the twentieth century – and shrink when trade dries up. Rome declined from a million inhabitants in 100 BC to less than 20,000 in the early Middle Ages. Since people have generally done more dying than procreating when in cities, big cities have always depended on rural immigrants to sustain their numbers.

Just as agriculture appeared in six or seven parts of the world simultaneously, suggesting an evolutionary determinism, so the same is true, a few thousand years later, of cities. Large urban settlements, with communal buildings, monuments and shared infrastructure, start popping up after seven thousand years ago in several fertile river valleys. The oldest cities were in southern Mesopotamia, in what is now Iraq. Their emergence signified that production was becoming more specialised, consumption more diversified.

It seems that farmers on the rich alluvial soils of the southern Euphrates valley began to grow sufficiently prosperous, in a period of high rainfall, to exchange their grain and woven wool for timber and precious stones from the people in the hills to the north. From about 7,500 years ago, a distinctive 'Ubaid' style of pottery, clay sickles and house design spread all across the

Near East, reaching up into the mountains of Iran, across to the Mediterranean and along the shores of the Arabian peninsula, where fishermen sold fish to Ubaid merchants in exchange for grain and nets. This was a trading diaspora, not an empire: the domestic habits of the distant people who adopted Ubaid style remained distinctive, showing that they were not colonists from Mesopotamia, but locals aping the Ubaid habits.

## The ur-city

So Ubaid Mesopotamia, by exporting grain and cloth, drew its neighbours into exporting timber and later metal. The Ubaids must have become rich enough to support chiefs and priests. Inevitably, these had ideas above their stations, for when, after 6,000 years ago, the Ubaid culture disappeared, it was replaced by something that looks much more like an empire – the 'Uruk expansion'. Uruk was a large city, probably the first the world had ever seen, housing more than 50,000 people within its six miles of wall (King Gilgamesh may have built the wall – having plundered his trading partners' lands and earned their enmity). All the signs are that Uruk, its agriculture made prosperous by sophisticated irrigation canals, had in the words of the archae-ologist Gil Stein 'developed centralized institutions to mobilize surplus labour and goods from the hinterlands in a meticulously administered political economy'. To put it more succinctly, a class of middlemen, of trade intermediaries, had emerged for the first time. These were people who lived not by production, nor by plunder and tribute, but by deals alone. Like traders ever since, they gathered as tightly together as possible to maximise information flow and minimise costs. Trade with the hills con-tinued, but increasingly it came to look like tribute as Uruk merchants' dwellings, complete with distinctive central halls, niched-façade temples and peculiar forms of pottery and stone tool, were plonked amid the rural settlements of trading partners in the hills. A cooperative trade network seems to have turned into something much more like colonialism. Tax and even slavery soon began to rear their ugly heads. Thus was set the pattern that would endure for the next 6,000 years – merchants make wealth; chiefs nationalise it.

The story of Ubaid and Uruk is familiar and modern. You can imagine the Ubaid merchants displaying their cloths and pots and groaning sacks of grain to the wide-eyed peasants of the hills. You can see the Uruk nabobs in their privileged enclaves, surrounded by subservient natives, like the British in India or Chinese in Singapore. It is with a start that you recall this is still essentially the Stone Age. Only towards the end of the Ubaid period is copper being smelted, and well into the Uruk times, sickles and knives are still made of stone or clay. Late in the Uruk period clay tablets appear with uniform marks on them meticulously accounting for merchants' stocks and profits. Those dull records, dug into the surfaces of clay tablets, are the ancestors of writing – accountancy was its first application. The message those tablets tell is that the market came long before the other appurtenances of civilisation. Exchange and trade were well established traditions before the first city, and record keeping may have played a crucial role in allowing cities to emerge full of strangers who could trust each other in transactions. It was the habit of exchange that enabled specialists to appear in Uruk, swelling the city with artisans and craftsmen who never went near the fields. For instance, there seems to have been almost mass production of bevelled-rim bowls that appear to have been disposable. Handed out at communal events like temple constructions, they were undoubtedly made in something like a factory, by workers paid to make them, not by moonlighting farmers.

Uruk did not last, because the climate dried out and the population collapsed, aided no doubt by soil erosion, salination, imperial overspending and uppity barbarians. But Uruk was followed by an endless series of empires on the same ground: Sumerian, Akkadian, Assyrian, Babylonian, neo-Assyrian, Persian, Hellenistic, Roman (briefly, under Trajan), Parthian, Abbasid, Mongol, Timurid, Ottoman, British, Saddamite, Bushite ... Each empire was the product of trading wealth and was itself the eventual cause of that wealth's destruction. Merchants and craftsmen make prosperity; chiefs, priests and thieves fritter it away.

## Cotton and fish

The urban revolution on the banks of the Euphrates was repeated on the banks of the Nile, Indus and Yellow rivers. Ancient Egypt could grow nearly two tonnes of wheat per hectare on land irrigated and replenished with nutrients by the annual flood of the Nile, providing an ample surplus of food, if peasants could be persuaded to produce one, to exchange for other goods, not excluding pyramids. Even more than in Mesopotamia, Egypt followed the path of irrigation, centralisation, monument building and eventual stagnation. Dependent on the flow of the Nile for their crops, the peasants became subject to whoever owned the boats and sluice gates, and he took most of the surplus. Unlike hunter-gatherers or herders, farmers faced with taxes have to stay put and pay, especially if surrounded by desert and dependent on irrigation ditches. So once Menes had unified the upper and lower valley and made himself the first pharaoh, the productive Egyptian economy found itself nationalised, monopolised, bureaucratised and eventually stifled by – in the words of two modern historians – the 'leaden authoritarianism' of its rulers.

On the banks of the Indus, an urban civilisation arose without spawning an emperor, at least not one whose name is known. Harappa and Mohenjo-Daro are known for the precisely standardised size of their bricks and their neat sanitary arrangements. The port of Lothal was distinguished by what appears to be a dock and tidal lock, and a factory for making beads. There is less sign of palaces or temples, let alone pyramids, but the anthropologist Gordon Childe's preliminary conclusion that the whole thing appears to have been rather egalitarian and peaceful turned out to be largely wishful thinking. Somebody was imposing a neat grid of streets and building a hefty 'citadel' of pillars, towers and walls. Smells like a monarch to me. As Sir Mortimer Wheeler wrote in his autobiography: 'I sat down and wrote to Gordon Childe in London that the bourgeois complacency of the Indus civilisation had dissolved into dust and that, instead, a thoroughly militaristic imperialism had raised its ugly head amongst the ruins.'

The Indus people were good at transport: bullock carts may have been used here for the first time and plank-built sailing boats. Transport allowed extensive trade. Some of the very earliest settlements in the region, such as Mehrgarh in Baluchistan, were importing lapis lazuli from north of the Hindu Kush mountains as early as 6,000 years ago. By the time of Harappa, copper came from Rajasthan, cotton from Gujarat, and lumber from the mountains. Even more remarkably, the archaeologist Shereen Ratnagar concluded that boats sent exports west to Mesopotamia, stopping at ports along the coast of what is now Iran – implying a seamanship that is surprising at such an early date. There can be little doubt that the great wealth of the Indus cities was generated by trade.

The Harappan people ate a lot of fish and grew a lot of cotton, things they had in common with citizens of another valley on the far side of the world. Caral in the desert of the Supe Valley in Peru was a large town with monuments, warehouses, temples and plazas. Discovered in the 1990s by Ruth Shady, it lies in a desert crossed by a river valley and was only the biggest of many towns in the area, some of which date from more than 5,000 years ago – the so-called Norte Chico civilisation. For archaeologists there are three baffling features of the ancient Peruvian towns. First, their people had no cereals in their diet. Maize was yet to be invented, and although there were several domesticated squashes and other foods, there was nothing so easily accumulated and stored as the grain which was the staple of Mesopotamia. The idea that cities are made possible by large-scale hoarding of grain thus takes a blow. Second, the Norte Chico towns have yielded no pottery of any kind: they were 'preceramic'. This surely made both the storage and the cooking of food more difficult, again undermining one of the favourite tenets of archaeologists trying to explain how cities began. And third, there is no evidence of warfare or defensive works. So the conventional wisdom that cereal stores made cities possible, that ceramic containers made them practical and that warfare made them necessary takes quite a knock from Norte Chico.

So what was driving people together into these South American towns? The answer, in a word, is trade. The settlements on the coast harvested fish in huge quantities, mainly anchovies and sardines, but also clams and mussels. For this they needed nets. The settlements in the interior grew huge quantities of cotton in fields irrigated with Andean snowmelt. They fashioned the cotton into nets, which they bartered for fish. There was not just mutual dependence, but mutual gain. A fisherman need only catch some more fish rather than spend time making his own nets; a cotton grower need only grow some more cotton rather than spend time fishing. Specialisation raised the standard of living for both. Caral lay at the centre of a large web of trade, reaching high into the Andes, over into the rainforest and far along the coast.

## The flag follows trade

To argue, therefore, that emperors or agricultural surpluses made the urban revolution is to get it backwards. Intensification of trade came first. Agricultural surpluses were summoned forth by trade, which offered farmers a way of turning their produce into valuable goods from elsewhere. Emperors, with their ziggurats and pyramids, were often made possible by trade. Throughout history, empires start as trade areas before they become the playthings of military plunderers from within or without. The urban revolution was an extension of the division of labour.

When a usurper named Sargon founded the Akkadian dynasty by conquest in the middle of the third millennium BC, he inherited the prosperity of the Syrian city of Ebla and its trading partners: a world in which grain, leather, textiles, silver and copper flowed easily between the Mediterranean and the Persian Gulf. Managing to resist the temptation of bureaucratic authoritarianism rather more than their Chinese and Egyptian contemporaries, the Akkadians allowed this trade to expand until it made fruitful contact with Lothal near the mouth of the Indus and bought the cotton and lapis lazuli of India with the wheat and bronze of Mesopotamia. A great free trade area stretched from the Nile to the Indus. An Akkadian merchant could handle Anatolian silver from a thousand miles to the west and Rajasthani copper from a thousand miles to the east. And that meant that he could raise the standard of living of the consumers he supplied, whether they were farmers or priests, by connecting them with distant producers of diverse goods.

Who was such a merchant? The economist Karl Polanyi argued in the 1950s that the concept of the market cannot be applied to any time before the fourth century BC, that until then instead of supply, demand and price, there was reciprocal exchange, state-sponsored redistribution of goods and top-down treaty trade in which agents were sent abroad to acquire things on behalf of the palace. Trade was administered, not spontaneous. But Polanyi's thesis or those of his fellow 'substantivists' has not stood the test of time well. It now seems that the state did not so much sponsor trade, as capture it. The more that comes to light about ancient trade, the more bottom-up it looks. While it is true that some Akkadian merchants may well have eventually seen themselves partly as civil servants sent abroad to acquire goods for their rulers, even they earned a living by trading for a profit themselves. Polanyi depicted a reflection of his own planning-obsessed times. The dirigiste mentality that dominated the second half of the twentieth century was always asking who is in charge, looking for who decided on a policy of trade. That is not how the world works. Trade emerged from the interactions of individuals. It evolved. Nobody was in charge.

So the typical Akkadian tamkarum or merchant was a businessman of the most surprisingly modern kind, who depended for his livelihood on freely exchanging goods for profit. Though there was no minted coinage, from the end of the fourth millennium BC there were silver-based prices, which fluctuated freely. The temple would act as a sort of bank, lending money at interest – and the Uruk word for high priest is the same as the word for accountant. By 2000 BC, under the Assyrian empire, merchants from Ashur operated in 'karum' enclaves in the independent states of Anatolia as thoroughly modern entrepreneurs with 'head offices, foreign branch-plants, corporate hierarchies, extra-territorial business law, and even a bit of foreign direct investment and value-added activity'. They bought gold, silver and copper in exchange for tin, goat-hair felt, woven textiles and perfumes shipped in on caravans of up to 300 donkeys. The profit margin was 100 per cent on tin and 200 per cent on textiles, but it had to be because the transport was unreliable and the risk of theft high. One such merchant, Pusu-Ken, operating in a tax-free zone in the Anatolian city of Kanesh, was to be found in 1900 BC lobbying the king, paying fines for evading textile import regulations imposed by the assembly, and sharing profits with his investor-partners, sounding in other words every inch the modern chief executive. Such merchants 'did not devote themselves to trading in copper and wool because Assyria needed them, but because that trade was a means of obtaining more gold and silver'. Profit ruled.

In these Bronze Age empires, commerce was the cause, not the symptom of prosperity. None the less, a free trade area lends itself easily to imperial domination. Soon, through tax, regulation and monopoly, the wealth generated by trade was being diverted into the luxury of the few and the oppression of the many. By 1500 BC you could argue that the richest parts of the world had sunk into the stagnation of palace socialism as the activities of merchants were progressively nationalised. Egyptian, Minoan, Babylonian and Shang dictators ruled over societies of rigid dirigisme, extravagant bureaucracy and feeble individual rights, stifling technological innovation, crowding out social innovation and punishing creativity. A Bronze Age empire stagnated for much the same reason that a nationalised industry stagnates: monopoly rewards caution and discourages experiment, the income is gradually captured by the interests of the producers at the expense of the interests of the consumers, and so on. The list of innovations achieved by the pharaohs is as thin as the list of innovations achieved by British Rail or the US Postal Service.

## The maritime revolution

Still, you cannot keep a good idea down. Around 1200 BC, the power of both Egypt and Assyria waned, the Minoans fell, the Myceneans fragmented and the Hittites came and went. It was a dark age for empires, and like the later Dark Ages that followed the fall of Rome, this political fragmentation, perhaps aided by a population decline, caused a burst of invention as demand rose among free people. The Philistines invented iron; the Canaanites the alphabet; and their coastal cousins, the Phoenicians, glass.

It was a different Phoenician invention, the bireme galley, that truly created the classical world. The people of Byblos, Tyre and Sidon lived close to great forests of magnificent cedars and cypresses, the hard, aromatic planks of which made especially durable boats. With decks of pine from Cyprus and oars of oak from Jordan (says Ezekiel), the Phoenician boat was greater than the sum of its far-flung parts. There was of course nothing new about the boat as a concept: boats had been plying the Nile, Euphrates, Indus and Yellow rivers for centuries, and the coasts of Asia and the Mediterranean for almost as long. But, realising their comparative advantage in timber, the Phoenicians built ships of greater capacity, finer trim and more seaworthy mortise joints than any people before them. Eventually they were able to build ships so large that they needed two banks of oars to propel them. Oars, though, were used only for manoeuvring close to shore. These were sailing ships and the larger they were, the more they could amplify the work of their human operators. Using the power of the wind, a comparatively small crew could transport a heavy cargo hundreds of miles further, and much more cheaply, than a caravan of donkeys could ever hope to manage.

Suddenly, for the first time, a large-scale seaborne division of labour became a possibility: wheat from Egypt could feed the Hittites in Anatolia; wool from Anatolia could clothe the Egyptians on the Nile; olive oil from Crete could enrich the diets of Assyrians in Mesopotamia. The ships of what is now Lebanon could trade for profit and scour the seas for tempting products. Grain, wine, honey, oil, resin, spices, ivory, ebony, leather, wool, cloth, tin, lead, iron, silver, horses, slaves, or a purple dye made from a gland in the body of the murex shellfish – there was little the Phoenicians could not find for an ambitious pharaoh with a harem to pamper, or a prosperous Assyrian farmer with a fiancée to impress.

All around the Mediterranean, markets grew into towns and ports into cities. Travelling farther afield, the Phoenicians' innovations multiplied: better keels, sails, navigational knowledge, accounting systems, log-keeping. Trade, once more, was the flywheel of the innovation machine. To the south, steeped in their religious obsessions, the Israelite pastoralists looked on in puritan horror at the explosion of wealth thus created. Isaiah cheerily anticipates Yahweh's destruction of Tyre, the 'market of the nations', to humble her pride. Ezekiel vents his *Schadenfreude* when Tyre is attacked: 'When thy wares went forth out of the seas, thou filledst many peoples; Thou didst enrich the kings of the earth with thy merchandise and thy riches ... Thou art become a terror; and thou shalt never be any more.' To the west, the warring island farmers of the Aegean looked down in warrior contempt on the bourgeois traders who were suddenly appearing in their midst. Throughout both the *Iliad* and the *Odyssey*, 'Homer' displays a relentlessly negative attitude to Phoenician traders and hints that they must be pirates. Greek trade in the age of Homer was supposed to handle precious reciprocal gifts between elites, not workaday goods in demand among ordinary people. The snobbery of the elite towards trade has ancient roots.

The effect of the Phoenicians must have been to create a burst of specialisation all around the Mediterranean. Villages, towns and regions would have discovered their comparative advantages in smelting metals, manufacturing pottery, tanning hides or growing grain. Mutual dependence and gains from trade would have emerged in unexpected places. Redressing the natural inequality in the location of metal ores, for example, benefits everybody. Cyprus may have lots of copper and Britain lots of tin, but put them together and bring them to Tyre and you can make the much more useful bronze. Tyrian traders founded Gadir, present-day Cadiz, around 750 BC not to settle the area but to trade with its inhabitants, in particular to exploit the silver ores of the Iberian hinterland – discovered, according to legend, when a forest fire caused rivulets of pure silver to pour from the hillsides. In doing so they must have turned the people of the region from largely self-sufficient peasants into producer-consumers. The Tartessian natives controlled the mining and smelting of the silver, selling it to the Tyrians at Gadir in exchange for oil, salt, wine and trinkets to charm the chiefs of the tribes farther into the interior. The Tyrians then took that silver (according to Diodorus, sometimes making silver anchors for their ships so as to squeeze a little more on board) east into the Mediterranean, exchanging it for staples and other luxuries.

No doubt just as the Tyrians could not believe their fortune at finding savages happy to give them so much silver for a little Cretan olive oil, so the Tartessians could not believe their luck at finding strange seagoing people prepared to give them such a convenient, storable, calorie-rich bounty for a mere metal. It is common to find that two traders both think their counterparts are idiotically overpaying: that is the beauty of Ricardo's magic trick. 'The English have no sense,' said a Montagnais trapper to a French missionary in seventeenth-century Canada. 'They give us twenty knives for this one beaver skin.' The contempt was mutual. When HMS *Dolphin*'s sailors found that a twenty-penny iron nail could buy a sexual encounter on Tahiti in 1767, neither sailors nor Tahitian men could believe their luck; whether the Tahitian women were as happy as their menfolk about this bargain goes unrecorded. Twelve days later, rampant inflation had set in and sex now cost a nine-inch marlinspike.

Traders from Gadir even worked their way south along the coast of Africa, acquiring gold from the inhabitants by 'silent trade': leaving goods on the shore and retreating. Comparative Ricardo ruled the Phoenician world. Tyre is the prototype of the trading port, the Genoa, Amsterdam, New York or Hong Kong of its day. The Phoenician diaspora is one of the great untold stories of history – untold because Tyre and its books were so utterly destroyed by thugs like Nebuchadnezzar, Cyrus and Alexander, and Carthage by the Scipios, so the story comes to us only through snippets from snobbish and envious neighbours. But in truth, was there ever a more admirable people than the Phoenicians? They knitted together not only the entire Mediterranean, but bits of the Atlantic, the Red Sea and the overland routes to Asia, yet they never had an emperor, had comparatively little time for religion and fought no memorable battles – unless you count Cannae, fought by a mercenary army paid by Carthage. I do not mean they were necessarily nice: they traded in slaves, sometimes resorted to war and did deals with the piratical Philistine 'sea peoples' who destroyed coastal cities around 1200 BC, but the Phoenicians seem to have managed to resist the temptations of turning into thieves, priests and chiefs better than most successful people in history. Through enter-prise they discovered social virtue.

## The virtue of fragmented government

The Phoenician diaspora teaches another important lesson, first advanced by David Hume: political fragmentation is often the friend, not the enemy, of economic advance, because of the stop which it gives 'both to power and authority'. There was no need for Tyre, Sidon, Carthage and Gadir to unite as a single political entity for them all to prosper. At most they were a federation. The extraordinary flowering of wealth and culture around the Aegean between 600 and 300 BC tells the same story. First the Milesians then the Athenians and their allies grew wealthy by trading among small, independent 'citizen states', not by uniting as an empire. Having copied the Phoenicians' ships and trading habits, Miletus, the most successful of the Ionian Greek cities, sat 'like a bloated spider' at the junction of four trade routes, east overland to Asia, north through the Hellespont to the Black Sea, south to Egypt and west to Italy. But though it established colonies all over the Black Sea, Miletus was not an imperial capital: it was first among equals. The city of Sybaris, a preferred trading partner of Miletus on a fertile plain in the toe of southern Italy, grew to perhaps several hundred thousand people and became a byword for opulence and refinement before it was destroyed by its enemies and buried under the diverted river Crathis in 510 BC.

The discovery of rich silver ores at Laurion in Attica in the 480s BC propelled the experimental democracy at Athens to the status of a regional economic superpower, not least by allowing it to finance a navy with which to defeat the Persians; but Athens too was *primus inter pares*. The Greek world depended crucially on finding gains from trade: grain from the Crimea, saffron from Libya and metals from Sicily swapped for olive oil from the Aegean itself. Modern philosophers who aspire to rise above the sordid economic reality of the world would do well to recall that this trade made possible the cross-fertilisation of ideas that led to great discoveries. Pythagoras probably got his theorem from a student of Thales the Milesian who learnt geometry on trade excursions to Egypt. We would never have heard of Pericles, Socrates or Aeschylus had there not been tens of thousands of slaves toiling underground at Laurion and tens of thousands of customers for Athenian goods all over the Mediterranean.

Yet as soon as Greece was unified into an empire by a thug – Philip of Macedon in 338 BC – it lost its edge. Had his son Alexander's empire lasted, it would undoubtedly have become as commercially and intellectually inert as its Persian predecessor. But because the empire fragmented on Alexander's death, parts of it were reborn as independent city states that lived off trade, most notably Alexandria in Egypt, which reached a third of a million people living in a state of famous wealth under the comparatively benign rule of the book-collecting Ptolemy III. That wealth was based on new roads to bring cash crops of cotton, wine, grain and papyrus within reach of the river Nile for export.

This is not to say that democratic city states are the only places where economic progress can occur, but it is to discern a pattern. Plainly, there is something beneficial to the growth of the division of labour when governments are limited (though not so weak that there is widespread piracy), republican or fragmented. The chief reason is surely that strong governments are, by definition, monopolies and monopolies always grow complacent, stagnant and self-serving. Monarchs love monopolies because where they cannot keep them to themselves, they can sell them, grant them to favourites and tax them. They also fall for the perpetual fallacy that they can make business work more efficiently if they plan it rather than allow and encourage it to evolve. The scientist and historian Terence Kealey points out that entrepreneurs are rational and if they find that wealth can more easily be stolen than created, then they will steal it: 'Humanity's great battle over the last 10,000 years has been the battle against monopoly.'

This is not disproved by the success of two empires from around the beginning of the Christian era: both Rome and India realised the benefits of economic unification before they managed to endure the disasters of political unification. The Mauryan empire in India seems to have harvested the prosperity of the Ganges valley to combine an imperial monarchy with expanding trade. It was ruled at its zenith in 250 BC by Asoka, a warrior who turned into a Buddhist pacifist once he had won (funny, that) and was as economically benign a head of state as you could wish. He built roads and waterways to encourage the movement of goods, established a common currency and opened maritime trade routes with China, south-east Asia and the Middle East, sparking an export-led boom in which cotton and silk textiles played a prominent part. Trade was carried on almost entirely by private firms (*sreni*) of a recognisably corporate kind; taxation, though extensive, was fairly administered. There were remarkable scientific advances, not least the invention of zero and the decimal system and the accurate calculation of pi. Asoka's empire disintegrated before it had become totalitarian, and its legacy was impressive: for the next few centuries the Indian subcontinent was both the most populous and the most prosperous part of the world, with a third of the world's people and a third of the world's GDP. It was without question the economic superpower of the day, dwarfing both China and Rome, and its capital city Pataliputra was the largest city in the world, famous for its gardens, luxuries and markets. Only later, under the Guptas, did the caste system ossify Indian commerce.

## From Ganges to Tiber

Asoka was a contemporary of Hannibal and Scipio, which brings me to Rome. Rome's particular speciality, from its very first days until the end of its empire, was simply to plunder its provinces to pay for bribes, luxuries, triumphs and soldiers' pensions nearer to home. There were four respectable ways for a prominent Roman to gain wealth: land-owning, booty from war, money lending and bribery. Cicero pocketed over two million sesterces (three times the sum he had previously quoted to illustrate 'luxury') from his governorship of Cilicia in 51 and 50 BC – and he had a reputation as an especially honest governor.

Yet there is no doubt that Rome's hegemony was built on trade. Rome was the final unification of Greece's and Carthage's trade zones, with a smattering of belligerent Etruscans and Latins in charge. 'The history of antiquity resounds with the sanguinary achievements of Aryan warrior elites,' wrote Thomas Carney, 'but it was the despised Levantines, Arameans, Syrians and Greeklings who constituted the economic heroes of antiquity.' The populous and prosperous cities of southern Italy, Sicily and points east that were the core of Rome's world were Greek-speaking; they did the hard work of keeping people rich while legionaries and consuls strutted their triumphs. The fact that standard histories of Rome barely mention the markets, merchants, ships and family firms that sustained the empire, preferring instead to bang on about battles, does not mean they did not exist. Ostia was a trading city as surely as Hong Kong is today, with 'a vast piazza housing the head offices of some five dozen companies'. Much of the Campanian country-side was devoted to slave-manned plantations growing wine and oil for export.

Moreover, Rome's continuing prosperity once the republic became an empire may be down at least partly to the 'discovery' of India. Following Augustus's absorption of Egypt, the Romans inherited the Egyptians' trade with the East, and soon the Red Sea was alive with massive Roman cargo ships carrying tin, lead, silver, glass and wine – the latter soon becoming an exciting novelty in India. Thanks to the discovery of the monsoon, which reliably blew ships eastward in summer and back westward in winter, the journey across the Arabian Sea was cut from years to months. At last Rome's ships made direct contact with the world's economic superpower. In the first century, the anonymous author of *The Periplus of the Erythrean Sea* described the navigation and trade of the Indian Ocean; Strabo wrote that 'now great fleets are sent as far as India'; and the emperor Tiberius complained of Indian luxuries draining the empire of its wealth. Peacocks from India became a favourite possession of Roman plutocrats. Indian ports like Barigaza (modern Bharuch in Gujarat) seem to have blossomed through exporting cotton cloth and other manufactures to the West. Soon, even within India, there were enclaves of Roman traders, whose hoards of amphorae and coins still sometimes come to light. Arikamedu, for example, on the east coast near modern Pondicherry, was exporting to China glass imported from Roman Syria (glass blowing was a new Roman invention and glass was suddenly much better and cheaper throughout the empire).

Think about this from the consumer's point of view. Nobody in China can blow glass; nobody in Europe can reel silk. Thanks to a middleman in India, however, the European can wear silk and the Chinese can use glass. The European may scoff at the ridiculous legend that this lovely cloth is made from the cocoons of caterpillars; and the Chinese may guffaw at the laughable fable that this transparent ceramic is made from sand. But both of them are better off and so is the Indian middleman. All three have acquired the labour of others. In Robert Wright's terms, this is a non-zero transaction. The collective brain has expanded across the entire Indian Ocean and lifted the standard of living at both ends.

# Ships of the desert

But the plundering, the lack of invention, the barbarians and above all Diocletian's red tape did for Rome in the end. As the empire disintegrated under this bureaucratic burden, at least in the west, money lending at interest stopped and coins ceased to circulate so freely. In the Dark Ages that followed, because free trade became impossible, cities shrank, markets atrophied, merchants disappeared, literacy declined and – crudely speaking – once Goth, Hun and Vandal plundering had run its course, everybody had to go back to being self-sufficient again. Europe de-urbanised. Even Rome and Constantinople fell to a fraction of their former populations. Trade with Egypt and India largely dried up, especially once the Arabs took control of Alexandria, so that not only did oriental imports such as papyrus, spices and silk cease to appear, but those export-oriented plantations in Campania became the plots of subsistence farmers instead. In that sense, the decline of the Roman empire turned consumer traders back into subsistence peasants. The Dark Ages were a massive experiment in the back-to-the-land hippy lifestyle (without the trust fund): you ground your own corn, sheared your own sheep, cured your own leather and cut your own wood. Any pathetic surplus you generated was confiscated to support a monk, or maybe you could occasionally sell something to buy a metal tool off a part-time blacksmith. Otherwise, subsistence replaced specialisation.

This was never, of course, absolutely true. Within each village or monastery there was a degree of specialisation, but it was not enough to support large towns. At least there were now, as there had not been in slave-powered Rome, incentives to improve technology. A steady trickle of innovations began to improve productivity in northern Europe long after the end of the western empire: the barrel, soap, spoked wheels, the overshot water wheel, the horseshoe and the horse collar. Fitfully, Byzantium prospered from what was left of the Mediterranean trade, but plague, war, politics and piracy kept getting in the way. The predatory expansion of the Carolingian Franks in the eighth century, caused by a modest revival of regional trade in grain and manufactures, began also to stimulate trade in spices and slaves across the Mediterranean. The Vikings, paddling their boats down the rivers of Russia to the Black Sea and the Mediterranean, partly revived the oriental trade (with a little pillage thrown in) – hence their sudden prosperity and power.

But meanwhile the torch passed east. As Europe sank back into self-sufficiency, Arabia was discovering gains from trade. The sudden emergence of an all-conquering prophet in the middle of a desert in the seventh century is rather baffling as the tale is usually told – one of religious inspiration and military leadership. What is missing from the story is the economic reason that Arabs were suddenly in a position to carry all before them. Thanks to a newly perfected technology, the camel, the people of the Arabian Peninsula found themselves well placed to profit from trade between East and West. The camel caravans of Arabia were the source of the wealth that carried Muhammad and his followers to power. The camel had been domesticated several thousand years earlier, but it was in the early centuries AD that it was at last made into a reliable beast of burden. It could carry far more than a donkey could, go to places a wheeled bullock cart could not, and because it could find its own forage en route, its fuel costs were essentially zero – like a sailing ship. For a while even the Byzantine sailing ships of the Red Sea, waiting for the right winds and running the gauntlet of increasingly numerous pirates, found themselves at a competitive disadvantage compared with 'ships of the desert'. With the route down the Euphrates disrupted by wars between Sassanid Persia and Byzantine Constantinople, the way was open for the people of Mecca, like dry Phoenicians, to become rich through trade. Spices, slaves and textiles went north and west; while metals, wine and glass went south and east.

Later, by adopting two Chinese inventions, the lateen sail and the sternpost rudder, the Arabs extended their commercial tentacles deep into Africa and the Far East. A dhow that sank off Belitung in Indonesia in AD 826 was carrying objects of gold, silver, lead, lacquer, bronze and 57,000 ceramics, including 40,000 Changsha bowls, 1,000 funerary urns and 800 inkpots – mass-produced exports from the kilns of Hunan for the well-heeled consumers of Basra and Baghdad. Not coincidentally, the free-trading Arabs exchanged ideas as well as goods and culture thrived. As they spilled out of their homeland, Arabs brought luxury and learning to an area stretching from Aden to Cordoba, before the inevitable imperial complacency and then severe priestly repression set in at home. Once the priesthood tightened its grip, books were burned, not read.

## The merchant of Pisa

In due course, these Muslim gains from trade began to lift Europe out of its self-sufficiency thanks largely to Jewish traders, who in the tenth century abandoned the increasingly oppressive court of the Abbasids in Baghdad for the more tolerant regime of the Egyptian Fatimids. Settling along the southern shores of the Mediterranean and in Sicily, these Maghribi traders developed their own rules of contract enforcement and punishment by ostracism, quite outside the official courts. Like all the best entrepreneurs, they thrived despite, rather than because of their government. And it was they who began to trade with the ports of Italy. Italian peasants started to discover that instead of dividing their land among impoverished heirs they could send sons to town to trade with Maghribi Jews.

Northern Italy, because of a stand-off between the Holy Roman emperor and the pope, was temporarily favoured by an absence of greedy rent-seeking kings. When Arab piracy and papal plunder paused under the influence of the first Otto, the towns of Lombardy and Tuscany found themselves free to set up their own governments, and since towns were there because of trade, these governments became dominated by the interests of merchants. Amalfi, Pisa, and above all Genoa began to flourish on the back of the Maghribi trade. It was a Pisan trader living in north Africa, Fibonacci, who brought Indian–Arabic decimals, fractions and the calculation of interest to Europe's notice in his book *Liber Abaci*, published in 1202. Genoa's trade with North Africa doubled after an agreement for the protection of merchants was reached in 1161, and by 1293 the city's trade exceeded the entire revenue of the king of France. Lucca acquired a strong position in the silk trade and then in banking. Florence became wealthy through weaving wool and silk. Milan, gateway to the Alpine passes, flourished as a market town. And Venice, long independent in the safety of its lagoon, gradually became the epitome of the trading state. Despite competing and often warring with each other, republican city states, run by merchants, not only took care not to tax or regulate trade into extinction, but did everything they could to encourage it: in Venice, for example, the government built and leased ships and arranged convoys.

Italy's prosperity was felt in northern Europe, too. Venetian merchants crossed the Brenner pass into Germany in search of silver and began to appear at Champagne fairs in Flanders – another no man's land between kingdoms – bringing silk, spices, sugar and lacquer in exchange for wool. In the early 1400s, for instance, Giovanni Arnolfini settled in Bruges as an agent for his family silk business in Lucca, and was immortalised in the famous painting by van Eyck. Although a small percentage of the European population in the Middle Ages would have even encountered silk and sugar, let alone regularly, and a tiny proportion of Europe's GDP came from such trade, none the less it is undeniable that Europe's reawakening was boosted by contact with the productivity of China, India, Arabia and Byzantium through Italian trade. Regions that participated in Asian trade grew richer than the regions that did not: by 1500 Italy's GDP per capita was 60 per cent higher than the European average. But historians often put too much emphasis on exotic trade with the Orient. As late as 1600, European trade with Asia, dominated because of transport costs by luxuries such as spices, was only half the value of the inter-regional European trade in cattle alone. Europe could trade with Asia because it traded so much with itself, not vice versa.

Inexorably, gains from trade could be rediscovered – people could become consumers again, which meant that they could also become producers of cash crops to sell to each other. If I grow a bit more wheat and you tan a bit more leather, then I can feed you and you can shoe me ... Eventually in the twelfth century towns started to grow at a rapid rate. By 1200, Europe was once again a place of markets, merchants and craftsmen, though heavily dependent on the 70 per cent who worked the land to produce food, fibre, fuel and housing material. In an unusually warm climate the continent was enjoying an economic boom. Living standards rose all across the continent of Europe, especially in the north, where the Hanseatic merchants from Lübeck and other cities, equipped with new, slow, but capacious sailing ships called cogs, did for the Baltic and the North Sea what the Genoese had done for the Mediterranean. They brought timber, fur, wax, herrings and resin west and south in exchange for cloth and grain. Like the Maghribis they developed their own *lex mercatoria*, merchant law, with sanctions against those who broke their contracts when abroad, quite independent of national laws. Through the rivers of Russia and the Black Sea, the merchants of Visby on Gotland even re-established contact with the Orient via Novgorod, bypassing the Arabs who controlled the Strait of Gibraltar.

## The Moloch state

China, meanwhile, was heading the other way, into stagnation and poverty. China went from a state of economic and technological exuberance in around AD 1000 to one of dense population, agrarian backwardness and desperate poverty in 1950. According to Angus Maddison's estimates, it was the only region in the world with a lower GDP per capita in 1950 than in 1000. The blame for this lies squarely with China's governments.

Pause, first, to admire the exuberance. China's best moments came when it was fragmented, not united. The economy first truly prospered in the unstable Zhou dynasty of the first millennium BC. Later, after the Han empire fell apart in AD 220, the Three Kingdoms period saw a flourishing of culture and technology. When the Tang empire came to an end in 907, and the 'Five Dynasties and Ten Kingdoms' fought each other incessantly, China experienced its most spectacular burst of invention and prosperity yet, which the Song dynasty inherited. Even the rebirth of China in the late twentieth century owes much to the fragmentation of government and to an explosion of local autonomy. The burst of economic activity in China after 1978 was driven by 'township and village enterprises', agencies of the government given local freedom to start companies. One of the paradoxical features of modern China is the weakness of a central, would-be authoritarian government.

By the late 1000s, the Chinese were masters of silk, tea, porcelain, paper and printing, not to mention the compass and gunpowder. They used multi-spindle cotton wheels, hydraulic trip hammers, as well as umbrellas, matches, toothbrushes and playing cards. They made coke from coal to smelt high-grade iron: they were making 125,000 tonnes of pig iron a year. They used water power to spin hemp yarn. They had magnificent water clocks. All across the Yangtze delta the Confucian dictum 'men plough; women weave' was obeyed with industrious efficiency so that peasants were working for cash as well as subsistence and were using that cash to consume goods. Art, science and engineering flourished. Bridges and pagodas sprang up everywhere. Woodblock printing quenched a raging thirst for literature. The Song era had, in short, a highly elaborate division of labour: many people were consuming what each other produced.

Then came the calamities of the 1200s and 1300s. First the Mongol invasion, then the Black Death, then a series of natural disasters, followed by the all too unnatural disaster of totalitarian Ming rule. The Black Death, as I shall argue in the next chapter, spurred Europe into further gains from trade and escaping the trap of self-sufficiency; why did it not have the same effect in China, where it left the country half as populous as before and therefore presumably rich in surplus land to support disposable income? The blame rests squarely with the Ming dynasty. Western Europe only bounced back from the Black Death because it had regions of independent city states run by and for merchants, notably in Italy and Flanders. This made it harder for landowners to reimpose serfdom and restrictions on peasant movement after the plague had briefly empowered the labouring classes. In Eastern Europe, Mamluk Egypt and Ming China, serfdom was effectively restored.

Empires, indeed governments generally, tend to be good things at first and bad things the longer they last. First they improve society's ability to flourish by providing central services and removing impediments to trade and specialisation; thus, even Genghis Khan's Pax Mongolica lubricated Asia's overland trade by exterminating brigands along the Silk Road, thus lowering the cost of oriental goods in European parlours. But then, as Peter Turchin argues following the lead of the medieval geographer Ibn Khaldun, governments gradually employ more and more ambitious elites who capture a greater and greater share of the society's income by interfering more and more in people's lives as they give themselves more and more rules to enforce, until they kill the goose that lays the golden eggs. There is a lesson for today. Economists are quick to speak of 'market failure', and rightly so, but a greater threat comes from 'government failure'. Because it is a monopoly, government brings inefficiency and stagnation to most things it runs; government agencies pursue the inflation of their budgets rather than the service of their customers; pressure groups form an unholy alliance with agencies to extract more money from taxpayers for their members. Yet despite all this, most clever people still call for government to run more things and assume that if it did so, it would somehow be more perfect, more selfless, next time.

Not only did the Ming emperors nationalise much of industry and trade, creating state monopolies in salt, iron, tea, alcohol, foreign trade and education, but they interfered with the everyday lives of their citizens and censored expression to a totalitarian degree. Ming officials had high social status and low salaries, a combination that inevitably bred corruption and rent-seeking. Like all bureaucrats they instinctively mistrusted innovation as a threat to their positions and spent more and more of their energy on looking after their own interests rather than the goals they were put there to pursue. As Etienne

Balazs put it:

> The reach of the Moloch-state, the omnipotence of the bureau-
> cracy, goes much further. There are clothing regulations, a
> regulation of public and private construction (dimensions of
> houses); the colours one wears, the music one hears, the festivals
> – all are regulated. There are rules for birth and rules for death;
> the providential State watches minutely over every step of its
> subjects, from cradle to grave. It is a regime of paperwork and
> harassment, endless paperwork and endless harassment.

Do not be fooled by the present tense: this is Ming, not
Maoist, China that Balazs is describing. The behaviour of
Hongwu, the first of the Ming emperors, is an object lesson in
how to stifle the economy: forbid all trade and travel without
government permission; force merchants to register an inven-
tory of their goods once a month; order peasants to grow for
their own consumption and not for the market; and allow
inflation to devalue the paper currency 10,000-fold. His son
Yong-Le added some more items to the list: move the capital at
vast expense; maintain a gigantic army; invade Vietnam unsuc-
cessfully; put your favourite eunuch in charge of a nationalised
fleet of monstrous ships with 27,000 passengers, five astrologers
and a giraffe aboard, then in a fit of pique at the failure of this
mission to make a profit, ban everybody else from building ships
or trading abroad.

Yet the Chinese people were bursting to trade with the world.
In the 1500s Portuguese carracks took silk from Macao to Japan
in exchange for silver. In the 1600s junks that had slipped un-
officially from the coast of Fujian arrived in Manila laden with
silk, cotton, porcelain, gunpowder, mercury, copper, walnuts
and tea. There they met a large Spanish galleon stuffed with
silver from the Potosi mine in Peru, which had crossed the
Pacific from Acapulco. It is no accident that when the Ming

dynasty fell, weakened by the silver drought caused by the loss of three Acapulco galleons in three years, it fell to Manchu traders who financed their conquest by the profitable exchange of goods with Korea and Japan.

Part of the problem was that a Chinese artisan could not flee to work under a more tolerant ruler or in a more congenial republic, as Europeans did routinely. Because of its peninsulas and mountain ranges, Europe is much harder to unify than China: ask Charles V, Louis XIV, Napoleon or Hitler. For a while the Romans achieved a sort of European unity, and the result was just like the Ming: stagnation and bureaucracy. Under the emperor Diocletian (just as under the emperor Yong-Le) 'tax collectors began to outnumber taxpayers', said Lactantius, and 'a multitude of governors and hordes of directors oppressed every region – almost every city; and to these were added count- less collectors and secretaries and assistants to the directors.'

Since then, Europe had been fragmented among warring states. So Europeans took to their heels all the time, sometimes fleeing from cruel rulers as French Huguenots and Spanish Jews did, sometimes drawn to ambitious ones, sometimes seeking republican freedom. The Italian Christopher Columbus gave up on Portugal and tried Spain instead. The Sforzas lured engineers to Milan; Louis XI enticed Italian silk makers to set up in Lyon; Johann Gutenberg moved from Mainz to Strasbourg in search of investors; Gustavus Adolphus started the Swedish iron industry by bringing in a Walloon named Louis de Geer; John Kay, English inventor of the flying shuttle, was paid 2,500 livres a year by the French authorities to tour Normandy demonstrating his machine. In one especially bizarre case of industrial poaching in the early 1700s, Augustus the Strong of Saxony got a monopoly on the manufacture of porcelain by the cunning ploy of imprisoning a passing charlatan who claimed to be able to make gold – lest any other state get him. The man in question, Johann Friedrich Bottger, made no gold, but perfected a colleague's technique for making fine porcelain in the hope that this would win him back his freedom. So Augustus locked him even more securely in a hilltop castle at Meissen and put him to work churning out teapots and vases. In short, competition was a grand incentive to European industrialisation, and a brake on bureaucratic suffocation, at the national as well as the corporate level.

## Repeal the corn laws again

The greatest beneficiaries of European political fragmentation were the Dutch. By 1670, uncommanded by emperors and even fragmented among themselves, the Dutch so dominated European international trade that their merchant marine was bigger than that of France, England, Scotland, the Holy Roman Empire, Spain and Portugal – combined. They brought grain from the Baltic, herrings from the North Sea, whale blubber from the Arctic, fruit and wine from southern Europe, spices from the Orient and of course their own manufactures to whoever wanted them. Through efficient ship construction (not least a new division of labour within the shipyard, as observed by a curious William Petty) they undercut all other shipping costs by more than one-third. It did not last long. Within a century Louis XIV and others had ended the Dutch golden age with a mixture of war, mercantilist retaliation and high taxes, imposed to fight wars. Yet another attempt to use free trade to lift living standards bit the dust. But because this was not monolithic China, the baton was picked up by others, especially the British.

Victorian Britain's great good fortune was that at the moment of industrial take-off Robert Peel embraced free trade, whereas Yong-Le had banned it. Between 1846 and 1860, Britain unilaterally adopted a string of measures to open its markets to free trade to a degree unprecedented in history. It abolished the corn laws, terminated the navigation acts, removed all tariffs and agreed trade treaties with France and others incorporating the 'most favoured nation' principle – that any liberalisation applied to all trading parties. This spread tariff reduction like a virus through the countries of the world and genuine global free trade arrived at last – a planetary Phoenician experiment. So at the crucial moment America could specialise in providing food and fibre to Britain and Europe, which could further specialise in providing manufactures for the consumers of the world. Both sides benefited. By 1920, for example, 80 per cent of all beef eaten in London was imported, mostly from Argentina, which was one of the richest countries in the world as a consequence. Both sides of the estuary of the River Plate became a vast slaughterhouse where beef was canned, salted and dried for export, the name of the Uruguayan town of Fray Bentos turning into a synonym for canned meat in Britain.

The message from history is so blatantly obvious – that free trade causes mutual prosperity while protectionism causes poverty – that it seems incredible that anybody ever thinks otherwise. There is not a single example of a country opening its borders to trade and ending up poorer (coerced trade in slaves or drugs may be a different matter). Free trade works for countries even if they do it and their neighbours do not. Imagine a situation in which your street is prepared to accept produce from other streets but they are only allowed what they produce: who loses? Yet in the aftermath of the First World War, one by one countries tried beggaring their neighbours in the twentieth century. As currencies devalued and unemployment rose in the 1930s, government after government sought self-sufficiency and import substitution: Greece under Ioannis Metaxas, Spain under Francisco Franco, America under Smoot-Hawley. Global trade fell by two-thirds between 1929 and 1934. In India in the 1930s, the British government imposed tariffs to protect wheat farmers, cotton manufacturers and sugar producers against cheap imports from Australia, Japan and Java respectively. These protectionist measures exacerbated the economic collapse. In five years from 1929, Japanese silk exports collapsed from 36 per cent of the total to 13 per cent. Little wonder that with a rapidly growing population but a shrinking opportunity to export both goods and people, the Japanese regime began seeking imperial space instead.

Then after the Second World War, the entire continent of Latin America broke with free trade under the influence of an Argentinian economist named Raul Prebisch, who thought he had found the flaw in Ricardo's logic, and achieved decades of stagnation. India, under Jawaharlal Nehru, went for autarky too, closing its borders to trade in the hope of sparking a boom in import substitution. It too found stagnation. Still they tried: North Korea under Kim Il Sung, Albania under Enver Hoxha, China under Mao Zedong, Cuba under Fidel Castro – every country that tried protectionism suffered. Countries that went the other way include Singapore, Hong Kong, Taiwan, South Korea and later Mauritius, bywords for miraculous growth. Countries that changed tack in the twentieth century include Japan, Germany, Chile, post-Mao China, India and more recently Uganda and Ghana. China's Open Door policy, which cut import tariffs from 55 per cent to 10 per cent in twenty years, transformed it from one of the most protected to one of the most open markets in the world. The result was the world's greatest economic boom. Trade, says Johann Norberg, is like a machine that turns potatoes into computers, or anything into anything: who would not want to have such a machine at their disposal?

Trade, for example, could transform Africa's prospects. China's purchases from Africa (not counting its direct invest-ments there) quintupled in the Nineties and quintupled again in the Noughties, yet they still account for just 2 per cent of China's foreign trade. China may be about to repeat some of Europe's colonial exploitative mistakes in Africa, but in terms of being open to trade from the continent it puts Europe and

America to shame. Farm subsidies and import tariffs on cotton, sugar, rice and other products cost Africa $500 billion a year in lost export opportunities – or twelve times the entire aid budget to the continent.

Yes, of course, trade is disruptive. Cheap imports can destroy jobs at home – though in doing so they always create far more both at home and abroad, by freeing up consumers' cash to buy other goods and services. If Europeans find their shoes made cheaply in Vietnam, then they have more to spend on getting their hair done and there are more nice jobs for Europeans in hair salons and fewer dull ones in shoe factories. Sure, manufacturers will and do seek out countries that tolerate lower wages and lower standards – though, prodded by Western activists, in practice their main effect is then to raise the wages and standards in such places, where they most need raising. It is less of a race to the bottom, more of a race to raise the bottom. Nike's sweatshops in Vietnam, for example, pay wages three times as high as local state owned factories and have far better facilities. That drives up wages and standards. During the period of most rapid expansion of trade and out-sourcing, child labour has halved since 1980: if that is driving down standards, let there be more of it.

## The apotheosis of the city

Trade draws people to cities and swells the slums. Is this not a bad thing? No. Satanic the mills of the industrial revolution may have looked to romantic poets, but they were also beacons of opportunity to young people facing the squalor and crowding of a country cottage on too small a plot of land. As Ford Madox Ford celebrated in his Edwardian novel *The Soul of London*, the city may have seemed dirty and squalid to the rich but it was seen by the working class as a place of liberation and enterprise. Ask a modern Indian woman why she wants to leave her rural village for a Mumbai slum. Because the city, for all its dangers and squalor, represents opportunity, the chance to escape from the village of her birth, where there is drudgery without wages, suffocating family control and where work happens in the merciless heat of the sun or the drenching downpour of the monsoon. Just as Henry Ford said he was driven to invent the gasoline buggy to escape the 'crushing boredom of life on a mid-west farm', so, says Suketa Mehta, 'for the young person in an Indian village, the call of Mumbai isn't just about money. It's also about freedom.'

All across Asia, Latin America and Africa, a tide of subsistence farmers is leaving the land to move to cities and find paid work. To many Westerners, suffused with *nostalgie de la boue* (nostalgia for mud), this is a regrettable trend. Many charities and aid agencies see their job as helping to prevent subsistence farmers having to move to the city by making life in the countryside more sustainable. 'Many of my contemporaries in the developed world,' writes Stewart Brand, 'regard subsistence farming as soulful and organic, but it is a poverty trap and an environmental disaster.' Surely a Nairobi slum or a São Paolo favela is a worse place to be than a tranquil rural village? Not for the people who move there. Given the chance they eloquently express their preference for the relative freedom and opportunity of the city, however poor the living conditions. 'I am better off in all facets of life compared to my peers left behind in the village,' says Deroi Kwesi Andrew, a teacher earning $4 a day in Accra. Rural self-sufficiency is a romantic mirage. Urban opportunity is what people want. In 2008 for the first time more than half the people in the world lived in cities. That is not a bad thing. It is a measure of economic progress that more than half the population can leave subsistence and seek the possibilities of a life based on the collective brain instead. Two-thirds of economic growth happens in cities.

Not long ago, demographers expected new technology to

hollow out cities as people began to telecommute from tranquil suburbs. But no – even in weightless industries like finance people prefer to press into ever closer contact with each other in glass towers to do their exchanging and specialising, and they are prepared to pay absurdly high rents to do so. By 2025, it looks as if there will be five billion people living in cities (and rural populations will actually be falling fast), and there will be eight cities with more than twenty million people each: Tokyo, Mumbai, Delhi, Dhaka, São Paolo, Mexico City, New York and Calcutta. As far as the planet is concerned, this is good news because city dwellers take up less space, use less energy and have less impact on natural ecosystems than country dwellers. The world's cities already contain half the world's people, but they occupy less than 3 per cent of the world's land area. 'Urban sprawl' may disgust some American environmentalists, but on a global scale, the very opposite is happening: as villages empty, people are living in denser and denser anthills. As Edward Glaeser put it, 'Thoreau was wrong. Living in the country is not the right way to care for the Earth. The best thing that we can do for the planet is build more skyscrapers.'

After a 'stinking hot' evening in a taxi in central Delhi in the 1960s, the ecologist Paul Ehrlich had an epiphany. 'The streets seemed alive with people. People eating, people washing, people sleeping. People visiting, arguing, and screaming. People thrusting their hands through the taxi window, begging. People defecating and urinating. People clinging to buses. People herding animals. People, people, people, people.' It was then that Ehrlich, like so many Westerners with culture shock, decided that the world had (to quote his chapter title) 'too many people'. However good life might get, perhaps in the end it is all in vain because of population growth. Was he right? It is time to understand old 'Population Malthus'.

## CHAPTER 6

# Escaping Malthus's trap: population after 1200

The great question is now at issue, whether man shall henceforth start
forwards with accelerated velocity towards illimitable, and hitherto
unconceived improvement; or be condemned to a perpetual oscillation
between happiness and misery.

T. R. MALTHUS
*Essay on Population*

## PERCENTAGE INCREASE IN WORLD POPULATION

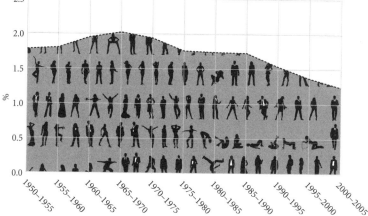

Since human beings are just another kind of animal, the story of population should be a simple one. Give us more food and we will have more babies until we reach the density at which starvation, predators and parasites crash the system. In some episodes of human history something like this has indeed happened. Yet often, after the crash, population density settles at a higher level than before. The subsistence level keeps on rising, erratically, but inexorably. In terms of power and relative wealth, modern Egypt may be a shadow of its pharaonic self, but it is much more heavily populated today than it was in Ramses II's day.

There is another odd feature. On the way up the graph, abundant food encourages some people to specialise in something other than growing or catching food, while others produce food for sale not for self-sufficiency. The division of labour increases. But when the food supply becomes tight, near the top of the graph, fewer people will be prepared to sell their food or will have a surplus to sell. They will feed it to their families and make do without the goods they were wont to buy from others. The non-farmers, finding both food and customers for their services harder to come by, will have to give up their jobs and return to growing their own food themselves. So there is a cycle of rising and falling specialisation in human populations. The economist Vernon Smith, in his memoirs, recalls how in the Depression his family moved in the 1930s from Wichita, Kansas, to a farm when his father was laid off as a machinist, because 'we could at least grow most of our own food and participate in a subsistence economy.' This return to subsistence happens often in human history.

In the animal world, this is unique. In no animal species do individuals become more specialised as population is rising, nor less specialised as population is stalling or falling. In fact, the whole notion of specialised individuals is rare outside the human race, and where specialisation does happen – in ants, for example – it does not wax and wane in this way.

This suggests that good old-fashioned Malthusian population limitation does not really apply to human beings, because of their habit of exchange and specialisation. That is to say, instead of dying from famine and pestilence when too numerous for their food supply, people can increase their specialisation, which allows more to subsist on the available resources. On the other hand, if exchange becomes harder, they will reduce their specialisation, which can lead to a population crisis even without an increase in population. The Malthusian crisis comes not as a result of population growth directly, but because of decreasing specialisation. Increasing self-sufficiency is the very signature of a civilisation under stress, the definition of a falling standard of living. Until 1800 this was how every economic boom ended: with a partial return to self-sufficiency driven by predation by elites, or diminishing returns from agriculture. It is hard to be sure given the patchy information that this is what happened to Mesopotamia and Egypt after 1500 BC, or India and Rome after AD 500, but it is pretty clear that it happened to China and to Japan in later centuries. As Greg Clark puts it, 'In the preindustrial world, sporadic technological advance produced people, not wealth.'

## The medieval collapse

Robert Malthus and David Ricardo, though they were good friends, disagreed on much. But in one respect they were entirely aligned – that unchecked population could drive down the standard of living.

Malthus: 'In some countries, the population appears to have been forced, that is, the people have been habituated by degrees to live upon the smallest possible quantity of food ... China seems to answer this description.'

Ricardo: 'The land being limited in quantity, and differing in quality, with every increased portion of capital employed on it there will be a decreased rate of production.'

At first glance, medieval England furnishes a tidy example of such diminishing returns. The thirteenth century, a time of mild weather across Europe, saw a prolonged expansion of the population, which then crashed in the following century as the weather deteriorated. The 1200s were the golden high-water mark of the Middle Ages. Courts were richly furnished; monasteries flourished; cathedrals rose towards the sky; troubadours strutted their stuff. Watermills, windmills, bridges and ports were built all over England. Fairs and markets proliferated and thrived: there was an unprecedented surge in commercial activity between 1150 and 1300. A good part of it was driven by the wool trade. As Flemish merchants sought out more and more English wool to supply the cloth makers of Flanders, so they provided livelihoods for ship owners, fullers and above all sheep farmers. The national sheep flock boomed to perhaps ten million animals, more than two sheep per person. The English had found a comparative advantage in their mild, wet, grass-growing climate – a gain from trade – supplying Europe's fibre. Specialisation and exchange fuelled population growth.

For example, in 1225, of the 124 people assessed for a survey in the Wiltshire village of Damerham, fifty-nine owned sheep, with a combined flock of 1,259 animals. That meant they sold wool for cash rather than strove for self-sufficiency. They presumably used that cash to buy bread from the baker who bought flour from the miller, who bought grain from other farmers, who therefore got cash too. Instead of self-sufficiency, everybody was now in the market and had disposable income. People wanted to travel to the market in nearby Salisbury to buy things: so the carter was doing well, too, and the merchants in Salisbury. In 1258 a spectacular cathedral began to take shape in Salisbury, on the back of the wool boom, because the Church was coining it in tithes and taxes. Put yourself in the shoes of a grain farmer in Damerham. The miller wants all you can grow, so you encourage both your sons to marry early and rent a few acres off you. The carter, the miller, the baker, the merchant and the shepherds are all doing the same: setting their children up in business. Family formation – which had always been as much an economic as a biological decision – increased markedly in the thirteenth century. The consequence of all this early and frequent marriage was fecundity. In the thirteenth century the population of England seems to have doubled, from over two million to something like five million people.

Inevitably, and gradually, the population boom overtook the economy's productivity. Rents inflated and wages deflated: the rich were bidding up land prices while the poor were bidding down wages. By 1315 real wages had halved in a century, although because of family formation, family income was probably not falling as fast as individual wages. For example, a miller in Feering in Essex in the 1290s agreed to halve his wage when his employer took on another employee. Chances are the new employee was the miller's son and they were simply sharing the same income within the family. None the less, as pay packets shrank, demand for the goods supplied by merchants must have begun to stall. To feed the growing population, marginal land was being ploughed, and was yielding fewer and fewer grains for each grain sown. Diminishing returns dominated. Predatory priests and chiefs did not help.

Before long hunger was a real risk. It came suddenly in the sodden summers of 1315 and 1317, when wheat yields more than halved all across the north of Europe. The crops rotted in the fields; some people were forced to eat their own seed corn. Mothers abandoned their babies. There were rumours of fresh corpses of criminals pulled from gallows for food. In the years that followed, with continuing poor harvests and unusually cold winters, a fatal murrain spread among hungry oxen, and that left some land unploughed, further exacerbating the food shortage. The population then stagnated for three decades until the Black Death arrived in the 1340s and caused a crash in human numbers. The plague returned in the 1360s, followed by more bad harvests and more plague outbreaks. By 1450, the population of England had been reduced to roughly where it had been in 1200.

Yet neither the boom of the thirteenth century, nor the bust of the fourteenth, can be described in simplistic Ricardian and Malthusian terms. The carrying capacity of the land was not much increased in the first period by Ricardian technological change, nor much diminished in the second by Malthusian falls in yield. What changed was the economy's, rather than the land's, capacity to support so many people. After all, the Black Death was not caused by overpopulation, but by a bacterium. Ironically, the plague may have been one of the sparks that lit the Renaissance, because the shortage of labour shifted income from rents to wages as landlords struggled to find both tenants and employees. With rising wages, some of the surviving peasantry could once more just afford the oriental luxuries and fine cloth that Lombard and Hanseatic merchants supplied. There was a rash of financial innovation: bills of credit to solve the problem of how to pay for goods without transporting silver through bandit country, double-entry book-keeping, insurance. Italian bankers began to appear all across the continent, financing kings and their wars, sometimes at a profit, sometimes at a disastrous loss. The wealth that the Italian trading towns had generated soon found its way into scholarship, art or science, or in the case of Leonardo da Vinci, all three. Per capita income in England was probably higher in 1450 than it would be again before 1820.

The point is this. In 1300, Europe was probably on a trajectory towards a labour-intensive 'industrious' revolution of diminishing returns. Remember the miller of Feering who halved his wage by sharing his job with his son in the 1290s? Or consider the women who were paid half what their menfolk earned when they carried water (for making mortar) to the site of a new windmill being constructed at Dover Castle in 1294.

No doubt they were delighted to have a job and earn a little cash, but they came so cheap they provided their employer with an incentive not to buy a cart and bullock. Yet by 1400, Europe had partly switched to a labour-saving 'industrial' trajectory instead, and the pattern was repeated after the cold and brutal seventeenth century, when famine, plague and war once more reduced the European population: in 1692–4, perhaps 15 per cent of all French people starved to death. Unlike Mesopotamia, Egypt, India, Mexico, Peru, China and Rome, early modern Europe became capital-intensive, not labour-intensive. That capital was used to get work out of animals, rivers and breezes, rather than people. Europe was, in Joel Mokyr's words, 'the first society to build an economy on non-human power rather than on the backs of slaves and coolies'.

## The industrious revolution

To imagine what would have happened to Europe without the Black Death, consider the case of Japan in the eighteenth century. In the 1600s Japan was a relatively prosperous and sophisticated country with a population the size of France and Spain combined, and a strong manufacturing industry, especially in paper products, cotton textiles and weapons – much of them for export. In 1592, the Japanese had conquered Korea carrying tens of thousands of home-made arquebuses copied from Portuguese designs. Japan was none the less mainly an agrarian economy with plentiful herds of sheep and goats, lots of pigs, some cattle and oxen and quite a few horses. The plough was in common use, both ox-drawn and horse-drawn.

By the 1800s, domestic farm animals had virtually disappeared. Sheep and goats were almost unknown, horses and cattle were very rare and even pigs were few in number. As the traveller Isabella Bird remarked in 1880, 'As animals are not used for milk, draught or food and there are no pasture lands, both the country and the farm-yards have a singular silence and an inanimate look.' Carriages, carts (and even wheelbarrows) were scarce. Instead the power needed for transport came from human beings carrying goods hung from poles on their shoulders and racks on their backs. Watermills, though the technology had been known for a long time, were little used; rice was threshed and ground by hand querns or stone-weighted trip hammers, powered by treadle. Human rice pounders could be heard toiling away, naked behind a curtain, for hours at a time, even in cities like Tokyo; the irrigation pumps needed for the rice fields were often driven by pedalling coolies. Above all, the plough was now virtually unknown in the entire country. Fields were cultivated by men and women with hoes. Where Europeans used animal, water and wind power, the Japanese did the work themselves.

What seems to have happened is that some time between 1700 and 1800, the Japanese collectively gave up the plough in favour of the hoe because people were cheaper to hire than draught animals. This was a time of rapid population expansion, made possible by the high productivity of paddy rice, naturally fertilised by nitrogen-fixing cyanobacteria in the water and therefore needing little manure (though human night soil was assiduously collected, carefully stored and diligently applied to the land). With abundant food and a fastidious approach to hygiene, the Japanese population boomed to the point where land was scarce, labour was cheap and it was literally more economic to use human labour to hoe the land than to set aside precious acres for pasture to support oxen or horses to draw a plough. So the Japanese, to a spectacular extent, retreated from technology and trade and reduced their demands on merchants as they became more self-sufficient. The market for technology of all kinds atrophied. They even gave up capital-intensive guns in favour of labour-intensive swords. A good Japanese sword had a blade of strong though soft steel, but with a brittle, hard edge made lethally sharp by incessant hammering.

Europe probably came close to going down the same path as Japan in the eighteenth century. Just as in the thirteenth century, the European population boomed in the 1700s, helped by wealth generated by local and oriental trade and agricultural improvements. New crops like the potato, though often treated with suspicion when urged on the populace by rulers (Marie-Antoinette's wearing of potato flowers put the French off eating them for decades), allowed the population of some countries such as Ireland to boom. Potatoes could be grown using a spade rather than a plough, and their fantastic productivity – more than thrice the calories per acre of wheat or rye – and high nutrient content encouraged a very dense population. An Irish acre in 1840 could yield six tonnes of potatoes, almost as much food as an acre of rice paddy in the Yangtze delta. (Sir William Petty, lamenting the idleness of the Irish in 1691, blamed the potato: 'What need have they of work, who can content themselves with potato's [sic], whereof the labour of one man can feed Forty?' Adam Smith begged to differ, crediting the potato for London having the 'strongest men and the most beautiful women perhaps in the British dominions'.) At the time, an English worker needed twenty acres to grow his bread and cheese. The subsistence farmers of Ireland, even into the 1800s, were not only dependent mainly on their own muscle power for cultivation and transport, but were 'out of the market', consuming very few manufactured goods for lack of disposable income. (Rapacious English landlords did not help.) As the size of each family potato plot shrank, Ireland was a Malthusian disaster waiting to happen even before the *Phytophthora* famine of 1845 killed a million people and drove a million more to America. In the Scottish Highlands too, the population boom of the 1700s caused a retreat to subsistence, or crofting as it was known there. Only a vast 'clearance' and emigration to America and Australia, highly coerced and highly resented to this day, relieved the Malthusian pressure.

Denmark followed Japan's path, too, for a while. The Danes responded to increasing ecological constraints in the eighteenth century by intensifying their agricultural labour. They banned cattle from forests to protect the supply of future fuel, which increased the price of manure. To maintain the fertility of their soil, they worked extraordinarily hard at ditching, clover growing and marling (laboriously digging up and spreading lime and clay subsoil to neutralise and release nutrients from sandy or acid soils). Hours of work increased by more than 50 per cent. By the 1800s, Denmark had become a country that was trapped by its own self-sufficiency. Its people were so busy farming that none could be spared for other industries and few could afford to consume manufactured products. Living standards stagnated, admittedly at a relatively decent level. Eventually in the late nineteenth century the industrialisation of its neighbours then created a market for Danish agricultural exports and these could slowly raise the living standards of Danes.

## British exceptionalism

It was Britain's fate to escape the quasi-Malthusian trap into which Japan, Ireland and Denmark fell. The reasons are many and debatable, but here it is worth noting a surprising demographic factor. Britain, more than any other country, had unintentionally prepared itself for industrial life in an elemental, human way. For centuries – leaving out the aristocrats (who left fewer heirs because they died from falling off horses) – the relatively rich had more children than the relatively poor. On average a merchant in Britain who left £1,000 in his will had four surviving children, while a labourer who left £10 had only two – this was in around 1600, but the differential was similar at other dates. Such differential reproduction happened in China, too, but to a much lesser extent. Because there was little or no increase in the standard of living between 1200 and 1700, this overbreeding by the rich meant there was constant downward mobility. Gregory Clark has shown from legal records that rare surnames of the poor survived much less well than rare surnames of the rich.

By 1700, therefore, in Britain most of the poor were actually the descendants of the rich. They had perhaps carried down with them into the working classes many of the habits and customs of the rich: literacy, for example, numeracy and perhaps industriousness or financial prudence. This theory accounts especially well for the otherwise puzzling rise in literacy during the early modern period. It may also account for the steady decline in violence. Your chances of being a victim of homicide in England fell from 0.3 per thousand in 1250 to 0.02 per thousand in 1800: you were ten times more likely to be killed in the earlier period.

Fascinating as this demographic discovery is, it cannot fully explain the industrial revolution. The same was not nearly as true of Holland in its golden age; and it would, for example, struggle to explain China's rapid and successful industrialisation after 1980 – in the wake of a policy of deliberate murder and humiliation of the literate and the bourgeois in the Cultural Revolution.

What Europe achieved after 1750 – uniquely, precariously, unexpectedly – was an increasing division of labour that meant that each person could produce more each year and therefore could consume more each year, which created the demand for still more production. Two things, says the historian Kenneth Pomeranz, were vital to Europe's achievement: coal and America. The ultimate reason that the British economic take-off kept on going where the Chinese – or for that matter, the Dutch, Italian, Arab, Roman, Mauryan, Phoenician or Mesopotamian – did not was because the British escaped the Malthusian fate. The acres they needed to provide themselves with corn, cotton, sugar, tea and fuel just kept on materialising elsewhere. Here are Pomeranz's numbers: in around 1830, Britain had seventeen million acres of arable land, twenty-five million acres of pastureland and less than two million acres of forest. But she consumed sugar from the West Indies equivalent (in calories) to the produce of at least another two million acres of wheat; timber from Canada equivalent to another one million acres of woodland, cotton from the Americas equivalent to the wool produced on an astonishing twenty-three million acres of pastureland, and coal from underground equivalent to fifteen million acres of forest. Without these vast 'ghost acres' Britain's industrial revolution, which was only just starting to raise living standards in 1830, would have already shuddered to a halt for lack of calories, cotton or coal.

Not only did the Americas ship back their produce; they also allowed a safety valve for emigration to relieve the Malthusian pressure of the population boom induced by industrialisation. Germany, in particular, as it industrialised rapidly in the nineteenth century, saw a huge increase in the birth rate, but a flood of emigrants to the United States prevented the division of land among multiple heirs and the return to poverty and self-sufficiency that had afflicted Japan two centuries before.

When Asia experienced a population boom in the early twentieth century, it had no such emigration safety valve. In fact, Western countries firmly and deliberately closed the door, terrified by the 'yellow peril' that might otherwise head their way. The result was a typical Malthusian increase in self-sufficiency. By 1950 China and India were bursting with the self-sufficient agrarian poor.

## The demographic transition

It is hard now to recall just how coercive were the population policies urged by experts in the mid-twentieth century. When President Lyndon Johnson's adviser Joseph Califano suggested that an increase in famine relief should be announced before a visit by Indira Gandhi to the United States, Johnson supposedly replied that he was not going to 'piss away foreign aid in nations where they refuse to deal with their own population problems'. Garrett Hardin, in his famous essay 'The Tragedy of the Commons' (remembered these days as being about collective action, but actually a long argument for coerced population control), found 'freedom to breed intolerable', coercion 'a necessity' and that 'the only way we can preserve and nurture other and more precious freedoms is by relinquishing the freedom to breed, and that very soon.' Hardin's view was nearly universal. 'Adding a sterilant to drinking water or staple foods is a suggestion that seems to horrify people more than most proposals for involuntary fertility control,' wrote John Holdren (now President Obama's science adviser) and Paul and Anne Ehrlich in 1977, but not to worry: 'It has been concluded that compulsory population-control laws, even including laws requiring compulsory abortion, could be sustained under the existing Constitution if the population crisis became sufficiently severe to endanger the society.' All right-thinking people agreed, as they so often do, that top-down government action was needed: people must be ordered or at least bribed to accept sterilisation and punished for refusing it.

Which is exactly what happened. Egged on by Western governments and pressure groups such as the International Planned Parenthood Foundation, coerced sterilisation became a pattern in many parts of Asia in the 1970s. 'Dalkon Shield' contraceptive devices, the subject of safety lawsuits in America, were bought in bulk by the American federal government and shipped to Asia. Chinese women were forcibly taken from their homes to be sterilised. Cheered on by Robert McNamara's World Bank, Sanjay Gandhi, the son of the Indian prime minister, ran a vast campaign of rewards and coercion to force eight million poor Indians to accept vasectomies. In one episode, recounted by the historian Matthew Connelly, the village of Uttawar was surrounded by police and every eligible male sterilised. In response, a crowd gathered to defend the nearby village of Pipli, but police fired on the crowd, killing four people. A government official was unapologetic. In this war against 'people pollution', force was justified: 'if some excesses appear, don't blame me ... Whether you like it or not, there will be a few dead people.' Eventually Sanjay Gandhi's policies proved so unpopular that his mother lost an election by a landslide in 1977, and family planning was treated with deep suspicion for many years thereafter.

Yet the tragedy is that this top-down coercion was not only counter-productive; it was unnecessary. Birth rates were already falling rapidly in the 1970s all across the continent of Asia quite voluntarily. They fell just as far and just as fast without coercion. They continue to fall today. As soon as it felt prosperity from trade, Asia experienced precisely the same transition to lower birth rates that Europe had experienced before.

Bangladesh today is the most densely populated large country in the world, with more than two thousand people living on every square mile; it has a population greater than Russia living on an area smaller than Florida. In 1955 Bangladesh had a birth rate of 6.8 children per woman. Today, fifty years later, that ratio has more than halved, to about 2.7 children per woman. On current trends Bangladesh's population will soon cease growing altogether. Its neighbour India has seen a similar collapse in fecundity, from 5.9 to 2.6 children per woman. In Pakistan the birth rate did not start dropping till the mid-1980s, but its decline has been catching up its neighbours: it has halved in just twenty years to 3.2 children per woman. Between them these three countries account for about a quarter of the world's population. If they had not seen their birth rates fall so fast, the world population boom would have become deafening.

Yet they are not alone. Throughout the world, birth rates are falling. There is no country in the world that has a higher birth rate than it had in 1960, and in the less developed world as a whole the birth rate has approximately halved. Until 2002, the United Nations, when projecting future world population density, assumed that birth rates would never fall below 2.1 children per woman in most countries: that is the 'replacement rate', at which a woman produces enough babies to replace her and her husband, with 0.1 babies added in to cover childhood deaths and a slightly male-biased sex ratio. But in 2002, the UN changed this assumption as it became clear that in country after country the decline in baby-making went straight through the 2.1 level and kept on dropping. If anything, the decline may accelerate as the effect of small family size compounds. Nearly half the world now has fertility below 2.1. Sri Lanka's birth rate, at 1.9, is already well below replacement level. Russia's population is falling so fast it will be one-third smaller in 2050 than it was at its peak in the early 1990s.

Do these statistics surprise you? Everybody knows the population of the world is growing. But remarkably few people seem to know that the rate of increase in world population has been falling since the early 1960s and that the raw number of new people added each year has been falling since the late 1980s. As the environmentalist Stewart Brand puts it, 'Most environmentalists still haven't got the word. Worldwide, birth-rates are in free fall … On every part of every continent and in every culture (even Mormon), birth rates are headed down. They reach replacement level and keep on dropping.' This is happening despite people living longer and thus swelling the ranks of the world population for longer, and despite the fact that babies are no longer dying as frequently as they did in the early twentieth century. Population growth is slowing even while death rates are falling.

Frankly, this is an extraordinary bit of luck. Had the human race continued to turn wealth into more babies as it did for so many centuries, it would come to grief eventually. When the world population looked like it would hit fifteen billion by 2050 and keep on rising after that, there was a genuine risk of not feeding or watering that number comfortably, at least not while hanging on to any natural habitats. But now that even the United Nations' best estimate is that world population will probably start falling once it peaks at 9.2 billion in 2075, there is every prospect of feeding the world for ever. After all, there are already 6.8 billion on the earth and they are still feeding better and better every decade. Only 2.4 billion to go.

Think of it this way. After the world population first hit a billion in (best guess) 1804, the human race had another 123 years to work out how to feed the next billion, the two billion milestone being reached in 1927. The next billions took thirty-three, fourteen, thirteen and twelve years respectively to arrive. Yet despite the accelerating pace, the world food supply in calories per head improved dramatically. The rate at which the billions are being added is now falling. The seven billionth person won't be born till 2013, fourteen years after the six billionth, the eight billionth will come fifteen years after that and the nine billionth another twenty-six years after that. The ten billionth, it is now officially forecast, will never come at all.

In technical jargon, the entire world is experiencing the second half of a 'demographic transition' from high mortality and high fertility to low mortality and low fertility. It is a process that has occurred in many countries, starting with France at the end of the eighteenth century then spreading to Scandinavia and Britain in the nineteenth century and to the rest of Europe in the early twentieth century. Asia began to follow the same path in the 1960s, Latin America in the 1970s and most of Africa in the 1980s. It is now a worldwide phenomenon: with the exception of Kazakhstan, there is no country where birth rate is high and rising. The pattern is always the same: mortality falls first, causing a population boom, then a few decades later, fecundity falls quite suddenly and quite rapidly. It usually takes about fifteen years for birth rate to fall by 40 per cent. Even Yemen, the country with the highest birth rate in the world for most of the 1970s with an average of nearly nine babies per woman, has halved the number. Once the demographic transition starts happening in a country it happens at all levels of society pretty well at the same time.

Not everybody saw the demographic transition coming, but some did. When the journalist John Maddox wrote a book in 1973 arguing that the demographic transition was already slowing Asian birth rates, he was treated to a condescending blast by Paul Ehrlich and John Holdren:

> The most serious of Maddox's many demographic errors is his invocation of a 'demographic transition' as the cure for population growth in Asia, Africa and Latin America. He expects that birth rates there will drop as they did in developed countries following the industrial revolution. Since most underdeveloped countries are unlikely to have an industrial revolution, this seems somewhat optimistic at best. But even if those nations should follow that course, starting immediately, their population growth would continue for well over a century – perhaps producing by the year 2100 a world population of twenty thousand million.

Rarely has a paragraph proved so wrong so soon.

## An unexplained phenomenon

Deliciously, nobody really knows how to explain this mysteriously predictable phenomenon. Demographic transition theory is a splendidly confused field. The birth-rate collapse seems to be largely a bottom-up thing that emerges by cultural evolution, spreads by word of mouth, and is not commanded by fiat from above. Neither governments nor churches can take much credit. After all, the European demographic transition happened in the nineteenth century without any official encouragement or even knowledge. In the case of France, it happened in the teeth of official encouragement to breed. Likewise, the modern transition began without any government family-planning policies in many countries, especially Latin America. China's highly coerced ('one child') birth-rate decline since 1955 (from 5.59 to 1.73 children, or 69 per cent) is almost exactly mirrored by Sri Lanka's largely voluntary one over the same time period (5.70 to 1.88, or 67 per cent). As for religion, Italy's plunging birth rate (now 1.3 children per woman) in the pope's backyard has always seemed moderately amusing to non-Catholics. Of course, the provision of family planning advice surely helps, and in parts of Asia may have accelerated the transition, but on the whole it seems to help women cheaply and easily achieve what they wish to achieve anyway. The onset of Britain's demographic transition in the 1870s coincided with the publication of bestsellers on contraception by Annie Besant and Charles Bradlaugh – but which caused which?

So what might be the cause of these episodes of quite extraordinary downward shift in human fecundity? Top of the list of explanations, paradoxically, comes falling child mortality. The more babies are likely to die, the more their parents bear. Only when women think their children will survive do they plan and complete their families rather than just keep breeding. This remarkable fact seems to be very poorly known. Most Western, educated people seem to think, rationally enough, that keeping babies alive in poor countries is only making the population problem worse and that ... well, the implication is usually left unspoken. Jeffrey Sachs recounts that on 'countless occasions' after a lecture a member of the audience has 'whispered' to him 'if we save all those children, won't they simply starve as adults?' Answer: no. If we save children from dying, people will have smaller families. In Niger or Afghanistan today, where more than fifteen of every 100 babies die before their first birthdays, the average woman will give birth seven times in her lifetime; in Nepal and Namibia, where less than five babies out of every 100 die, the average woman gives birth three times. But the correlation is not exact. Burma has twice the infant mortality and half the birth rate of Guatemala, for instance.

Another factor is wealth. Having more income means you can afford more babies, but it also means you can afford more luxuries to divert you from constant breeding. Children are consumer goods, but rather time-consuming and demanding ones compared with, say, cars. The transition seems to kick in as countries grow richer, but there is no exact level of income at which it happens, and the poor and the rich within any country start reducing their birth rate about the same time. Once again, there are exceptions: Yemen has almost twice the birth rate and almost twice the income per head of Laos.

Is it female emancipation? Certainly, the correlation between widespread female education and low birth rate is pretty tight, and the high fecundity of many Arab countries must in part reflect women's relative lack of control over their own lives. Probably by far the best policy for reducing population is to encourage female education. It is evolutionarily plausible that in the human species, females want to have relatively few children and give them high-quality upbringing, whereas males like to have lots of children and care less about the quality of their upbringing. So the empowerment of women through education gives them the upper hand. But there are exceptions here too: 90 per cent of girls complete primary school in Kenya, which has twice the birth rate of Morocco, where only 72 per cent of girls complete primary school.

Is it urbanisation? Certainly, as people move from farms, where children can help in the fields, to cities where housing is expensive and jobs are outside the home, they find large families to be a drawback. Most cities are – and always have been – places where death rates exceed birth rates. Immigration sustains their numbers. Yet this cannot be the whole story: Nigeria is twice as urbanised and twice as fecund as Bangladesh.

In other words, the best that can be said for sure about the demographic transition is that countries lower their birth rates as they grow healthier, wealthier, better educated, more urbanised and more emancipated. A typical woman probably reasons thus: now I know my children will probably not die of disease, I do not need to have so many; now I can get a job to support those children, I do not want to interrupt my career too often; now I have an education and a pay cheque, I can take control of contraception; now education can get my children non-farming jobs, I shall have only as many as I can support through school; now I can buy consumer goods, I shall be careful not to spread my income across too large a family; now I live in a city I will plan my family. Or some combination of such thoughts. And she will be encouraged by the examples of others, and by family-planning clinics.

To argue that the demographic transition is a mysterious, evolutionary, natural phenomenon, rather than a successful government policy, is not to say that it cannot be given a push. If Africa's slow fall in birth rates could be accelerated, there would be great dividends in terms of welfare. A bold programme, driven by philanthropy or even government aid, but not tied to teaching sexual abstinence, to cut child mortality in countries like Niger, and hence bring forward the fall in family size, and to spread the news of family planning out to rural villages, could mean that Africa has 300 million fewer mouths to feed in 2050 than it otherwise would. However, politicians should be careful not to repeat in Africa the high-minded brutality that Asia experienced in the 1970s.

It is somewhat distasteful to the intelligentsia to accept that consumption and commerce could be the friend of population control, or that it is when they 'enter the market' as consumers that people plan their families – this is not what most market-phobic professors, preaching anticapitalist asceticism, want to hear. Yet the relationship is there, and it is strong. Seth Norton found that the birth rate was more than twice as high in countries with little economic freedom (average 4.27 children per woman) compared with countries with high economic freedom (average 1.82 children per woman). Besides, there is quite a neat exception which proves this rule. The Anabaptist sects in North America, the Hutterites and Amish, have largely resisted the demographic transition; that is to say, they have large families. This has been achieved despite – or rather because of – an ascetic emphasis on family roles, which immunises them against the spread of time-consuming hobbies (including higher education) and a taste for expensive gadgets.

What a happy conclusion. Human beings are a species that stops its own population expansions once the division of labour reaches the point at which individuals are all trading goods and services with each other, rather than trying to be self-sufficient. The more interdependent and well-off we all become, the more population will stabilise well within the resources of the planet. As Ron Bailey puts it, in complete contradiction of Garrett Hardin: 'There is no need to impose coercive population control measures; economic freedom actually generates a benign invisible hand of population control.'

Most economists are now more worried about the effects of imploding populations than they are about exploding ones. Countries with very low birth rates have rapidly ageing work-forces. This means more and more old people eating the savings and taxes of fewer and fewer people of working age. They are right to be concerned, though they would be wrong to be apocalyptic, after all, today's 40-year-olds will surely be happier to continue operating computers in their seventies than today's 70-year-olds are to continue operating machine tools. And once again, the rational optimist can bring a measure of comfort. The latest research uncovers a second demographic transition in which the very richest countries see a slight increase in their birth rate once they pass a certain level of prosperity. The United States, for example, saw its birth rate bottom out at 1.74 children per woman in about 1976; since then it has risen to 2.05. Birth rates have risen in eighteen of the twenty-four countries that have a Human Development Index greater than 0.94. The puzzling exceptions are ones such as Japan and South Korea, which see a continuing decline. Hans-Peter Kohler of the University of Pennsylvania, who co-authored the new study, believes that these countries lag in providing women with better opportunities for work–life balance as they get richer.

So, all in all, the news on global population could hardly be better, though it would be nice if the improvements were coming faster. The explosions are petering out; and the declines are bottoming out. The more prosperous and free that people become, the more their birth rate settles at around two children per woman with no coercion necessary. Now, is that not good news?

# CHAPTER 7

# The release of slaves: energy after 1700

With coal almost any feat is possible or easy; without it we are thrown back in the laborious poverty of earlier times.

STANLEY JEVONS
*The Coal Question*

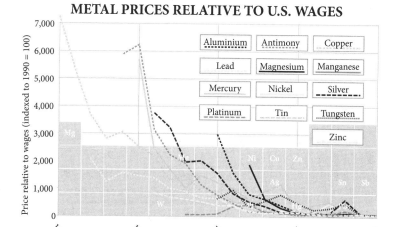

**METAL PRICES RELATIVE TO U.S. WAGES**

In 1807, as Parliament in London was preparing to pass at last William Wilberforce's bill to abolish the slave trade, the largest factory complex in the world had just opened at Ancoats in Manchester. Powered by steam and lit by gas, both generated by coal, Murrays' Mills drew curious visitors from all over country and beyond to marvel at their modern machinery. There is a connection between these two events. The Lancashire cotton industry was rapidly converting from water power to coal. The world would follow suit and by the late twentieth century, 85 per cent of all the energy used by humankind would come from fossil fuels. It was fossil fuels that eventually made slavery – along with animal power, and wood, wind and water – uneconomic. Wilberforce's ambition would have been harder to obtain without fossil fuels. 'History supports this truth,' writes the economist Don Boudreaux: 'Capitalism exterminated slavery.'

The story of energy is simple. Once upon a time all work was done by people for themselves using their own muscles. Then there came a time when some people got other people to do the work for them, and the result was pyramids and leisure for a few, drudgery and exhaustion for the many. Then there was a gradual progression from one source of energy to another: human to animal to water to wind to fossil fuel. In each case, the amount of work one man could do for another was ampli-fied by the animal or the machine. The Roman empire was built largely on human muscle power, in the shape of slaves. It was Spartacus and his friends who built the roads and houses, who tilled the ground and trampled the grapes. There were horses, forges and sailing ships as well, but the chief source of watts in Rome was people. The period that followed the Roman empire, especially in Europe, saw the widespread replacement of that human muscle power by animal muscle power. The European early Middle Ages were the age of the ox. The invention of dried-grass hay enabled northern Europeans to feed oxen through the winter. Slaves were replaced by beasts, more out of practicality than compassion one suspects. Oxen eat simpler food, complain less and are stronger than slaves. Oxen need to graze, so this civilisation had to be based on villages rather than cities. With the invention of the horse collar, oxen then gave way to horses, which can plough at nearly twice the speed of an ox, thus doubling the productivity of a man and enabling each farmer either to feed more people or to spend more time consuming other's work. In England, horses were 20 per cent of draught animals in 1086, and 60 per cent by 1574.

In turn oxen and horses were soon being replaced by inanimate power. The watermill, known to the Romans but comparatively little used, became so common in the Dark Ages that by the time of the Domesday Book (1086), there was one for every fifty people in southern England. Two hundred years later, the number of watermills had doubled again. By 1300 there were sixty-eight watermills on a single mile of the Seine in Paris, and others floating on barges.

The Cistercian monastic order took the watermill to its technical zenith, not only improving and perfecting it, but aggressively suppressing rival animal-powered mills by legal action. With gears, cams and trip hammers, they used the water to achieve multiple ends. At Clairvaux, for example, the water from the river first turned the mill wheel to crush the grain, then shook the sieve to separate flour from bran, then topped up the vats to make beer, then moved on to work the fullers' hammers against the raw cloth, then trickled into the tannery and was finally directed to where it could wash away waste.

The windmill appeared first in the twelfth century and spread rapidly throughout the Low Countries, where water power was not an option. But it was peat, rather than wind, that gave the Dutch the power to become the world's workshop in the 1600s. Peat dug on a vast scale from freshly drained bogs fuelled the brick, ceramic, beer, soap, salt and sugar industries. Harlem bleached linen for the whole of Germany. At a time when timber was scarce and expensive, peat gave the Dutch their chance.

Hay, water and wind are ways of drawing upon the sun's energy: the sun powers plants, rain and the wind. Timber is a way of drawing on a store of the sun's energy laid down in previous decades – on solar capital, as it were. Peat is an older store of the sunlight – solar capital laid down over millennia. And coal, whose high energy content enabled the British to overtake the Dutch, is still older sunlight, mostly captured around 300 million years before. The secret of the industrial revolution was shifting from current solar power to stored solar power. Not that human muscle power disappeared: slavery continued, in Russia, the Caribbean and America as well as many other places. But gradually, erratically, more and more of the goods people made were made with fossil energy.

Fossil fuels cannot explain the start of the industrial revolution. But they do explain why it did not end. Once fossil fuels joined in, economic growth truly took off, and became almost infinitely capable of bursting through the Malthusian ceiling and raising living standards. Only then did growth become, in a word, sustainable. This leads to a shocking irony. I am about to argue that economic growth only became sustainable when it began to rely on non-renewable, non-green, non-clean power. Every economic boom in history, from Uruk onwards, had ended in bust because renewable sources of energy ran out: timber, cropland, pasture, labour, water, peat. All self-replenishing, but far too slowly, and easily exhausted by a swelling populace. Coal not only did not run out, no matter how much was used: it actually became cheaper and more abundant as time went by, in marked contrast to charcoal, which always grew more expensive once its use expanded beyond a certain point, for the simple reason that people had to go further in search of timber. Had England never used coal, it could still have had an industrial miracle of sorts, because it could have (and did) use water power to drive the frames and looms that turned Lancashire into the cotton capital of the world. But water power, though renewable, is very much finite, and Britain's industrial boom would have petered out as expansion became impossible, population pressure overtook income and wages fell, depressing demand.

This is not to imply that non-renewable resources are infinite – of course not. The Atlantic Ocean is not infinite, but that does not mean you have to worry about bumping into Newfoundland if you row a dinghy out of a harbour in Ireland. Some things are finite but vast; some things are infinitely renewable, but very limited. Non-renewable resources such as coal are sufficiently abundant to allow an expansion of both economic activity and population to the point where they can generate sustainable wealth for all the people of the planet without hitting a Malthusian ceiling, and can then hand the baton to some other form of energy. The blinding brightness of this realisation still amazes me: we can build a civilisation in which everybody lives the life of the Sun King, because everybody is served by (and serves) a thousand servants, each of whose service is amplified by extraordinary amounts of inanimate energy and each of whom is also living like the Sun King. I will deal in later chapters with the many objections that pessimistic environmentalists will raise, including the question of the atmosphere's non-renewable capacity for absorbing carbon dioxide.

## Wealthier yet and wealthier

Before I make the case that fossil fuels, by driving pistons and dynamos, made modern living standards possible, first, a digression about living standards. Did industrialisation really improve them? There are still people about, including it seems those who write the textbooks from which my children learn history, who follow Karl Marx in believing that the industrial revolution drove down most living standards, by cramming carefree and merrie yokels into satanic mills and polluted tenements, where they were worked till they broke and then coughed their way to early deaths. Is it really necessary to point out that poverty, inequality, child labour, disease and pollution existed before there were factories? In the case of poverty, the rural pauper of 1700 was markedly worse off than the urban pauper of 1850 and there were many more of him. In Gregory King's survey of the British population in 1688, 1.2 million labourers lived on just £4 a year and 1.3 million 'cottagers' – peasants – on just £2 a year. That is to say, half the entire nation lived in abject poverty; without charity they would starve. During the industrial revolution, there was plenty of poverty but not nearly as much as this nor nearly as severe. Even farm labourers' income rose during the industrial revolution. As for inequality, in terms of both physical stature and number of surviving children, the gap narrowed between the richest and the poorest during industrialisation. That could not have happened if economic inequality increased. As for child labour, a patent for a hand-driven linen-spinning machine from 1678, long before powered mills, happily boasts that 'a child three or four years of age may do as much as a child of seven or eight years old.' As for disease, deaths from infectious disease fell steadily throughout the period. As for pollution, smog undoubtedly increased in industrial cities, but the sewage-filled streets of Samuel Pepys's London were more noisome than anything in Elizabeth Gaskell's Manchester of the 1850s.

The plain fact is that the mechanisation of production in the industrial revolution raised incomes across all classes. The average Englishman's income, having apparently stagnated for three centuries, began to rise around 1800 and by 1850 was 50 per cent above its 1750 level, despite a trebling of population. The rise was steepest for unskilled workers: the wage premium for skilled building workers fell steadily. Income inequality fell, and gender inequality, too. The share of national income captured by labour rose, while the share captured by land fell: the rent of an acre of English farmland buys as many goods now as it did in the 1760s, while the real wage of an hour of work buys immensely more. Real wages rose faster than real output throughout the nineteenth century, meaning that the benefit of cheaper goods was being garnered chiefly by the workers as consumers, not by bosses or landlords. That is to say, the people who produced manufactured goods could also increasingly afford to consume them.

While it is undoubtedly true that by modern standards the workers who manned the factories and mills of 1800 in England laboured for inhuman hours from an early age in conditions of terrible danger, noise and dirt, returning to crowded and insanitary homes through polluted streets, and had dreadful job security, diet, health care and education, it is none the less just as undeniably true that they lived better lives than their farm-labourer grandfathers and wool-spinning grandmothers had done. That was why they flocked to the factories from the farms – and would do so again in New England in the 1870s, in the American South in the 1900s, in Japan in the 1920s, in Taiwan in the 1960s, in Hong Kong in the 1970s and in China today. That was why the jobs in the mills were denied to the Irish in New England and the blacks in North Carolina.

Here are three anecdotes to illustrate the notion that factory jobs are often preferable to farm ones. A farm worker named William Turnbull, born in 1870, told my grandmother that he started work at thirteen, for sixpence a day, working six days a week, from 6 a.m. to 6 p.m., usually outdoors whatever the weather, with just Good Friday, Christmas Day and half of New Year's day as his only holidays. On market days he started herding sheep or cattle to town, carrying a lantern, at 1 or 2 a.m. A cotton picker from North Carolina in the 1920s explained to a different historian why the mill was so much better than the farm: 'Once we went to work in the mill after we moved here from the farm, we had more clothes and more kinds of food than we did when we was a-farmin'. And we had a better house. So yes when we came to the mill life was easier.' And in the 1990s Liang Ying was delighted to run away from the family rubber farm in southern China, where she had daily to cut the bark of hundreds of rubber trees in pre-dawn darkness, to get a job at a textile factory in Shenzhen: 'If you were me, what would you prefer, the factory or the farm?' The economist Pietra Rivoli writes, 'As generations of mill girls and seamstresses from Europe, America and Asia are bound together by this common sweatshop experience – controlled, exploited, overworked, and underpaid – they are bound together too by one absolute certainty, shared across both oceans and centuries: this beats the hell out of life on the farm.'

The reason that the poverty of early industrial England strikes us so forcibly is that this was the first time writers and politicians took notice of it and took exception to it, not because it had not existed before. Mrs Gaskell and Mr Dickens had no equivalents in previous centuries; factory acts and child labour restrictions were unaffordable before. The industrial revolution caused a leap in the wealth-generating capacity of the population that greatly outstripped its breeding potential but it thereby also caused an increase in compassion, much of which was expressed through the actions of charities and governments.

## The metal Midlands

The burst of innovation which Britain experienced quite suddenly in the late 1700s was both the cause and consequence of mechanisation, of the amplification of one person's labour by machinery and fuel. The tiny nation of Britain, with just eight million people in 1750, compared with twenty-five million in far more sophisticated France, thirty-one million in far more populous Japan and 270 million in far more productive China, embarked upon a phenomenal economic expansion that would propel it to world domination within a century. Between 1750 and 1850 British men (some of them immigrants) invented an astonishing range of labour-saving and labour-amplifying devices, which allowed them to produce more, sell more, earn more, spend more, live better and have more surviving children. A famous print entitled 'The Distinguished Men of Science of Great Britain Living in the Year 1807–8', the year that Parliament abolished the slave trade, depicts fifty-one great engineers and scientists all alive at the time – as if they were gathered together by an artist in the library of the Royal Institution. Here are the men who made canals (Thomas Telford), tunnels (Marc Brunel), steam engines (James Watt), locomotives (Richard Trevithick), rockets (William Congreve), hydraulic presses (Joseph Bramah); men who invented the machine tool (Henry Maudslay), the power loom (Edmund Cartwright), the factory (Matthew Boulton), the miner's lamp (Humphry Davy) and the smallpox vaccine (Edward Jenner). Here are astronomers like Nevil Maskelyne and William Herschel, physicists like Henry Cavendish and Count Rumford, chemists like John Dalton and William Henry, botanists like Joseph Banks, polymaths like Thomas Young, and many more. You look at such a picture and wonder, 'How did any one country have so much talent in the same place?'

The premise is false, of course, because it was the aura of the time and place that drew forth (and attracted from abroad – Brunel was French, Rumford American) such talent. For all their brilliance, there are Watts, Davys, Jenners and Youngs galore in every country at every time. But only rarely do sufficient capital, freedom, education, culture and opportunity come together in such a way as to draw them out. Two centuries later, somebody could paint a picture of the great men of Silicon Valley and posterity will stand amazed at the thought that giants like Gordon Moore and Robert Noyce, Steve Jobs and Sergey Brin,

Herb Boyer and Leroy Hood all lived at the same time and in the same place.

Just as the Californian is today, so in 1700 the British manufacturing entrepreneur was unusually free, compared with both European and Asian equivalents, to invest, invent, expand and reap the profits. His huge capital city was unusual in being dominated by merchants, not the government, and always had been. He also had a world market thanks to the British ships that were plying the tropics of the world. His rural hinterland was filled with people free to sell their labour to the highest bidder. Most of the continent was still dominated by change-resistant lords and serfs neither of whom had an incentive to be more productive. In much of central and eastern Europe, serfdom gained a new lease of life in the eighteenth century following the wars and famines of the 1600s. Peasants owed much of their labour or their produce to seigneurs (plus a tithe to the Church) and had little freedom to move, so they had little incentive to be more productive or commercial. Lords, meanwhile, fiercely resisted the attempts by reforming monarchs to free their vassals. 'The landlord looks on the serf as a tool necessary to cultivate his lands,' explained one Hungarian liberal, 'and as a chattel which he inherited from his parents, or purchased, or acquired as a reward.'

Even where freedom to trade and prosper was obtained – around Toulouse, in Silesia, in Bohemia – enforcers of rules and extorters of bribes were legion, while frequent wars played havoc with commerce. Seventeen tolls were exacted in sixteen leagues in the Limousin valley. France, three times as populous as England, was 'cut up by internal customs barriers into three major trade areas and by informal custom, obsolete tolls and charges, and above all poor communications into a mosaic of semi-autarkic cells'. Internal smuggling was rife. Spain was 'an archipelago, islands of local production and consumption, isolated from each other by centuries of internal tariffs'. The

Englishman, by contrast, did not have to answer to petty bureaucrats and pesky tax collectors to the same extent. For this he could partly thank the upheavals of the previous century, including a civil war and a 'glorious revolution' against James II's arbitrary government. The latter event was more than a king-swap; it was in effect a semi-hostile management buy-in of the entire country by Dutch venture capitalists, which resulted in a rush of Dutch capital investment, a lurch towards foreign trade as the engine of state policy in emulation of Holland, and a constitutional shake-up that empowered a parliament of merchants. William III had to settle for respecting his people's property rights if he was to keep the throne. Add in that Britain did not support a standing army, that a heavily indented coastline allowed seaborne trade to reach most parts of the country, and that the administrative capital of the country was also its commercial capital, and it becomes clear that this was not a bad place to start or expand a business in say 1700. 'Nowhere else,' says David Landes, 'was the countryside so infused with manufacture; nowhere else, the pressures and incentives to change greater, the forces of tradition weaker.'

The small town of Birmingham, with no restrictive guilds and no civic charter, had begun to thrive as a centre of the metal-working trade in the early 1600s. By 1683 it had over 200 forges producing iron using coal. The heady combination of available skills and freedom of enterprise created a boom in an industry known as the 'toy trade', though the items made were mostly buckles, pins, nails, buttons and small utensils, rather than toys themselves. More patents were issued in Birmingham than in any other city than London in the eighteenth century, though few would count as major 'inventions': this was incremental exploration of the possibilities of iron, brass, tinplate and copper. The work was done in small workshops, with little new-fangled machinery, but it was split into skilled, specialised trades and organised along increasingly sophisticated lines.

Manufacturers spun out of each other's firms and started business on their own account, just as they would do around San Francisco Bay in the 1980s.

## Demand it and they will supply

People do not start businesses unless there is demand from consumers. One root cause of England's miracle was that thanks to trade enough Britons were rich enough after 1700 to buy the goods and services supplied by manufacturers so that it paid manufacturers to go out and find more productive technologies, and in doing so they stumbled upon something close to an economic perpetual motion machine. 'One of the most extraordinary facts of the [eighteenth] century was the enlargement of the consuming classes,' says Robert Friedel. 'There was a consumer revolution in eighteenth century England,' writes Neil McKendrick: 'more men and women than ever before in human history enjoyed the experience of acquiring material possessions.' Compared with mainland Europeans they were wearing wool cloth (as opposed to linen), eating beef (as opposed to cheese) and white wheat bread (as opposed to rye). For Daniel Defoe, writing in 1728, a low level of demand from the masses was far more important than a rich demand from a few:

Poor People, Journey-Men, working and pains-taking people ... These are the People that carry off the Gross of your Consumption; 'tis for these your markets are kept open late on Saturday nights ... Their Numbers are not Hundreds or Thousands, or Hundreds of Thousands, but Millions; 'tis by their Multitude, I say, that all the Wheels of Trade are set on Foot, the Manufacture and Produce of the Land and Sea, finished, cur'd, and fitted for the Markets Abroad; 'tis by the Largeness of their Gettings, that they are supported, and by the Largeness of their number that the whole country is supported.

Initially, it was the cost of luxuries that fell fastest. If you could afford only to buy food, fuel and fibre, you were not much better off than your medieval predecessor; but if you could afford spices, wine, silk, books, sugar, candles, buckles and the like, then you were three times better off, not because your income had gone up, but because the price of these goods was coming down thanks to the efforts of traders in the East India Company and their ilk. There was a mania for Indian cotton and Chinese porcelain and it was by copying these Oriental imports that the industrialists got started. Josiah Wedgwood, for instance, was not technically better at making pottery and porcelain than many others, but he was supremely good at making sure it was affordable, by dividing labour among skilled workers and applying steam to the process. He was also very good at marketing porcelain to the consuming classes by making it seem to be both posh and affordable – the holy grail of marketing ever since.

Cotton tells the tale best, though. In the 1600s, English people wore wool, linen and – if they were rich – silk. Cotton was almost unknown, though some refugees from Spanish per-secution in Antwerp settled in Norwich as cotton weavers. But trade with India was bringing more and more 'calico' cotton cloth into the country, where its light, soft, washable character, and the way it could be colourfully printed and dyed, attracted demand from the well off. The weavers of wool and silk resented this upstart rival, and pressed Parliament for protection against it. In 1699, all judges and students were told to wear gowns of wool; in 1700 all corpses were ordered to wear shrouds of sheep's wool; and from 1701 it was decreed that 'all calicoes painted, dyed, printed or stained … shall not be worn'. So ladies of fashion bought plain muslin and had it dyed. Riots broke out and women seen wearing cotton were even attacked by gangs of silk or wool weavers. Cotton was considered unpatriotic. By 1722 Parliament had bowed to the wishes of these weavers and on Christmas day that year, when the Calico Act took effect, it became illegal to wear cotton of any kind, or even to use it in home furnishings. Not for the last time, the narrow interest of producers triumphed over the broader interest of consumers in an act of trade protectionism.

And not for the last time, protectionism would fail, even backfire. To get round the law, East India merchants began to import raw cotton instead, and entrepreneurs started 'putting out' cotton to the cottages of rural spinsters and weavers to be made into cloth for export or even, mixed with a little linen or wool to keep it legal (the Calico Act was eventually repealed in 1774), for domestic sale. They had already been 'putting out' wool for decades, stealing a march on the Low Countries, where powerful craft guilds had prevented 'putting out' by smashing looms in rural cottages. The putters-out were clothiers with a reputation for loan sharking who made a living mostly by supplying raw wool to workers in their cottage homes and paying to collect finished cloth later, minus any interest on loans. The wives and daughters of farmers, and their menfolk in certain seasons, were in effect prepared to add to family income by selling labour as well as produce. Sometimes they found themselves in debt, because they borrowed money from the putters-out to equip themselves.

You can see these folk as desperate wage slaves driven off communal land by enclosure acts, the division of common land into private plots that gradually spread across most of England between about 1550 and 1800. But this is misleading. It is more accurate to see the rural textile workers as taking the first step on the ladder of producing and consuming, of specialisation and exchange. They were escaping self-sufficiency into the cash economy. It is true that some people were dispossessed of their livelihoods by enclosure, but enclosure actually increased paid employment for farm labourers, so it was for most a shift from low-grade self-sufficiency to slightly better production and consumption. Besides, Irish and Scottish as well as English migrants flocked to the textile districts to join the cottage industries. These were people giving up peasant drudgery for the chance of joining the cash economy, albeit at a low wage and for hard work. People were marrying younger and consequently giving birth to more children.

The result was that the very people who were joining the industry as workers would soon begin to be its customers. Suddenly the rising income of the average British worker met the falling cost of cotton cloth and suddenly everybody could afford to wear (and wash) cotton underwear. The historian Edward Baines noted in 1835 that the 'wonderful cheapness of cotton goods' was now benefiting the 'bulk of the people': 'a country-wake in the nineteenth century may display as much finery as a drawing room in the eighteenth.' The capitalist achievement, reflected Joseph Schumpeter a century later, 'does not typically consist of providing more silk stockings for queens but in bringing them within reach of factory girls in return for steadily decreasing amounts of effort.'

But increasing supply was not easy, because even the remotest Pennine valleys and Welsh marches were now thickly settled with the cottages of weavers and spinsters, transport was dear and some of the workers were earning good enough wages to take weekend holidays, occasionally even drinking their pay away till Monday night, preferring consumption to extra income. As the twentieth-century economist Colin Clark put it, 'Leisure has a real value even to very poor people.'

So, stuck between booming demand and stalling supply, the putters-out and their suppliers were ripe customers for any kind of productivity-enhancing invention, and with such an incentive, the inventors soon obliged. John Kay's flying shuttle, James Hargreaves's spinning jenny, Richard Arkwright's water frame, Samuel Crompton's mule – these were all just milestones on a continuous road of incrementally improving productivity.

The jenny worked up to twenty times as fast as a spinning wheel and produced a more consistent yarn, but it was still operated entirely by human muscle power. Yet by 1800 the jenny was already obsolete, because the frame was several hundred times as fast. Frames were increasingly powered by watermills. Ten years after that the 'mule', a machine that combined features of both the jenny and the frame, already outnumbered the frame by more than ten spindles to one. And mules would soon be powered by steam. The result was a vast expansion in the amount of cotton worked and a steep fall in the price of woven cloth. British exports of cotton goods quintupled in the 1780s and quintupled again in the 1790s. The price of a pound of fine-spun cotton yarn fell from 38 shillings in 1786 to just 3 shillings in 1832.

Until 1800 most of the raw cotton spun in England came from Asia. But Chinese and Indian cotton growers either could not or would not increase their output. They had little fresh land to exploit and little incentive either: the zemindar landlord or the imperial bureaucrat took the profit of any productivity increase. Instead it was the southern states of America that took up the opportunity. From producing an insignificant quantity of cotton in 1790, America became the world's biggest producer by the 1820s and by 1860 was growing two-thirds of the world's cotton. Cotton accounted for half of all American exports by value between 1815 and 1860.

Slaves did the work. Cotton was a labour-intensive crop, in which a single man could sow, weed (again and again), harvest and clean the product of just eighteen acres, and there were few economies of scale. In land-rich, thinly populated America, the only way to expand production was to kill the market in labour altogether: to force the workers to work for no wage. As the economist Pietra Rivoli puts it: 'It was not the perils of the labour market, but the suppression of the market that doomed the lives of slaves.' The affordability of cloth for the English working class was made possible by the buying and selling of captured Africans.

## King coal

So far fossil fuels have played only a small part. Now imagine what would have happened next if Britain had possessed no accessible coal reserves. Coal exists all over the world, but some British coal fields were close to the surface and close enough to navigable waterways to be cheaply transported. The cost of transporting coal overland was prohibitive until the railway came along. It was not that coal was a cheaper source of power than the alternative – coal took a century to compete on price with water power in factories – but that it was effectively limitless in supply. The harnessing of water power soon experienced diminishing returns as it reached saturation point in the Pennines. Nor was there any other, renewable fuel that could supply the need. In the first half of the eighteenth century, even the relatively tiny English iron industry was close to moribund for lack of charcoal fuel on a largely deforested island. What timber there was in the south of England was in demand for ship building, which bid up its price. So in search of charcoal to feed their forges, the iron masters left the Sussex Weald and moved to the West Midlands, then to the Welsh Marches, to South Yorkshire and eventually to Cumberland. Imports of wrought iron from well-forested Sweden and Russia met the growing demand from the mechanisation of the textile industry, but even these imports could not meet the needs of the industrial revolution. Only coal could do that. There was never going to be enough wind, water or wood in England to power the factories, let alone in the right place.

This was the position in which China found itself. In 1700 it had a vibrant textile industry, perhaps equally ripe for mechanisation, but it was a long way from coalfields, and its domestic iron industry was dependent entirely on charcoal, whose price was rising as forests retreated. Part of the problem was that Shanxi and Inner Mongolia, where the coal was, had been depopulated by barbarians and plague in the three centuries after 1100, so the country's demographic and economic centre of gravity shifted south to the Yangtze valley. Because none of the coal reserves were close to navigable water, China's iron industry gave up its early experiment with fossil fuel. The price of iron rose in China, discouraging inventors from using it for machinery. So industrial activity in China experienced diminishing returns and as the population grew, people had less and less incentive both to consume and to invent. Besides, the imperial bureaucracy would have had an attack of the vapours if asked to allow independent entrepreneurs to 'put out' work, unregulated, in the countryside, let alone build factories.

Efficiency in the coal industry did not itself contribute significantly to rising productivity in Britain even in the nineteenth century. Cotton contributed thirty-four times as much as coal to productivity growth in industrialising Britain. Coal's cost per tonne at the pithead in Newcastle rose slightly between the 1740s and 1860s, though the price in London fell because of lower taxes and falling transport costs. The miner's safety lamp aside, coal used few new technologies after the steam-driven pump. Well into the twentieth century, the equipment of the typical miner consisted of a lamp, a pick-axe, wooden pit props and a pony. The great increase in coal consumption (five-fold in the eighteenth century, fourteen-fold in the nineteenth century) was the result mainly of more investment, not more productivity. Contrast this with the iron industry, where the amount of coal needed to smelt a tonne of pig iron and then refine it into wrought iron halved every thirty years. It took almost as much human muscle power to mine a tonne of coal in 1900 as it did in 1800. Not until opencast (strip) mining began in the second half of the twentieth century did the tonnage produced per miner really begin to rise steeply.

This is one reason that the coal industry, like all mining industries before and since, was characterised by dreadful working conditions tolerated only because of somewhat higher wages than could be got in farm labour. They were higher, at least initially, or else Scottish and Irish people would not have flocked to Tyneside in the nineteenth century. The wages of a coal hewer in the North-east of England were twice as high, and rising twice as fast, as those of a farm worker in the nineteenth century.

Without coal, innovation in England's textile, iron and transport industry would have had to stagnate after 1800, when all that ferment of invention had as of yet had almost no effect on living standards. As the historian Tony Wrigley has put it: 'Until almost the middle of the nineteenth century it was still reasonable to fear a fate for England similar to that which had overtaken Holland. Hence the prominence of the stationary state in the prognostications of the classical economists.' Wrigley made the case that it was the transition from an organic economy, which grew its own fuel, to a mineral economy, which mined it, that enabled Britain to escape stagnation. It was coal that gave the industrial revolution its surprising second wind, that kept the mills, forges and locomotives running, and that eventually fuelled the so-called second industrial revolution of the 1860s, when electricity, chemicals and telegraphs brought Europe unprecedented prosperity and global power. Coal gave Britain fuel equivalent to the output of fifteen million extra acres of forest to burn, an area nearly the size of Scotland. By 1870, the burning of coal in Britain was generating as many calories as would have been expended by 850 million labourers. It was as if each worker had twenty servants at his beck and call. The capacity of the country's steam engines alone was equivalent to six million horses or forty million men, who would otherwise have eaten three times the entire wheat harvest. That is how much energy had been harnessed to the application of the division of labour. That is how impossible the task of Britain's nineteenth-century miracle would have been without fossil fuels.

Now Lancashire could beat the world for both quality and price. In 1750, India's muslins and calicoes were the envy of weavers everywhere. A century later, despite wages that were four or five times higher than in India, Lancashire was able to flood even India with cheap cotton cloth, some of it manufactured from Indian raw cotton that had made a 13,000-mile round trip. This was thanks entirely to the productivity of Lancashire's mechanised mills. That is how much difference having fossil fuels made. No matter how low his wages, an Indian weaver could not compete with the operator of a steam-driven Manchester mule. By 1900, 40 per cent of the world's cotton goods were produced within thirty miles of Manchester.

Industrialisation became contagious: the increased productivity of cotton mills encouraged demand from the chemical industry, which invented chlorine for bleaching, and from the printing industry, which turned to drum printing to print coloured cloth. By cutting the price of cotton, it also released consumer expenditure for other goods, which stimulated other manufacturing inventions. And of course to make the new machines, it demanded high-quality iron, which was made possible by cheap coal.

The crucial thing about coal was that, unlike forests and streams, it did not experience diminishing returns and rising prices. The price of coal may not have fallen much in the 1800s, but nor did it rise despite an enormous increase in the volume of consumption. In 1800 Britain was consuming over twelve million tonnes of coal a year, three times what it had used in 1750. The coal was still being used for two purposes only: domestic heating and general manufacturing, which at that date meant mostly bricks, glass, salt and metals. By 1830, consumption of coal had doubled, with iron manufacture taking 16 per cent and collieries themselves 5 per cent. By 1860, the country had consumed a billion tonnes and was now using it to drive the wheels of locomotives and the paddle wheels of ships. By 1930 Britain was using sixty-eight times as much coal as it had in 1750 and was now making electricity and gas with it as well. Today most coal is used for generating electricity.

## Dynamo

Electricity's contribution to human welfare can hardly be exaggerated. To my generation it is a dull utility, as inevitable, ubiquitous and mundane as water or air. Its pylons and wires are ugly, its plugs tiresome, its failures infuriating, its fire risks frightening, its bills annoying and its power stations monstrous symbols of man-made climate change (complete with Al Gore hurricanes coming from their stacks). But try to see its magic. Try to see it through the eyes of somebody who has never known power that was invisible and weightless, that could be transmitted miles through a slender wire, that can do almost anything, from lighting to toasting, from propulsion to music playing. Two billion people alive today have never turned on a light switch.

Imagine yourself at the Vienna exhibition of 1873. There is a stand exhibiting the work of the splendidly named semi-literate Belgian inventor Zénobe Théophile Gramme, and it is manned by his business partner, the equally euphonious French engineer Hippolyte Fontaine. They are showing off the Gramme dynamo, the first electricity generator that can produce a smooth current, and a steady light, when set spinning by hand or by a steam engine. Over the next five years, their dynamos will power hundreds of new industrial lighting installations all over Paris. In the Vienna exhibition, one of the workmen makes a careless mistake. He connects the wires from the spinning dynamo accidentally to the spare dynamo that is there to provide a back-up in case the first one fails. The reserve dynamo immediately begins to spin all by itself, in effect it becomes a motor. Fontaine's mind starts spinning too. He calls for the longest wire that can be found and connects the two dynamos by a wire that is 250 metres long. The reserve dynamo springs to life as soon as it is connected. Suddenly it became clear that electricity can transmit power over a distance far greater than belts, chains or cogs could.

By 1878, Gramme dynamos, turned by water in the river Marne, were transmitting power to two other Gramme dynamos working as motors three miles away, which in turn were pulling ploughs by cable through a field at the Menier estate near Paris, watched by wide-eyed grandees of the London Institute of Mechanical Engineers. A cascade of inventions followed: electric railways from William Siemens, better light bulbs from Joseph Swan and Thomas Edison, alternating current from George Westinghouse, Nikola Tesla and Sebastian de Ferranti, turbine generators from Charles Parsons. The electrification of the world began, and although like the computer it took decades to show up in the productivity statistics, its triumph was inexorable and its effect far-reaching. Today, 130 years later, electricity is still transforming people's lives when it first reaches them, bringing colourless, smokeless, weightless energy into the home. One recent study in the Philippines estimated that the average household derives $108 a month in benefits from connecting to the electricity grid – cheaper lighting ($37), cheaper radio and television ($19), more years in education ($20), time saving ($24) and business productivity ($8). Heck, it even affects the birth rate as television replaces procreation as an evening activity.

The earth receives 174 million billion watts of sunlight, about 10,000 times as much as the fossil-fuel output that human beings use. Or, to put it another way, a patch of ground roughly five yards by five yards receives as much sunlight as you need to run your techno life. So why pay for electricity, when there is power all about you? Because, even allowing for inconveniently timed winter, night, clouds and the shade of trees, this drenching rain of photons is all but useless. It does not come in the form of electricity, let alone car fuel or plastic. Joule for joule, wood is less convenient than coal, which is less convenient than natural gas, which is less convenient than electricity, which is less convenient than the electricity currently trickling through my mobile telephone. I am prepared to pay good money for somebody to deliver me refined and applied electrons on demand, just as I am for steaks or shirts.

Suppose you had said to my hypothetical family of 1800, eating their gristly stew in front of a log fire, that in two centuries their descendants would need to fetch no logs or water, and carry out no sewage, because water, gas and a magic form of invisible power called electricity would come into their home through pipes and wires. They would jump at the chance to have such a home, but they would warily ask how they could possibly afford it. Suppose that you then told them that to earn such a home, they need only ensure that father and mother both have to go to work for eight hours in an office, travelling roughly forty minutes each way in a horseless carriage, and that the children need not work at all, but should go to school to be sure of getting such jobs when they started work at twenty. They would be more than dumbfounded; they would be delirious with excitement. Where, they would cry, is the catch?

## Heat is work and work is heat

Can I stretch the industrial revolution upon the Procrustean bed of my hypothesis, as I have done for the upper Palaeolithic, Neolithic, urban and commercial revolutions, too? Thanks mainly to new energy technologies, what took a textile worker twenty minutes in 1750 took just one minute in 1850. He could therefore either supply twenty times as many people in a day's work, or supply each customer with twenty times as much cloth, or free his customer to spend $^{19}\!/_{20}$ ths of his income on something other than shirts. That was in essence why the second half of the industrial revolution made Britain rich. It made it possible for fewer people to supply more people with more goods and more services – in Adam Smith's words, to make 'a smaller quantity of labour produce a greater quantity of work'. There was a step change in the number of people that could be served or supplied by one person, a great leap in the specialisation of production and the diversification of consumption. Coal had made everybody into a little Louis XIV.

Today, the average person on the planet consumes power at the rate of about 2,500 watts, or to put it a different way, uses 600 calories per second. About 85 per cent of that comes from burning coal, oil and gas, the rest from nuclear and hydro (wind, solar and biomass are mere asterisks on the chart, as is the food you eat). Since a reasonably fit person on an exercise bicycle can generate about fifty watts, this means that it would take 150 slaves, working eight-hour shifts each, to pedal you to your current lifestyle. (Americans would need 660 slaves, French 360 and Nigerians 16.) Next time you lament human dependence on fossil fuels, pause to imagine that for every family of four you see in the street, there should be 600 unpaid slaves back home, living in abject poverty: if they had any better lifestyle they would need their own slaves. That is close to a trillion people.

You can take this *reductio ad absurdum* two ways. You can regret the sinful profligacy of the modern world, which is the conventional reaction, or you can conclude that were it not for fossil fuels, 99 per cent of people would have to live in slavery for the rest to have a decent standard of living, as indeed they did in Bronze Age empires. This is not to try to make you love coal and oil, but to drive home how much your *Louis Quatorze* standard of living is made possible by the invention of energy-substitutes for slaves. Let me repeat a declaration of interest here: I am descended from a long line of people who profited from the mining of coal, and I still do. Coal has huge drawbacks – it emits carbon dioxide, radioactivity and mercury; but my point here is to note how it contributes to human prosperity as well. Coal makes the electricity that lights your house, spins your washing machine and smelts the aluminium from which your aeroplane was made; oil fuels the ships, trucks and planes that filled your supermarket and makes the plastic from which your children's toys are made; gas heats your home, bakes your bread and makes the fertiliser that grows your food. These are your slaves.

But can it last? That fossil fuels will run out soon is an anxiety as old as fossil fuels themselves. Predicting an imminent increase in the price of coal as demand expanded and supplies ran short, the economist Stanley Jevons opined in 1865: 'It is thence simply inferred that we cannot long continue our present rate of progress', adding: 'it is useless to think of substituting any other kind of fuel for coal' and so his fellow Britons 'must either leave the country in a vast body or remain here to create painful pressure and poverty'. So influential was Jevons's jeremiad about what would now be called 'peak coal' that it led to a newspaper-led 'coal panic' of 1866, to William Gladstone's budgetary promise of that year to start paying down the national debt while coal lasted and to a Royal Commission on the coal supply. Ironically, this was the very decade when vast coal reserves were discovered all over the world and petroleum drilling began in earnest in the Caucasus and North America.

In the twentieth century oil has been the chief cause of anxiety. In 1914, the United States Bureau of Mines predicted that American oil reserves would last ten years. In 1939 the Department of the Interior said American oil would last thirteen years. Twelve years later it said the oil would last another thirteen years. President Jimmy Carter announced in the 1970s that: 'We could use up all of the proven reserves of oil in the entire world by the end of the next decade.' In 1970, there were 550 billion barrels of oil reserves in the world and between 1970 and 1990 the world used 600 billion barrels of oil. So reserves should have been overdrawn by fifty billion barrels by 1990. In fact, by 1990 unexploited reserves amounted to 900 billion barrels – not counting the Athabasca tar sands of Alberta, the Orinoco tar shales of Venezuela and the oil shale of the Rocky Mountains, which between them contain about six trillion barrels of heavy oil, or twenty times the proven oil reserves in Saudi Arabia. These heavy oil reserves are costly to exploit, but it is possible that bacterial refining will soon make them competitive with conventional oil even at 'normal' prices. The same false predictions of the imminent exhaustion of the natural gas supply have recurred throughout recent decades. Shale gas finds have recently doubled America's gas resources to nearly three centuries' worth.

Oil, coal and gas are finite. But between them they will last decades, perhaps centuries, and people will find alternatives long before they run out. Fuel can be synthesised from water using any source of power, such as nuclear or solar. At the moment, it costs too much to do so, but as efficiency increases and oil prices rise, then the equation will look different.

Moreover, it is an undeniable if surprising fact, often overlooked, that fossil fuels have spared much of the landscape from industrialisation. Before fossil fuels, energy was grown on land and it needed lots of land to grow it. Where I live, streams flow free; timber grows and rots in the woods; pasture supports cows; skylines are not scarred by windmills – where, were it not for fossil fuels, these acres would be desperately needed to power human lives. If America were to grow all its own transport fuel as biofuel it would need 30 per cent more farmland than it currently uses to grow food. Where would it grow food then? To get an idea of just how landscape-eating the renewable alternatives are, consider that to supply just the current 300 million inhabitants of the United States with their current power demand of roughly 10,000 watts each (2,400 calories per second) would require:

- solar panels the size of Spain
- or wind farms the size of Kazakhstan
- or woodland the size of India and Pakistan
- or hayfields for horses the size of Russia and Canada combined
- or hydroelectric dams with catchments one-third larger than all the continents put together

As it is, a clutch of coal and nuclear power stations and a handful of oil refineries and gas pipelines supply the 300 million Americans with nearly all their energy from an almost laughably small footprint – even taking into account the land despoiled by strip mines. For example, in the Appalachian coal region where strip mining happens, roughly 7 per cent of twelve million acres is being affected over twenty years, or an area two-thirds the size of Delaware. That's a big area, but nothing like the numbers above. Wind turbines require five to ten times as much concrete and steel per watt as nuclear power plants, not to mention miles of paved roads and overhead cables. To label the land-devouring monsters of renewable energy 'green', virtuous or clean strikes me as bizarre. If you like wilderness, as I do, the last thing you want is to go back to the medieval habit of using the landscape surrounding us to make power. Just one wind farm at Altamont in California kills twenty-four golden eagles every year: if an oil firm did that it would be in court. Hundreds of orang-utans are killed a year because they get in the way of oil-palm bio-fuel plantations. 'Let's stop sanctifying false and minor gods,' says the energy expert Jesse Ausubel, 'and heretically chant "Renewables are not green".'

The truth is, it was western Europe's incredible good fortune that just when humankind began to bear down on its landscapes and habitats most heavily, instead of ecological disaster as happened in Babylon, there appeared from underground a near-magical substance so that the landscape could be partly spared. Today you do not have to use acres to grow your transport fuel (oil has replaced hay for horses), your heating fuel (natural gas for timber), your power (coal for water), or your lighting (nuclear and coal for beeswax and tallow). You still have to grow much of your clothing, although 'fleeces' now come from oil. More's the pity: if cotton could be replaced by a synthetic substance of the same quality, the Aral Sea could be restored and parts of India and China given back to tigers. The one thing nobody has yet figured out how to make in factories using coal or oil is food – thank goodness – though even here natural gas provided the energy to fix about half the nitrogen atoms in your average meal.

## The mad world of biofuels

This is what makes the ethanol and biofuel boondoggle so enraging. Not even Jonathan Swift would dare to write a satire in which politicians argued that – in a world where species are vanishing and more than a billion people are barely able to afford to eat – it would somehow be good for the planet to clear rainforests to grow palm oil, or give up food-crop land to grow biofuels, solely so that people could burn fuel derived from carbohydrate rather than hydrocarbons in their cars, thus driving up the price of food for the poor. Ludicrous is too weak a word for this heinous crime. But I will calm myself just long enough to go through the numbers in case nobody has heard them.

In 2005, the world made roughly ten billion tonnes of ethanol, 45 per cent of it from Brazilian sugar cane and 45 per cent from American maize. Add in a billion tonnes of biodiesel made from European rape seed and the result is that roughly 5 per cent of the world's crop land has been taken out of growing food and put into growing fuel (20 per cent in the United States). Together with drought in Australia and more meat eating in China, this was the key factor that helped push world food supply below world food demand in 2008 and cause food riots all over the world. Between 2004 and 2007 the world maize harvest increased by fifty-one million tonnes, but fifty million tonnes went into ethanol, leaving nothing to meet the increase of demand for all other uses of thirty-three million tonnes: hence the price rose. The poor, remember, spend 70 per cent of their incomes on food. In effect, American car drivers were taking carbohydrates out of the mouths of the poor to fill their tanks.

Which might just be acceptable if either biofuel had a big environmental benefit, or it saved Americans money so they could afford to buy more goods and services from the poor and help them out of poverty that way. But since Americans are in effect being taxed thrice over to pay for the ethanol industry – they subsidise the growing of maize, they subsidise the manufacture of ethanol and they pay more for their food – the ability of American consumers to contribute to demand for manufactured goods is actually hurt by ethanol, not helped. Meanwhile, the environmental benefits of biofuels are not just illusory; they are negative. Fermenting carbohydrate is an inefficient business compared with burning hydrocarbon. Every acre of maize or sugar cane requires tractor fuel, fertilisers, pesticides, truck fuel and distillation fuel – all of which are fuel. So the question is: how much fuel does it take to grow fuel? Answer: about the same amount. The US Department of Agriculture estimated in 2002 that each unit of energy put into growing maize ethanol produces 1.34 units of output, but only by counting the energy of dried distillers' grain, a by-product of the production process that can go into cattle feed. Without that, the gain was just 9 per cent. Other studies, though, came to less positive conclusions, including one estimate that there was a 29 per cent loss of energy in the process. Drilling for and refining oil, by contrast, gets you a 600 per cent energy return or more on your energy used. Which sounds the better investment?

Even if you grant a net energy gain from ethanol – and Brazilian sugar cane is rather better, but only thanks to the fact it employs armies of underpaid human labour – that does not translate into environmental benefits. Using oil to drive cars releases carbon dioxide, which is a greenhouse gas. Using tractors to grow crops also releases nitrous oxide from soil, which is a stronger greenhouse gas with nearly 300 times the warming potential of carbon dioxide. And every increment in the price of grain that the biofuel industry causes means more pressure on rainforests, the destruction of which is the single most cost-effective way of adding carbon dioxide to the atmosphere. Converting the cerrado soils of Brazil to soybean diesel, or the peat lands of Malaysia to palm-oil diesel, says Joseph Fargione of the Nature Conservancy, releases '17–420 times more $CO_2$ than the annual greenhouse gas reductions that these biofuels would provide by displacing fossil fuels'. Or, to put it another way, it would take decades or centuries for the investment to pay back in climate terms. If you want to reduce carbon dioxide in the atmosphere, replant a forest on former farmland.

Moreover, it takes about 130 gallons of water to grow, and five gallons of water to distil a single gallon of maize ethanol – assuming that only 15 per cent of the crop is irrigated. By contrast, it takes less than three gallons of water to extract, and two gallons to refine, a gallon of gasoline. To meet America's stated aim of growing thirty-five billion gallons of ethanol a year would require using as much water as is consumed each year by the entire population of California. Be in no doubt: the biofuel industry is not just bad for the economy. It is bad for the planet, too. The chief reason it gained such a stranglehold on American politicians is because of the lobbying and political funding supplied by big companies.

Now, given that I am a fan of the future, I must not dismiss the first generation of biofuels prematurely. There are better crops coming along, whose ability to shoot themselves in the ecological footprint may not be so marked. Tropical sugar beet can generate huge yields using less water, and plants like jatropha may yet prove good at getting fuel from waste ground – if genetically engineered. And surely, algae, grown in water, have a chance to outyield them all without requiring irrigation, of course.

But do not forget the single most important problem with biofuels, the one that makes them so capable of making environmental problems worse – they need land. A sustainable future for nine billion people on one planet is going to come from using as little land as possible for each of people's needs. And if food yields from land continue to increase at the current rate, the current acreage of farmland will – just – feed the world in 2050, so the extra land for growing fuel will have to come from rainforests and other wild habitats. Another way of putting the same point is to borrow the familiar environmentalist lament that the human race is already, to quote the ecologist E.O. Wilson, 'appropriating between 20 and 40 per cent of the solar energy captured in organic material'. Why would you want to increase that percentage, leaving still less for other species? Ruining habitats and landscapes and extinguishing species to fuel a civilisation is a medieval mistake that surely need not be repeated, when there are coal seams and tar shales and nuclear reactors to hand.

Ah, for one good reason, you reply: climate change. I will address that issue in chapter 10. For now, simply note that if it were not for the climate-change argument, you could not begin to justify the claim that renewable energy is green and fossil energy is not.

# Efficiency and demand

Civilisation, like life itself, has always been about capturing energy. That is to say, just as a successful species is one that converts the sun's energy into offspring more rapidly than another species, so the same is true of a nation. Progressively, as the aeons passed, life as a whole has grown gradually more and more efficient at doing this, at locally cheating the second law of thermodynamics. The plants and animals that dominate the earth today channel more of the sun's energy through their bodies than their ancestors of the Cambrian period (when, for example, there were no plants on land). Likewise, human history is a tale of progressively discovering and diverting sources of energy to support human lifestyle. Domesticated crops captured more solar energy for the first farmers; draught animals channelled more plant energy into raising human living standards; watermills took the sun's evaporation engine and used it to enrich medieval monks. 'Civilisation, like life, is a Sisyphean flight from chaos,' as Peter Huber and Mark Mills put it. 'The chaos will prevail in the end, but it is our mission to postpone that day for as long as we can and to push things in the opposite direction with all the ingenuity and determination we can muster. Energy isn't the problem. Energy is the solution.'

The Newcomen steam engine worked at 1 per cent efficiency – that is to say it converted 1 per cent of the heat from burning coal into useful work. Watt's engine was 10 per cent efficient and rotated much faster. Otto's internal combustion engine was about 20 per cent efficient and faster still. A modern combined-cycle turbine is about 60 per cent efficient at making electricity from natural gas and runs at 1,000 rpm. Modern civilisation therefore gets more and more work out of each tonne of fossil fuel. This increasing efficiency would, you might think, gradually reduce the need to burn so much coal, oil and gas. As a country goes through an industrial revolution, at first more and more people join the fossil-fuel system – i.e., they start to use fossil fuels in both their work and their home – so more and more gets used. The 'energy intensity' (watts per dollar of GDP) actually rises. This happened in China in the 1990s, for example. Then later, once most people are in the system, efficiency does start to bite and energy intensity starts to fall. This is happening in India today. The United States now uses one-half as much energy per unit of GDP as it did in 1950. The world is using 1.6 per cent less energy for each dollar of GDP growth every year. Surely now energy usage will eventually also start to fall?

That is what I thought, until one day I tried to have an unnecessary conversation on a mobile telephone while a man was using a leaf-blower nearby. Even if everybody lags his loft and switches to compact fluorescent light bulbs, and throws out his patio heaters and gets his power from more efficient power stations, and loses his job in a steel plant but gets a new one in a call centre, the falling energy intensity of the economy will be offset by the new opportunities wealth brings to use energy in new ways. Cheap light bulbs let people plug in more lights. Silicon chips use so little power that they are everywhere and in aggregate their effect mounts up. A search engine may not use as much energy as a steam engine, but lots of them soon add up. Energy efficiency has been rising for a very long time and so has energy consumption. This is known as the Jevons paradox after the Victorian economist Stanley Jevons, who put it thus: 'It is wholly a confusion of ideas to suppose that the economical use of fuel is equivalent to a diminished consumption. The very contrary is the truth. As a rule, new modes of economy will lead to an increase of consumption.'

I am not saying fossil fuels are irreplaceable. I can easily envisage a world in 2050 in which fossil fuels have declined in importance relative to other forms of energy. I can envisage plug-in hybrid cars that use cheap off-peak (nuclear) electricity for their first twenty miles; I can imagine vast solar-power farms exporting electricity from sunny deserts in Algeria or Arizona; I can imagine hot-dry-rock geothermal plants; above all, I foresee pebble-bed, passive-safe, modular nuclear reactors everywhere. I can even imagine wind, tide, wave and biomass energy making small contributions, though these should be a last resort because they are so expensive and environmentally destructive. But this I know: we will need the watts from somewhere. They are our slaves. Thomas Edison deserves the last word: 'I am ashamed at the number of things around my house and shops that are done by animals – human beings, I mean – and ought to be done by a motor without any sense of fatigue or pain. Hereafter a motor must do all the chores.'

## CHAPTER 8

# The invention of invention: increasing returns after 1800

He who receives an idea from me, receives instruction himself without lessening mine; as he who lights his taper at mine, receives light without darkening me.

THOMAS JEFFERSON
*Letter to Isaac McPherson*

**WORLD PRODUCT**

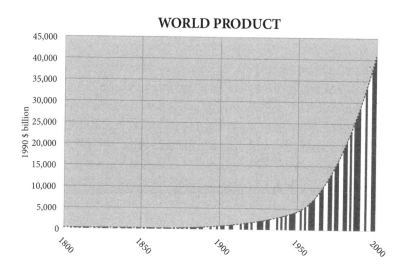

The phrase 'diminishing returns' is such a cliché that few give it much thought. Picking out the pecans from a bowl of salted nuts (a vice of mine) gives diminishing returns: the pieces of pecan in the bowl get rarer and smaller. The fingers keep finding almonds, hazelnuts, cashews or even – God forbid – Brazil nuts. Gradually the bowl, like a moribund gold mine, ceases to yield decent returns of pecan. Now imagine a bowl of nuts that had the opposite character. The more pecans you took, the larger and more numerous they grew. Implausible, I admit. Yet that is precisely the character of the human experience since 100,000 years ago. Inexorably, the global nut bowl has yielded ever more pecans, however many get used. The pace of acceleration of returns lurched upwards around 10,000 years ago in the agricultural revolution. It then lurched upwards again in AD 1800 and the acceleration continued in the twentieth century. The most fundamental feature of the modern world since 1800 – more profound than flight, radio, nuclear weapons or websites, more momentous than science, health, or material well-being – has been the continuing discovery of 'increasing returns' so rapid that they outpaced even the population explosion.

The more you prosper, the more you can prosper. The more you invent, the more inventions become possible. How can this be possible? The world of things – of pecans or power stations – is indeed often subject to diminishing returns. But the world of ideas is not. The more knowledge you generate, the more you can generate. And the engine that is driving prosperity in the modern world is the accelerating generation of useful knowledge. So, for example, a bicycle is a thing and is subject to diminishing returns. One bicycle is very useful, but there is not much extra gain in having two, let alone three. But the idea 'bicycle' does not diminish in value. No matter how many times you tell somebody how to make or ride a bicycle, the idea will not grow stale or useless or fray at the edges. Like Thomas

Jefferson's candle flame, it gives without losing. Indeed, the very opposite happens. The more people you tell about bicycles, the more people will come back with useful new features for bicycles – mudguards, lighter frames, racing tyres, child seats, electric motors. The dissemination of useful knowledge causes that useful knowledge to breed more useful knowledge.

Nobody predicted this. The pioneers of political economy expected eventual stagnation. Adam Smith, David Ricardo and Robert Malthus all foresaw that diminishing returns would eventually set in, that the improvement in living standards they were seeing would peter out. 'The discovery, and useful application of machinery, always leads to the increase of the net produce of the country, although it may not, and will not, after an inconsiderable interval, increase the value of that net produce,' said Ricardo: all tends towards what he called a 'stationary state'. Even John Stuart Mill, conceding that returns were showing no signs of diminishing in the 1840s, put it down to a miracle, innovation, he said, was an external factor, a cause but not an effect of economic growth, an inexplicable slice of luck. And Mill's optimism was not shared by his successors. As discovery began to slow, so competition would drive the profits of enterprise out of the increasingly perfect market till all that was left was rent and monopoly. With Smith's invisible hand guiding infinite market participants possessed of perfect information to profitless equilibria and vanishing returns, neo-classical economics gloomily forecast the end of growth.

It was a description of an entirely fictional world. The concept of a steady final state, applied to a dynamic system like the economy, is as wrong as any philosophical abstraction can be. It is Pareto piffle. As the economist Eamonn Butler puts it, the 'perfect market is not just an abstraction; it's plain daft ... Whenever you see the word equilibrium in a textbook, blot it out.' It is wrong because it assumes perfect competition, perfect knowledge and perfect rationality, none of which do or can exist. It is the planned economy, not the market, that requires perfect knowledge.

The possibility of new knowledge makes the steady state impossible. Somewhere somebody will have a new idea and that idea will enable him to invent a new combination of atoms both to create and to exploit imperfections in the market. As Friedrich Hayek argued, knowledge is dispersed throughout society, because each person has a special perspective. Knowledge can never be gathered together in one place. It is collective, not individual. Yet the failure of any particular market to match the perfect market no more constitutes 'market failure' than the failure of a particular marriage to match the perfect marriage constitutes 'marriage failure'.

In an exactly analogous way, the science of ecology has an enduring fallacy that in the natural world there is some perfect state of balance to which an ecosystem will return after disturbance. This obsession with 'the balance of nature' runs right through Western science, since even before Aristotle, and sees its recent expression in concepts like ecological climax, the natural vegetation that will clothe an area if it is left for long enough. But it is bunk. Take the place where I am sitting. Supposedly, its climax vegetation is oak forest, but the oaks only arrived a few thousand years ago, replacing the pines, the birch and before that the tundra. Just 18,000 years ago, where I sit was under a mile of ice, and 120,000 years ago it was a steaming swamp complete with hippos. Which of these is its 'natural' state? Besides, even if the climate settled down to an unvarying stability (something it has never done), oak saplings cannot thrive under oaks (oak-eating pests rain down on them), so after a few thousand years of oak domination an oak forest gives way to something else. Lake Victoria was bone-dry 15,000 years ago. The Great Barrier Reef was partly a range of coastal hills 20,000 years ago. The Amazon rainforest is in a state of constant perturbation: from tree falls to fires and floods, its diversity requires it to be constantly changing. There is no equilibrium in nature; there is only constant dynamism. As Heraclitus put it, 'Nothing endures but change.'

## Innovation is like a bush fire

To explain the modern global economy, then, you have to explain where this perpetual innovation machine came from. What kick-started the increasing returns? They were not planned, directed or ordered: they emerged, evolved, bottom-up, from specialisation and exchange. The accelerated exchange of ideas and people made possible by technology fuelled the accelerating growth of wealth that has characterised the past century. Politicians, capitalists and officials are flotsam bobbing upriver on the tidal bore of invention.

Even so, the generation of new useful knowledge is very far from routine, uniform, steady or continuous. Although the human race as a whole has experienced incessant change, individual peoples saw a much more intermittent flickering progress because the pace and place of that change was itself always changing. Innovation is like a bush fire that burns brightly for a short time, then dies down before flaring up somewhere else. At 50,000 years ago, the hottest hot-spot was west Asia (ovens, bows-and-arrows), at 10,000 the Fertile Crescent (farming, pottery), at 5,000 Mesopotamia (metal, cities), at 2,000 India (textiles, zero), at 1,000 China (porcelain, printing), at 500 Italy (double-entry book-keeping, Leonardo), at 400 the Low Countries (the Amsterdam Exchange Bank), at 300 France (Canal du Midi), at 200 England (steam), at 100 Germany (fertiliser); at 75 America (mass production), at 50 California (credit card), at 25 Japan (Walkman). No country remains for long the leader in knowledge creation.

At first blush, this is surprising, especially if increasing returns to innovation are possible. Why must the torch be passed elsewhere at all? As I have argued in the previous three chapters, the answer lies in two phenomena: institutions and population. In the past, when societies gorged on innovation, they soon allowed their babies to grow too numerous for their land, reducing the leisure, wealth and market that inventors needed (in effect, the merchant's sons became struggling peasants again). Or they allowed their bureaucrats to write too many rules, their chiefs to wage too many wars, or their priests to build too many monasteries (in effect, the merchants' sons became soldiers, sybarites or monks). Or they sank into finance and became parasitic rentiers. As Joel Mokyr puts it: 'Prosperity and success led to the emergence of predators and parasites in various forms and guises who eventually slaughtered the geese that laid the golden eggs.' Again and again, the flame of invention would splutter and die ... only to flare up elsewhere. The good news is that there is always a new torch lit. So far.

Just as it is true that the bush fire breaks out in different parts of the world at different times, so it leaps from technology to technology. Today, just as during the printing revolution of 500 years ago, communications is aflame with increasing returns, but transport is spluttering with diminishing returns. That is to say, the speed and efficiency of cars and aeroplanes are only very slowly improving and each improvement is incrementally more expensive. A greater and greater amount of effort is needed to squeeze the next few miles per gallon out of vehicles of any kind, whereas each tranche of extra megabits comes more cheaply for now. Very roughly, the best industry to be in as an innovator was: 1800 – textiles; 1830 – railways; 1860 – chemicals; 1890 – electricity; 1920 – cars; 1950 – aeroplanes; 1980 – computers; 2010 – the web. Whereas the nineteenth century saw a rash of new ways to move people about (railways, bicycles, cars, steamships), the twentieth century saw a rash of new ways to move information about (telephones, radio, television, satellites, fax, the internet, mobile telephones). Admittedly, the telegraph came long before the aeroplane, but the general point stands. The satellite is a neat example of a technology invented as a by-product of a transport project (space travel), which found a use in communications instead. Increasing returns would indeed peter out if innovators did not have a new wave to catch every thirty years, it seems.

Note that the greatest impact of an increasing-return wave comes long after the technology is first invented. It comes when the technology is democratised. Gutenberg's printing press took decades to generate the Reformation. Today's container ships go not much faster than a nineteenth-century steamship and today's internet sends each pulse little quicker than a nineteenth-century telegraph – but everybody is using them, not just the rich. Jets travel at the same speeds as they did in the 1970s, but budget airlines are new. As long ago as 1944, George Orwell was tired of the way the world appeared to be shrinking, supposedly a modern event. After reading what he called a 'batch of rather shallowly optimistic "progressive" books', he was struck by the repetition of certain phrases which had been fashionable before 1914. The phrases included the 'abolition of distance' and the 'disappearance of frontiers'.

But Orwell's scepticism misses the point. It is not the speed but the cost – in terms of hours of work – that counts. The death of distance may not be new, but it has been made affordable to all. Speed was once a luxury. In Orwell's day only the richest or most politically powerful could afford to travel by air or to import exotic goods or make an international telephone call. Now almost everybody can afford the cheap goods carried by container ships; almost everybody can afford the internet; almost everybody can afford to travel by jet. When I was young a transatlantic telephone call was absurdly expensive; now a transpacific email is absurdly cheap. The story of the twentieth century was the story of giving everybody access to the privileges of the rich, both by making people richer and by making services cheaper.

Likewise, when the credit card took off in California in the 1960s, driven by Joseph Williams of Bank of America, there was nothing new about buying on credit. It was as old as Babylon. There was not even anything new about charge cards. Diner's Club had been issuing cards for the convenience of restaurant users since the early 1950s and department stores for longer than that. What the BankAmericard achieved, especially once it emerged as Visa from the chaos of the mass mailings in the late 1960s, under Dee Hock's reinvention, was the democratisation of credit. The electronic possibility that your card could be authorised for a purchase anywhere in the country or even the world was a powerful lubricant to specialisation and exchange in the economy of the late twentieth century, allowing consumers to express their choice to borrow against future earnings when it made sense. There was, of course, irresponsibility, but the credit card did not lead, as most intellectual grandees had feared, to financial chaos. In the early 1970s, when credit cards were new, politicians of all stripes denounced them as unsound, unsafe and predatory, a view widely shared even by those who used the cards themselves: Lewis Mandell discovered that Americans were 'far more likely to use credit cards than to approve of them'.

This nicely captures the paradox of the modern world, that people embrace technological change and hate it at the same time. 'People don't like change,' Michael Crichton once told me, 'and the notion that technology is exciting is true for only a handful of people. The rest are depressed or annoyed by the changes.' Pity the inventor's lot then. He is the source of society's enrichment and yet nobody likes what he does. 'When a new invention is first propounded,' said William Petty in 1679, 'in the beginning every man objects and the poor inventor runs the gauntloop of all petulant wits.'

What is the flywheel of the perpetual innovation machine that drives the modern world? Why has innovation become routine and how was it that in Alfred North Whitehead's words, 'the greatest invention of the nineteenth century was the invention of the method of invention'? Is it down to the expansion of science, the application of money, the granting of intellectual property or is it something else, something much more bottom-up?

## Driven by science?

Much as I love science for its own sake, I find it hard to argue that discovery necessarily precedes invention and that most new practical applications flow from the minting of esoteric insights by natural philosophers. Francis Bacon was the first to make the case that inventors are applying the work of discoverers, and that science is the father of invention. As the scientist Terence Kealey has observed, modern politicians are in thrall to Bacon. They believe that the recipe for making new ideas is easy: pour public money into science, which is a public good, because nobody will pay for the generation of ideas if the taxpayer does not, and watch new technologies emerge from the downstream end of the pipe. Trouble is, there are two false premises here: first, science is much more like the daughter than the mother of technology; and second, it does not follow that only the taxpayer will pay for ideas in science.

It used to be popular to argue that the European scientific revolution of the seventeenth century unleashed the rational curiosity of the educated classes, whose theories were then applied in the form of new technologies, which in turn allowed standards of living to rise. China, on this theory, somehow lacked this leap to scientific curiosity and philosophical discipline, so it failed to build on its technological lead. But history shows that this is back-to-front. Few of the inventions that made the industrial revolution owed anything to scientific theory.

It is, of course, true that England had a scientific revolution in the late 1600s, personified in people like Harvey, Hooke and Halley, not to mention Boyle, Petty and Newton, but their influence on what happened in England's manufacturing industry in the following century was negligible. Newton had more influence on Voltaire than he did on James Hargreaves. The industry that was transformed first and most, cotton spinning and weaving, was of little interest to scientists and vice versa. The jennies, gins, frames, mules and looms that revolutionised the working of cotton were invented by tinkering businessmen, not thinking boffins: by 'hard heads and clever fingers'. It has been said that nothing in their designs would have puzzled Archimedes.

Likewise, of the four men who made the biggest advances in the steam engine – Thomas Newcomen, James Watt, Richard Trevithick and George Stephenson – three were utterly ignorant of scientific theories, and historians disagree about whether the fourth, Watt, derived any influence from theory at all. It was they who made possible the theories of the vacuum and the laws of thermodynamics, not vice versa. Denis Papin, their French-born forerunner, was a scientist, but he got his insights from building an engine rather than the other way round. Heroic efforts by eighteenth-century scientists to prove that Newcomen got his chief insights from Papin's theories proved wholly unsuccessful.

Throughout the industrial revolution, scientists were the beneficiaries of new technology, much more than they were the benefactors. Even at the famous Lunar Society, where the industrial entrepreneur Josiah Wedgwood liked to rub shoulders with natural philosophers like Erasmus Darwin and Joseph Priestley, he got his best idea – the 'rose-turning' lathe – from a fellow factory owner, Matthew Boulton. And although Benjamin Franklin's fertile mind generated many inventions based on principles, from lightning rods to bifocal spectacles, none led to the founding of industries.

So top-down science played little part in the early years of the industrial revolution. In any case, English scientific virtuosity dries up at the key moment. Can you name a single great English scientific discovery of the first half of the eighteenth century? It was an especially barren time for natural philosophers, even in Britain. No, the industrial revolution was not sparked by some *deus ex machina* of scientific inspiration. Later science did contribute to the gathering pace of invention and the line between discovery and invention became increasingly blurred as the nineteenth century wore on. Thus only when the principles of electrical transmission were understood could the telegraph be perfected; once coal miners understood the succession of geological strata, they knew better where to sink new mines; once benzene's ring structure was known, manufacturers could design dyes rather than serendipitously stumble on them. And so on. But even most of this was, in Joel Mokyr's words, 'a semi-directed, groping, bumbling process of trial and error by clever, dexterous professionals with a vague but gradually clearer notion of the processes at work'. It is a stretch to call most of this science, however. It is what happens today in the garages and cafés of Silicon Valley, but not in the labs of Stanford University.

The twentieth century, too, is replete with technologies that owe just as little to philosophy and to universities as the cotton industry did: flight, solid-state electronics, software. To which scientist would you give credit for the mobile telephone or the search engine or the blog? In a lecture on serendipity in 2007, the Cambridge physicist Sir Richard Friend, citing the example of high-temperature superconductivity – which was stumbled upon in the 1980s and explained afterwards – admitted that even today scientists' job is really to come along and explain the empirical findings of technological tinkerers after they have discovered something.

The inescapable fact is that most technological change comes from attempts to improve existing technology. It happens on the shop floor among apprentices and mechanicals, or in the workplace among the users of computer programs, and only rarely as a result of the application and transfer of knowledge from the ivory towers of the intelligentsia. This is not to condemn science as useless. The seventeenth-century discoveries of gravity and the circulation of the blood were splendid additions to the sum of human knowledge. But they did less to raise standards of living than the cotton gin and the steam engine. And even the later stages of the industrial revolution are replete with examples of technologies that were developed in remarkable ignorance of why they worked. This was especially true in the biological world. Aspirin was curing headaches for more than a century before anybody had the faintest idea of how. Penicillin's ability to kill bacteria was finally understood around the time bacteria learnt to defeat it. Lime juice was preventing scurvy centuries before the discovery of vitamin C. Food was being preserved by canning long before anybody had any germ theory to explain why it helped.

## Capital?

Perhaps money is the answer to the question of what drives the innovation engine. The way to incentivise innovation, as any Silicon Valley venture capitalist will tell you, is to bring capital and talent together. For most of history, people have been adept at keeping them apart. Inventors will always go where the money can be found to back them. One of Britain's advantages in the eighteenth century was that it was accumulating a collective fortune, made from foreign trade, and a comparatively efficient capital market to distribute funds to innovators. More specifically, the industrial revolution required long-term investment in capital equipment that could not easily be liquidated – factories and machines, for the most part. More than other countries, Britain's capital markets were in a position to supply this investment in the eighteenth century. London had managed to borrow from Amsterdam and nurture in the eighteenth century joint-stock, limited liability companies, liquid markets in bonds and shares, and a banking system capable of generating credit. These helped to give inventors the wherewithal to turn their ideas into products. By contrast in France capital markets were haunted by John Law's failure, banks haunted by Louis XIV's defaults, and corporate law haunted by the arbitrary extortions of tax farmers.

In an eerie repetition of the same pattern, Silicon Valley owes much of its explosion of novelty to its venture capitalists on Sandhill Road. Where would Amazon, Compaq, Genentech, Google, Netscape and Sun be without Kleiner Perkins Caulfield? It is no coincidence that the growth of technology industries took off after the mid-1970s when Congress freed pension funds and non-profits to invest some of their assets in venture funds. California is not the birthplace of entrepreneurs; it is the place they go to do their enterprising; fully one-third of successful start-ups in California between 1980 and 2000 had Indian- or Chinese-born founders.

In imperial Rome, no doubt scores of unknown slaves knew how to make better olive presses, better watermills and better wool looms, while scores of plutocrats knew how to save, invest and consume. But the two lived miles apart, separated by venal middlemen who had no desire to bring them together. A telling anecdote about glass repeated by several Roman authors rather drives home the point. A man demonstrates to the emperor Tiberius his invention of an unbreakable form of glass, hoping for a reward. Tiberius asks if anybody else knows his secret and is assured nobody does. So Tiberius beheads the man to prevent the new material reducing the relative value of gold to that of mud. The moral of the tale – whether it is true or not – is not just that Roman inventors receive negative reward for their pains, but that venture capital was so scarce, the only way to get a new idea funded was to go to the emperor. Imperial China, too, sent strong signals of discouragement to anybody whose inventiveness challenged the status quo. A Christian missionary in Ming China wrote: 'Any man of genius is paralysed immediately by the thought that his efforts will bring him punishment rather than rewards.'

The financing of innovation gradually moved inside firms in the twentieth century. Private sector companies, haunted by the Schumpeterian fear that innovation can pull their whole market from them, and equally dazzled by dreams that they can pull the whole market from under their rivals, had gradually learnt to sew innovation into their culture and to set aside budgets for it. Corporate research and development budgets are only a century old and they have been growing pretty steadily all that time. The proportion of GDP spent by firms on research and development in America has more than doubled, to nearly 3 per cent, over the past half-century. Little wonder that there has been a corresponding increase in invention and application.

Delve beneath the statistical surface though, and the picture changes. Far from being able to spend their way into novelty and growth, companies are perpetually discovering that their R&D budgets get captured by increasingly defensive and complacent corporate bureaucrats, who spend them on low-risk, dull projects and fail to notice gigantic new opportunities, which thereby turn into threats. The pharmaceutical industry, having tried again and again to instil a sense of radical thinking into its research departments, has largely given up the attempt and now simply buys up small firms that have developed big ideas. The history of the computer industry is littered with examples of big opportunities missed by dominant players, which thereby find themselves challenged by fast-growing new rivals – IBM, Digital Equipment, Apple, Microsoft. Even Google will suffer this fate. The great innovators are still usually outsiders.

Though they may start out full of entrepreneurial zeal, once firms or bureaucracies grow large, they become risk-averse to the point of Luddism. The pioneer venture capitalist Georges Doriot said that the most dangerous moment in the life of a company was when it had succeeded, for then it stopped innovating. 'This telephone has too many shortcomings to be considered as a means of communication. The device is of inherently no value to us,' read a Western Union internal memo in 1876. That is why Apple, not IBM, perfected the personal computer, why the Wright brothers, not the French army, invented powered flight, why Jonas Salk, not the British National Health Service, invented a polio vaccine, why Amazon, not the Post Office, invented one-click ordering and why a Finnish lumber-supply company, not a national telephone monopoly, became the world leader in mobile telephony.

One solution is for companies to try to set their employees free to behave like entrepreneurs. Sony did this after it discovered in the 1990s that its famously pioneering technologists had succumbed to a 'not-invented-here' mentality. General Electric under Jack Welch managed it for a while by fragmenting the company into smaller competing units. 3M – flush with success after its employee Art Fry dreamed up the idea of non-stick sticky notes (Post-its) while trying to mark the place in his hymn book in church in 1980 – told its technologists to spend 15 per cent of their time working on their own projects and by harvesting customers' ideas.

Another solution is to out-source problems to be solved by a virtual market of inventors with the promise of a prize, as the British government did with the problem of measuring longitude at sea in the eighteenth century. The internet has revived this possibility in recent years. Sites like Innocentive and yet2.com allow companies both to post problems they cannot solve, promising rewards for their solution, and to post technologies they have invented that are looking for applications.

Retired engineers can make good money and have good fun pitting their wits on a freelance basis through such sites. The old model of in-house R&D will surely rapidly give way to this marketplace in innovation, or 'idea-agora' as Don Tapscott and Anthony Williams call it.

Money is certainly important in driving innovation, but it is by no means paramount. Even in the most entrepreneurial of economies, very little saving finds its way to innovators. Victorian British inventors lived under a regime that spent a large proportion of its outgoings on interest payments, in effect sending a signal that the safest thing for rich folk to do with their money was to collect rent on it from taxes on trade. Today, plenty of money is wasted on research that does not develop, and plenty of discoveries are made without the application of much money. When Mark Zuckerberg invented Facebook in 2004 as a Harvard student, he needed very little R&D expenditure. Even when expanding it into a business, his first investment of $500,000 from Peter Thiel, founder of Paypal, was tiny compared with what entrepreneurs needed in the age of steam or railways.

## Intellectual property?

Perhaps property is the answer. Inventors will not invent unless they can keep at least some of the proceeds of their inventions. After all, somebody will not invest time and effort in planting a crop in his field if he cannot expect to harvest it and keep the profit for himself – a fact Stalin, Mao and Robert Mugabe learned the hard way – so surely nobody will invest time and effort in developing a new tool or building a new kind of organisation if he cannot keep at least some of the rewards for himself.

Yet intellectual property is very different from real property, because it is useless if you keep it to yourself. The abstract concept can be infinitely shared. This creates an apparent dilemma for those who would encourage inventors. People get rich by selling each other things (and services), not ideas. Manufacture the best bicycles and you profit handsomely; come up with the idea of the bicycle and you get nothing because it is soon copied. If innovators are people who make ideas, rather than things, how can they profit from them? Does society need to invent a special mechanism to surround new ideas with fences, to make them more like houses and fields? If so, how are ideas to spread?

There are several ways to turn ideas into property. You can keep the recipe secret, as John Pemberton did for Coca-Cola in 1886. This works well where it is hard for rivals to 'reverse-engineer' your secrets by dismantling your products. Machinery, by contrast, betrays its secrets too easily. The British pioneers of industrial textile manufacture largely failed in their attempts to use trade secrecy laws to protect themselves. Though customs officers searched foreigners' possessions for plans of machinery, New Englanders like Francis Cabot Lowell sauntered innocently about the mills of Lancashire and Scotland ostensibly for his health while frantically memorising the details of Cartwright power looms, which he promptly copied on his return to Massachusetts. The dye industry relied mostly on secrecy till the 1860s when analytical chemistry reached the point where rivals could find out how dyes were made; it then turned to patents.

Or, second, you can capture the first-mover advantage, as Sam Walton, the founder of Wal-Mart, did throughout his career. Even as his retailing rivals were catching up, he was forging ahead with new cost-cutting tactics. Intel's dominance of the microchip industry, and 3M's of the diversified technology industry, were based not on protecting their inventions so much as on improving them faster than everyone else. Packet switching was the invention that made the internet possible, yet nobody made any royalties out of it. The way to keep your customers, if you are Michael Dell, Steve Jobs or Bill Gates, is to keep making your own products obsolete.

The third way to profit from invention is a patent, a copyright or a trademark. The various mechanisms of intellectual property are eerily echoed in the apparently lawless and highly competitive world of real recipes, recipes devised by French chefs for their restaurants. There is no legal protection for recipes: they cannot be patented, copyrighted or trademarked. But try setting up a new restaurant in Paris and pinching the best recipes from your rivals and you will rapidly find that this is not common land. As Emmanuelle Fauchart discovered by interviewing ten *chefs de cuisine* who had restaurants near Paris, seven with Michelin stars, the world of haute cuisine operates according to three norms, unwritten and unenforceable by law, but no less real for that. First, no chef may copy another chef's recipe exactly; second, if a chef tells a recipe to another chef, the second chef may not pass it on without permission; third, chefs must give credit to the original inventor of a technique or idea. In effect, these norms correspond to patents, trade secrecy contracts and copyright.

Yet there is little evidence that patents are really what drive inventors to invent. Most innovations are never patented. In the second half of the nineteenth century neither Holland nor Switzerland had a patent system, yet both countries flourished and attracted inventors. And the list of significant twentieth-century inventions that were never patented is a long one. It includes automatic transmission, Bakelite, ballpoint pens, cellophane, cyclotrons, gyrocompasses, jet engines, magnetic recording, power steering, safety razors and zippers. By contrast, the Wright brothers effectively grounded the nascent aircraft industry in the United States by enthusiastically defending their 1906 patent on powered flying machines. In 1920, there was a logjam in the manufacture of radios caused by the blocking patents held by four firms (RCA, GE, AT&T

and Westinghouse), which prevented each firm making the best possible radios.

In the 1990s the US Patent Office flirted with the idea of allowing the patenting of gene fragments, segments of sequenced genes that could be used to find faulty or normal genes. Had this happened, the human genome sequence would have become an impossible landscape in which to innovate. Even so, modern biotechnology firms frequently encounter what Carl Shapiro has called a 'patent thicket' when they try to develop a treatment for a new disease. If each step in a metabolic pathway is subject to a patent, a medical inventor can find himself negotiating away all his rewards before he even tests his idea. And the last patent holder to yield commands the highest potential pay-off.

Something similar happens in mobile telephony, where the big mobile firms have to fight their way through patent thickets to bring any innovation to market. At any one moment these firms are involved in scores of lawsuits as plaintiffs, defendants or interested third parties. The result, says one observer, is that 'lobbying and litigating may be a more profitable way to win market share than innovating or investing'. Today, the biggest generators of new patents in the US system are 'patent trolls' – firms that buy up weak patent applications with no intention of making the products in question, but with every intention of making money by suing those who infringe them. Research in Motion, the Canadian company that manufactures BlackBerries, had to pay $600m to a small patent troll called NTP that did no manufacturing itself but had acquired contested patents with the aim of profiting from their defence.

Michael Heller's analogy for the patent trolls is to the state of the river Rhine between the decay of Holy Roman imperial power and the emergence of modern states. Hundreds of castles grew up all along the Rhine, one every few miles, each occupied by a little robber baron princeling living off tolls exacted from boats travelling along the river. The collective effect was to stifle trade on the Rhine, and repeated attempts to form a league to lift the burden from the trade to the benefit of all came to naught. In the twentieth century there was a possibility in the early days of flight that every landowner would extract a toll from every aircraft that crossed his 'searchlight' of vertical ownership of the air just like the Rhine robber barons. In this case, good sense prevailed and the courts quickly extinguished such property rights in the sky.

Modern patent systems, despite attempts at reform, are all too often a gauntlet of phantom tollbooths, raising fees from passing inventors and thus damaging enterprise as surely as real toll booths damage trade. Yet, of course, some intellectual property does help. A patent can be a godsend to a small firm trying to break into the market of an established giant. In the pharmaceutical industry, where government insists on a massively expensive regime of testing for safety and efficacy before a product launch, innovation without some form of patent would be impossible. In one survey of 650 R&D executives from 130 different industries, only those in the chemical and pharmaceutical industries judged patents to be effective at stimulating innovation. Yet even here there are questions to be raised. Even when such firms spend their patent profits on research rather than on marketing to exploit the temporary monopoly, most of the money goes towards me-too drugs for diseases of Westerners.

Copyright law, too, is becoming a thicket. Zealous enforcement, especially in the music and film industry has made it increasingly hard for people to share, borrow and build upon even small snippets of invented art. Smaller and smaller fragments of songs are copyrighted, and the US courts have made an attempt to lengthen the lives of copyrights to the life of the author plus seventy years (it is fifty today). Yet in the eighteenth century when composers had no copyright in their music,

Mozart was not discouraged: only one country had allowed the copyrighting of music – Britain – and the result was a decline in Britain's already dismal ability to produce composers. Just as newspapers have derived little of their income from licensing copyrights, so there will be ways to charge people for music and film in the digital world.

Intellectual property is an important ingredient of innovation, when innovation is happening, but it does very little to explain why some times and places are more innovative than others.

# Government?

The government can take credit for a list of big inventions, from nuclear weapons to the internet, from radar to satellite navigation. Yet government is also notorious for its ability to misread technical change. When I was a journalist in the 1980s, European government bodies bombarded me with boastful claims for their latest initiatives in supporting various parts of the computer industry. The programmes had catchy names like Alvey, or Esprit or 'fifth-generation' computing, and they were going to help push European industry into the lead. Usually modelled on some equally abortive idea from MITI, the then fashionable but flat-footed Japanese ministry, they invariably picked losers and encouraged companies down cul-de-sacs. Mobile phones and search engines were not among their possible futures.

In America there was a truly breathtaking outburst of government-led idiocy at the same time that went under the name of Sematech. Based on the premise that the future lay in big companies making memory chips (which were increasingly being made in Asia) it poured $100 million into chip manufacturers on condition that they stopped competing with each other and pooled their efforts to stay in what was fast becoming a commodity business. An 1890 anti-trust act had to be revised to allow it. Even as late as 1988 dirigistes were still criticising the fragmented companies of Silicon Valley as 'chronically entrepreneurial' and incapable of long-term investing. This was when Microsoft, Apple, Intel and (later) Dell, Cisco, Yahoo, Google and Facebook – chronically entrepreneurial all, in their garage or bedroom beginnings – were just laying the foundations for their global dominance at the expense of precisely the big companies dirigistes admired.

Not that any lessons were learned. In the 1990s, governments poured their efforts into such dead-ends as high-definition television standards, interactive television, telecommuting villages and virtual reality, while technology quietly got on with exploring the possibilities of wi-fi, broadband and mobile instead. Innovation is not a predictable business and it responds poorly to dirigisme from civil servants.

So although government can pay people to stumble upon new technologies – satnav and the internet were by-products of other projects – it is hardly the source of most innovation. During the late twentieth century, as companies sewed innovation into their culture and as industrial behemoths repeatedly fell prey to upstarts, most public-sector agencies just trundled on as before, neither trying to become especially innovative themselves, nor dying to make way for new versions of themselves. The idea of a government agency that fears having its mission pinched by another government agency is so peculiar as to be unimaginable. If food retailing in Britain had been left to a National Food Service after the Second World War, one suspects that supermarkets would now be selling slightly better spam at slightly higher prices from behind Formica counters.

Of course, there are some things, like large hadron colliders and moon missions, that no private company would be allowed by its shareholders to provide, but are we so sure that even these would not catch the fancy of a Buffett, a Gates or Mittal if they were not already being paid for by taxpayers? Can you doubt that if NASA had not existed some rich man would by now have spent his fortune on a man-on-the-moon programme for the prestige alone? Public funding crowds out the possibility of knowing an answer to that question. A large study by the Organisation for Economic Co-operation and Development concluded that government spending on R&D has no observable effect on economic growth, despite what governments fondly believe. Indeed it 'crowds out resources that could be alternatively used by the private sector, including private R&D'. This rather astonishing conclusion has been almost completely ignored by governments.

## Exchange!

The perpetual innovation machine that drives the modern economy owes its existence not mainly to science (which is its beneficiary more than its benefactor); nor to money (which is not always a limiting factor); nor to patents (which often get in the way); nor to government (which is bad at innovation). It is not a top-down process at all. Instead, I am going to try now to persuade you that one word will suffice to explain this conundrum: exchange. It is the ever-increasing exchange of ideas that causes the ever-increasing rate of innovation in the modern world.

Go back to that word 'spillover'. The characteristic feature of a piece of new knowledge, whether practical or esoteric, whether technical or social, is that you can give it away and still keep it. You can light your taper at Jefferson's candle without darkening him. You cannot give away your bicycle and still ride it. But you can give away the idea of the bicycle and still retain it. As the economist Paul Romer has argued, human progress consists largely in accumulating recipes for rearranging atoms in ways that raise living standards. The recipe for a bicycle, greatly abridged, might read like this: take some iron, chromium and aluminium ore from the earth, some sap from a tropical tree, some oil from beneath the ground, some hide from a cow. Smelt the ores into metals, and cast into various shapes. Vulcanise the sap into rubber and mould into hollow circular rings. Fractionate the oil to make plastic and mould. Set aside to cool. Mould the hide into the shape of a seat. Combine the ingredients in the form of a bicycle, add the startlingly counter-intuitive discovery that things don't fall over so easily when they are moving forwards, and ride.

Innovators are therefore in the business of sharing. It is the most important thing they do, for unless they share their innovation it can have no benefit for them or for anybody else. And the one activity that got much easier to do after about 1800, and has got dramatically easier recently, is sharing. Travel and communication disseminated information much faster and much further. Newspapers, technical journals and telegraphs spread ideas as fast as they spread gossip. In a recent survey of forty-six major inventions, the time it took for the first competing copy to appear fell steadily from thirty-three years in 1895 to three years in 1975.

When Hero of Alexandria invented an 'aeolipile' or steam engine in the first century AD, and employed it in opening temple doors, the chances are that news of his invention spread so slowly and to so few people that it may never have reached the ears of cart designers. Ptolemaic astronomy was ingenious and precise, if not quite accurate, but it was never used for navigation, because astronomers and sailors did not meet. The secret of the modern world is its gigantic interconnectedness. Ideas are having sex with other ideas from all over the planet with ever-increasing promiscuity. The telephone had sex with the computer and spawned the internet. The first motor cars looked as though they were 'sired by the bicycle out of the horse carriage'. The idea for plastics came from photographic chemistry. The camera pill is an idea that came from a conversation between a gastroenterologist and a guided-missile designer. Almost every technology is a hybrid.

This is one area in which cultural evolution has an unfair advantage over genetic evolution. For insuperable practical reasons connected with the pairing of chromosomes during meiosis, cross fertilisation cannot happen between different species of animal. (It can, indeed does, happen between species of bacteria, 80 per cent of whose genes have been borrowed from other species on average – one reason bacteria are so darned good at evolving resistance to antibiotics, for example.) As soon as two races of animals have diverged substantially, they find themselves able to produce only sterile offspring – like mules – or none at all. That is the very definition of a species.

Technologies emerge from the coming together of existing technologies into wholes that are greater than the sum of their parts. Henry Ford once candidly admitted that he had invented nothing new. He had 'simply assembled into a car the discoveries of other men behind whom were centuries of work'. So objects betray in their design their descent from other objects: ideas that have given birth to other ideas. The first copper axes of 5,000 years ago were the same shape as the polished stone tools then in common use. Only later did they become much thinner as the properties of metals became better understood. Joseph Henry's first electric motor bore an uncanny resemblance to a rotative-beam Watt steam engine. Even the first transistor of the 1940s was a direct descendant of the crystal rectifiers invented by Ferdinand Braun in the 1870s and used to make 'cat's whisker' radio receivers in the early twentieth century. This is not always obvious in the history of technology because inventors like to deny their ancestors, exaggerating the revolutionary and unfathered nature of their breakthroughs, the better to claim the full glory (and sometimes the patents) for themselves. Thus Britons rightly celebrate Michael Faraday's genius in devising an electric motor and a dynamo – he was even recently on a banknote for a while – but forget that he got at least half the concept from the Dane Hans Christian Oersted. Americans learn that Edison invented the incandescent light bulb out of thin air, when his less commercially slick forerunners, Joseph Swan in Britain and Alexander Lodygin in Russia, deserve at least to share the credit, if not rather more. Samuel Morse, when applying for his patent on the telegraph, in the historian George Basalla's words, 'stoutly and falsely denied' that he had learned anything from Joseph Henry. Technologies reproduce, and they do so sexually.

It follows that spillover – the fact that others pinch your ideas – is not an accidental and tiresome drawback for the inventor. It is the whole point of the exercise. By spilling over, an innovation meets other innovations and mates with them. The history of the modern world is a history of ideas meeting, mixing, mating and mutating. And the reason that economic growth has accelerated so in the past two centuries is down to the fact that ideas have been mixing more than ever before. The result is gloriously unpredictable. When Charles Townes invented the laser in the 1950s, it was dismissed as 'an invention looking for a job'. Well, it has now found an astonishing range of jobs nobody could have imagined, from sending telephone messages down fibreglass wires to reading music off discs to printing documents, to curing short sight.

End users, too, have joined in the mating frenzy. Adam Smith recounted the tale of a boy whose job was to open and close the valve on a steam engine and who, to save time, rigged up a device to do it for him. He no doubt went to his grave without imparting the idea to others, or would have done if not immortalised by the Scottish sage, but today he would have shared his 'patch' with like-minded others on a chat site. Today, the open-source software industry, with products such as Linux and Apache, is booming on the back of a massive wave of selflessness – programmers who share their improvements with each other freely. Even Microsoft is being forced to embrace open-source systems and 'cloud computing' – shared on the net – blurring the line between free and proprietary computing. After all, even the cleverest in-house programmer is unlikely to be as smart as the collective efforts of ten thousand users at the 'bleeding edge' of a new idea. Wikipedia is written by people who never expect to profit from what they do. The computer-game industry is increasingly being taken over by its players. In product after product on the internet, innovation is driven by what Eric von Hippel calls 'free-revealing lead users': customers who are happy to tell manufacturers of incremental improvements they can suggest, and of unexpected things they have found they can do with new products. Lead users are often happy to free-reveal, because they enjoy basking in the reputation of their peers. (Eric von Hippel, incidentally, practices what he preaches: you can read his books on his websites for free.)

This is not confined to software. When a windsurfer named Larry Stanley first modified his surfboard to make jumping possible without parting company from the board, he never dreamed of selling the idea, but he told everybody how to do it including the manufacturers of boards and now his innovations can be bought in the form of new surfboards. The greatest lead-user innovation of all was probably the World Wide Web, devised by Sir Tim Berners-Lee in 1991 to solve the problem of sharing particle physics data between computers. Incidentally, nobody has yet suggested that research in software and surfboards must be government-funded because innovation in them would not happen without subsidy.

In other words, we may soon be living in a post-capitalist, post-corporate world, where individuals are free to come together in temporary aggregations to share, collaborate and innovate, where websites enable people to find employers, employees, customers and clients anywhere in the world. This is also, as

Geoffrey Miller reminds us, a world that will put 'infinite production ability in the service of infinite human lust, gluttony, sloth, wrath, greed, envy and pride'. But that is roughly what the elite said about cars, cotton factories, and – I'm guessing now – wheat and hand axes too. The world is turning bottom-up again; the top-down years are coming to an end.

## Infinite possibility

Were it not for this inexhaustible river of invention and discovery irrigating the fragile crop of human welfare, living standards would undoubtedly stagnate. Even with population tamed, fossil energy tapped and trade free, the human race could quickly discover the limits to growth if knowledge stopped growing. Trade would sort out who was best at making what; exchange could spread the division of labour to best effect, and fuel could amplify the efforts of every factory hand, but eventually there would be a slowing of growth. A menacing equilibrium would loom. In that sense, Ricardo and Mill were right. But so long as it can hop from country to country and from industry to industry, discovery is a fast-breeder chain reaction; innovation is a feedback loop; invention is a self-fulfilling prophecy. So equilibrium and stagnation are not only avoidable in a free-exchanging world; they are impossible.

Throughout history, though living standards might rise and fall, though population might boom and crash, knowledge was one thing that has showed inexorable upward progress. Fire, once invented, was never forgotten. The wheel came and never left. The bow and arrow has not been disinvented even though it is obsolete except in sport – it is better than ever. How to make a cup of coffee, why insulin cures diabetes and whether continental drift happens – it is a fair bet that somebody will know these things or be able to look them up for as long as there are people on the planet. We may have forgotten a few things along the way: nobody really knows how to use an Acheulean hand axe, and until recently nobody knew how to build a medieval siege catapult known as a trebuchet. (Trial-and-error by a Shropshire squire in the 1980s eventually produced full-scale trebuchets capable of tossing pianos more than 150 yards; only rock bands have since found a profitable application.) But these forgettings are dwarfed by the additions to knowledge. We have accumulated far more knowledge than we have lost. Not even the most determined pessimist would deny that his species collectively adds more and more to the aggregate store of human knowledge as each year passes.

Knowledge is not the same thing as material wealth. It is possible to mint new knowledge and yet do nothing for prosperity. The knowledge of how to fly a man to the moon, now nearly two generations old, has yet to enrich humankind much, urban myths about non-stick frying pans notwithstanding. The knowledge that Fermat's Last Theorem is true, that quasars are distant galaxies – these may never increase gross domestic product, though contemplating them may enhance the quality of someone's life. It is also possible to get rich without adding to the store of human knowledge, as many an African dictator, Russian kleptocrat or financial fraudster can tell you.

On the other hand, a piece of new knowledge lies behind every net advance in human economic welfare: the knowledge that electrons could be deputed to carry both energy and information makes possible almost everything I do, from boiling a kettle to sending a text message. The knowledge of how to pack pre-washed salad and save everybody time; the knowledge of how to vaccinate children against polio; the knowledge that insecticide-impregnated mosquito nets can prevent malaria; the knowledge that different-sized paper cups in coffee bars can still have the same-sized lids, saving cost in manufacture and confusion in the shop – a billion such pages of knowledge make up the book of human prosperity.

It was Paul Romer's great achievement in the 1990s to rescue the discipline of economics from the century-long cul-de-sac into which it had driven by failing to incorporate innovation. From time to time its practitioners had tried to escape into theorems of increasing returns – Mill in the 1840s, Allyn Young in the 1920s, Joseph Schumpeter in the 1940s, Robert Solow in the 1950s – but not until Romer's 'new growth theory' in the 1990s was economics fully back in the real world: a world where perpetual innovation brings brief bursts of profit through temporary monopoly to whoever can commandeer demand for new products or services, and long bursts of growth to everybody else who eventually gets to share the spilled-over idea. Robert Solow had concluded that innovation accounted for growth that could not be explained by an increase in labour, land or capital, but he saw innovation as an external force, a slice of luck that some economies had more of than others – his was Mill's theory with calculus. Things like climate, geography and political institutions determined the rate of innovation – which is bad luck for land-locked tropical dictatorships – and not much could be done about them. Romer saw that innovation itself was an item of investment, that new, applied knowledge was itself a product. So long as people who are spending money on trying to find new ideas can profit from them before they pass them on, then increasing returns are possible.

The wonderful thing about knowledge is that it is genuinely limitless. There is not even a theoretical possibility of exhausting the supply of ideas, discoveries and inventions. This is the biggest cause of all for my optimism. It is a beautiful feature of information systems that they are far vaster than physical systems: the combinatorial vastness of the universe of possible ideas dwarfs the puny universe of physical things. As Paul Romer puts it, the number of different software programs that can be put on one-gigabyte hard disks is twenty-seven million times greater than the number of atoms in the universe. Or if you were to combine any four of the 100 chemical elements into different alloys and compounds in different proportions ranging from one to ten, you would have 330 billion possible chemical compounds and alloys to test, or enough to keep a team of researchers busy testing a thousand a day for a million years.

Yet if innovation is limitless, why is everybody so pessimistic about the future?

# CHAPTER 9

# Turning points: pessimism after 1900

I have observed that not the man who hopes when others despair, but the man who despairs when others hope, is admired by a large class of persons as a sage.

JOHN STUART MILL
*Speech on 'perfectibility'*

## U.S. AIR POLLUTANT EMISSIONS

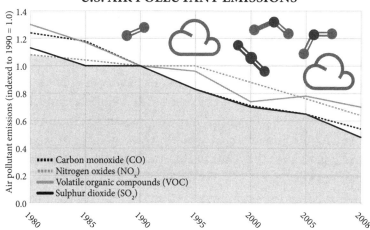

Air pollutant emissions (indexed to 1990 = 1.0)

- ▪▪▪▪ Carbon monoxide (CO)
- ▪▪▪▪ Nitrogen oxides (NO$_x$)
- Volatile organic compounds (VOC)
- Sulphur dioxide (SO$_2$)

A constant drumbeat of pessimism usually drowns out any triumphalist song of the kind I have vented in this book so far. If you say the world has been getting better you may get away with being called naïve and insensitive. If you say the world is going to go on getting better, you are considered embarrassingly mad. When the economist Julian Simon tried it in the 1990s, he was called everything from imbecile and Marxist to flat-earther and criminal. Yet no significant error came to light in Simon's book. When Bjørn Lomborg tried it in the 2000s, he was temporarily 'convicted' of scientific dishonesty by the Danish National Academy of Sciences, with no substantive examples given nor an opportunity to defend himself, on the basis of an error-strewn review in *Scientific American*. Yet no significant error has come to light in Lomborg's book. 'Implicit confidence in the beneficence of progress' said Hayek, 'has come to be regarded as the sign of a shallow mind.'

If, on the other hand, you say catastrophe is imminent, you may expect a McArthur genius award or even the Nobel Peace Prize. The bookshops are groaning under ziggurats of pessimism. The airwaves are crammed with doom. In my own adult lifetime, I have listened to implacable predictions of growing poverty, coming famines, expanding deserts, imminent plagues, impending water wars, inevitable oil exhaustion, mineral shortages, falling sperm counts, thinning ozone, acidifying rain, nuclear winters, mad-cow epidemics, Y2K computer bugs, killer bees, sex-change fish, global warming, ocean acidification and even asteroid impacts that would presently bring this happy interlude to a terrible end. I cannot recall a time when one or other of these scares was not solemnly espoused by sober, distinguished and serious elites and hysterically echoed by the media. I cannot recall a time when I was not being urged by somebody that the world could only survive if it abandoned the foolish goal of economic growth.

The fashionable reason for pessimism changed, but the pessimism was constant. In the 1960s the population explosion and global famine were top of the charts, in the 1970s the exhaustion of resources, in the 1980s acid rain, in the 1990s pandemics, in the 2000s global warming. One by one these scares came and (all but the last) went. Were we just lucky? Are we, in the memorable image of the old joke, like the man who falls past the first floor of the skyscraper and thinks 'So far so good!'? Or was it the pessimism that was unrealistic?

Let me make a square concession at the start: the pessimists are right when they say that, if the world continues as it is, it will end in disaster for all humanity. If all transport depends on oil, and oil runs out, then transport will cease. If agriculture continues to depend on irrigation and aquifers are depleted, then starvation will ensue. But notice the conditional: if. The world will not continue as it is. That is the whole point of human progress, the whole message of cultural evolution, the whole import of dynamic change – the whole thrust of this book. The real danger comes from slowing down change. It is my proposition that the human race has become a collective problem-solving machine and it solves problems by changing its ways. It does so through invention driven often by the market: scarcity drives up price; that encourages the development of alternatives and of efficiencies. It has happened often in history. When whales grew scarce, petroleum was used instead as a source of oil. (As Warren Meyer has put it, a poster of John D. Rockefeller should be on the wall of every Greenpeace office.) The pessimists' mistake is extrapolationism: assuming that the future is just a bigger version of the past. As Herb Stein once said, 'If something cannot go on forever, then it will not.'

So, for example, the environmentalist Lester Brown, writing in 2008, was pessimistic about what will happen if the Chinese are by 2030 as rich as the Americans are now:

If, for example, each person in China consumes paper at the current American rate, then in 2030 China's 1.46 billion people will need twice as much paper as is produced worldwide today. There go the world's forests. If we assume that in 2030 there are three cars for every four people in China, as there now are in the United States, China will have 1.1 billion cars. The world currently has 860 million cars. To provide the needed roads, highways, and parking lots, China would have to pave an area comparable to what it now plants in rice. By 2030 China would need 98 million barrels of oil a day. The world is currently producing 85 million barrels a day and may never produce much more than that. There go the world's oil reserves.

Brown is dead right with his extrapolations, but so was the man who (probably apocryphally) predicted ten feet of horse manure in the streets of London by 1950. So was IBM's founder Thomas Watson when he said in 1943 that there was a world market for five computers, and Ken Olson, the founder of Digital Equipment Corporation, when he said in 1977: 'There is no reason anyone would want a computer in their home.' Both remarks were true enough when computers weighed a tonne and cost a fortune. Even when the British astronomer royal and the British government space adviser said that space travel was respectively 'bunk' and 'utter bilge' – just before Sputnik flew – they were not wrong when they said it; just the world changed rather soon after they said it. It is the same with modern predictions of impossibility, like Lester Brown's. Paper and oil will all have to be used more frugally, or replaced by something else, by 2030, and land will have to be used more productively. What is the alternative? Banning Chinese prosperity? The question is not 'Can we go on as we are?' because of course the answer is 'No', but how best can we encourage the necessary torrent of change that will enable the Chinese and the Indians and even the Africans to live as prosperously as Americans do today.

# A brief history of bad news

There is a tendency to believe that pessimism is new, that our current dyspeptic view of technology and progress has emerged since Hiroshima and got worse since Chernobyl. History contradicts this. Pessimists have always been ubiquitous and have always been feted. 'Five years have seldom passed away in which some book or pamphlet has not been published,' wrote Adam Smith at the start of the industrial revolution, 'pretending to demonstrate that the wealth of the nation was fast declining, that the country was depopulated, agriculture neglected, manufactures decaying, and trade undone.'

Take the year 1830. Northern Europe and North America were much richer than they had ever been. They had enjoyed more than a decade of peace for the first time in more than a generation and they were brimming with novel inventions, discoveries and technology (a word which was coined that year): steam boats, cotton looms, suspension bridges, the Erie Canal, Portland cement, the electric motor, the first photograph, Fourier analysis. It was a world, in retrospect, pregnant with possibility, ready to explode into modernity. To be born then you would see a life of ever-increasing wealth, health, wisdom and safety.

Yet was the mood of 1830 optimistic? No, it was just like today: fashionable gloom was everywhere. Campaigners who went under the pseudonym of 'Captain Swing' took precisely the same approach to threshing machines in 1830 as their 1990s equivalents would take to genetically modified crops: they vandalised them. Some of the vociferous and numerous opponents of the Liverpool to Manchester Railway, which opened that year, forecast that passing trains would cause horses to abort their foals. Others mocked its pretensions to speed: 'What can be more palpably absurd and ridiculous than the prospect held out of locomotives travelling twice as fast as stagecoaches!' cried the

*Quarterly Review.* 'We trust that Parliament will, in all railways it may sanction, limit the speed to eight or nine miles an hour.' (Dr Arnold was more enlightened about the first steam train: 'I rejoice to see it, and think that feudality is gone forever.')

In that year, 1830, the British Poet Laureate Robert Southey had just published a book (*Thomas More; or, Colloquies on the Progress and Prospects of Society*) in which he imagined his alter ego escorting the ghost of Thomas More, the Tudor author of *Utopia*, round the English Lake District. Through the ghost of More, Southey rails against the condition of the people of England, and especially those who have left their rose-fringed cottages for the soulless tenements and factories of the industrial cities. He complains that their condition is worse than in the days of Henry VIII or even than in the days of Caesar and the Druids:

Look, for example, at the great mass of your populace in town and country – a tremendous proportion of the whole community! Are their bodily wants better, or more easily supplied? Are they subject to fewer calamities? Are they happier in childhood, youth, and manhood, and more comfortably or carefully provided for in old age, than when the land was unenclosed, and half covered with woods? ... Their condition is greatly worsened ... [They] have lost rather than gained by the alterations which have taken place during the last thousand years.

Not content with denigrating the present, Southey castigates the future. He – in the form of his fictional ghost of More – forecasts imminent misery, famine, plague and a decline of religion. The timing of this jeremiad was, in retrospect, hilarious. Not only technology, but living standards themselves, had begun their extraordinary break-out, their two centuries of unprecedented explosion. For the first time people's life expectancy was rapidly rising, child mortality rapidly falling, purchasing power burgeoning and options expanding. The rise of living standards over the next few decades would be especially marked among the unskilled working poor. British working-class real earnings were about to double in thirty years, an unprecedented occurrence. All across the world countries were looking enviously at Britain and saying 'I want some of that.' But for the reactionary, Tory, nostalgic Robert Southey, the future could only get worse. He would have been at home in the modern environmental movement, lamenting world trade, tutting at consumerism, despairing of technology, longing to return to the golden age of Merrie England when people ate their local, organic veg, danced round their maypoles, sheared their own sheep and did not clog up the airports on the way to their ghastly package holidays. As the modern philosopher John Gray puts it, echoing Southey, open-ended economic growth is 'the most vulgar ideal ever put before suffering mankind'.

Thomas Babington Macaulay was a poet, too, the author of 'Horatius' and other such well remembered ditties. In the *Edinburgh Review* of January 1830 he reviewed Southey's *Colloquies* and did not pull his punches. Far from idyllic, the life of the rural peasant was one of hellish poverty, he said; the factory towns were better off, which was why people were flocking to them. The poor rate was twenty shillings a head in rural Sussex and only five shillings in the industrial West Riding of Yorkshire.

> As to the effect of the manufacturing system on the bodily health, we must beg leave to estimate it by a standard far too low and vulgar for a mind so imaginative as that of Mr. Southey, the proportion of births and deaths. We know that, during the growth of this atrocious system, this new misery, to use the phrases of Mr. Southey, this new enormity, this birth of a portentous age, this pest which no man can approve whose heart is not seared or whose understanding has not been darkened, there has been a great diminution of mortality, and that this

diminution has been greater in the manufacturing towns than anywhere else.

As for the notion that life was better in the past, Macaulay warmed to his theme:

If any person had told the Parliament which met in perplexity and terror after the crash in 1720 that in 1830 the wealth of England would surpass all their wildest dreams …, that the rate of mortality would have diminished to one half of what it then was …, that stage-coaches would run from London to York in twenty-four hours, that men would be in the habit of sailing without wind, and would be beginning to ride without horses, our ancestors would have given as much credit to the prediction as they gave to Gulliver's Travels. Yet the prediction would have been true.

He went on (twenty-five years later, in his *History of England*):

We too shall, in our turn, be outstripped, and in our turn be envied. It may well be, in the twentieth century, that the peasant of Dorsetshire may think himself miserably paid with twenty shillings a week; that the carpenter at Greenwich may receive ten shillings a day; that labouring men may be as little used to dine without meat as they now are to eat rye bread; that sanitary police and medical discoveries may have added several more years to the average length of human life; that numerous comforts and luxuries which are now unknown, or confined to a few, may be within the reach of every diligent and thrifty working man.

The extraordinary thing about Macaulay's predictions is not that they were too barmy in their optimism but that they were far too cautious. Last week I took a stagecoach (well, a train) from London to York in two hours, not twenty-four, and ate a take-away salad of mango and crayfish (£3.60) that I bought at the station before I boarded. The week before I sailed without wind (at 37,000 feet) from London to New York in seven hours watching Daniel Day Lewis cover himself in oil. Today I rode my trusty Toyota without horses ten miles in fifteen minutes, listening to Schubert. A 'peasant' in Dorsetshire would indeed think himself miserably paid at twenty shillings (£70 in today's money) a week. Sanitation and medicine have not added several years to life expectancy, as Macaulay rashly predicted, they have doubled it. And as for comforts and luxuries, even the indolent and spendthrift working man has a television and a refrigerator, let alone the diligent and thrifty one.

## Turning-point-itis

'We cannot absolutely prove,' said Macaulay in 1830, 'that those are in error who tell us that society has reached a turning point, that we have seen our best days. But so said all who came before us, and with just as much apparent reason.' So, too, would say all that came after him. Defining moments, tipping points, thresholds and points of no return have been encountered, it seems, by pessimists in every generation since. A fresh crop of pessimists springs up each decade, unabashed in its certainty that it stands balanced upon the fulcrum of history. Throughout the half-century between 1875 and 1925, while European living standards shot up to unimaginable levels, while electricity and cars, typewriters and movies, friendly societies and universities, indoor toilets and vaccines pressed their ameliorating influence out into the lives of so many, intellectuals were obsessed with imminent decline, degeneration and disaster. Again and again, just as Macaulay had said, they wailed that society had reached a turning point; we had seen our best days.

The runaway bestseller of the 1890s was a book called *Degeneration*, by the German Max Nordau, which painted a picture of a society morally collapsing because of crime, immigration and urbanisation: 'we stand in the midst of an epidemic, a sort of Black Death of degeneration and hysteria.' An American bestseller of 1901 was Charles Wagner's *The Simple Life*, which argued that people had had enough of materialism and were about to migrate back to the farm. In 1914, Britain's Robert Tressell's posthumous *The Ragged Trousered Philanthropists* called his country 'a nation of ignorant, unintelligent, half-starved, broken-spirited degenerates'. The craze for eugenics that swept the world, embraced by left and right with equal fervour, after 1900 and caused the passage of illiberal and cruel laws in democracies like America as well as autocracies like Germany, took as its premise the deterioration of the blood lines caused by the overbreeding of the poor and the less intelligent. A huge intellectual consensus gathered around the idea that a distant catastrophe must be averted by harsh measures today (sound familiar?). 'The multiplication of the feeble-minded' said Winston Churchill in a memo to the prime minister in 1910, 'is a very terrible danger to the race.' Theodore Roosevelt was even more explicit: 'I wish very much that the wrong people could be prevented entirely from breeding; and when the evil nature of these people is sufficiently flagrant, this should be done. Criminals should be sterilized and feeble-minded persons forbidden to leave offspring behind them.' In the end, eugenics did far more harm to members of the human race than the evil it was intended to combat would ever have done. Or, as Isaiah Berlin put it, 'disregard for the preferences and interests of individuals alive today in order to pursue some distant social goal that their rulers have claimed is their duty to promote has been a common cause of misery for people throughout the ages.'

It was the thing intellectuals said they needed more of –

government – that did for the golden Edwardian afternoon, by declaring world war over a trivial issue. After it, what with inflation, unemployment, depression and fascism, there were plenty of excuses for pessimism between the two world wars. In 1918, in *The Education of Henry Adams*, Henry Adams, famously contrasting the spiritual energy of the Virgin Mary with the material energy of a huge dynamo seen at an exhibition, foresaw the 'ultimate, colossal, cosmic collapse' of civilisation. The drone of woe from pessimistic intellectuals was now a constant background hum: from T.S. Eliot, James Joyce, Ezra Pound, W.B. Yeats and Aldous Huxley. They were mostly looking the wrong way – at money and technology, not idealism and nationalism. 'Optimism is cowardice' scolded Oswald Spengler in 1923 in his bestselling polemic *The Decline of the West*, telling a generation of attentive readers of his mystical prose that the Western, Faustian world was about to follow Babylon and Rome into progressive decline as authoritarian 'Caesarism' at last came to rule, and blood triumphed over money. Caesarism did indeed rise from the ruins of capitalism in Italy, Germany, Russia and Spain, and proceeded to murder millions. By 1940, only a dozen nations remained democratic. Yet, dreadful as it was, the double war of 1914–45 did little to interrupt the improvement of lifespan and health of those who managed to survive. Despite the wars, in the half-century to 1950, the longevity, wealth and health of Europeans improved faster than ever before.

## Worse and worse

After the Second World War, led by Konrad Adenauer's West Germans, Europeans enthusiastically followed America down the path of free enterprise. There dawned a golden age after 1950 of peace (for most), prosperity (for many), leisure (for the young) and progress (in the form of accelerating technological change). Did the pessimists disappear? Was everybody cheerful? The heck they were. George Orwell kicked it off in 1942 with an essay complaining about the spiritual emptiness of the machine age and a book in 1948 warning of a totalitarian future. The torrent of gloomy prognostication that characterised the second half of the twentieth century was, like everything else from that time, unprecedented in its magnitude. Doom after doom was promised: nuclear war, pollution, overpopulation, famine, disease, violence, grey goo, vengeful technology – culminating in the eruption of civil chaos that would undoubtedly follow the inability of computers to cope with the year 2000. Remember that?

Consider the opening words of Agenda 21, the 600-page dirge signed by world leaders at a United Nations conference in Rio de Janeiro in 1992: 'Humanity stands at a defining moment in history. We are confronted with a perpetuation of disparities within and between nations, a worsening of poverty, hunger, ill health and illiteracy, and the continued deterioration of the ecosystems on which we depend for our well-being.' The following decade saw the sharpest decrease in poverty, hunger, ill health and illiteracy in human history. In the 1990s numbers in poverty fell in absolute as well as relative terms. Yet even the 1990s were marked by (in the words of Charles Leadbetter) 'an outpouring of self-doubt and even self-loathing from the intelligentsia of developed liberal societies'. An unspoken alliance, Leadbetter argued, developed between reactionaries and radicals, between nostalgic aristocrats, religious conservatives, eco-fundamentalists and angry anarchists, to persuade people that they should be anxious and alarmed. Their common theme was that individualism, technology and globalisation were leading us headlong into hell. Horrified by the rate of change, and the undermining of the status of noble intellectuals relative to brash tradesmen, 'the stasis-craving social critics who have shaped the western zeitgeist for decades' (in Virginia

Postrel's words) lashed out at the new and yearned for stability. 'It is the failure of modern man to observe the constraints necessary for maintaining the integrity and stability of the various social and ecological systems of which he is a part that is giving rise to their disintegration and destabilization' groaned the wealthy environmentalist Edward Goldsmith. The price of prosperity, in the words of the Prince of Wales, has been 'a progressive loss of harmony with the flow and rhythm of the natural world'.

Today, the drumbeat has become a cacophony. The generation that has experienced more peace, freedom, leisure time, education, medicine, travel, movies, mobile phones and massages than any generation in history is lapping up gloom at every opportunity. In an airport bookshop recently, I paused at the Current Affairs section and looked down the shelves. There were books by Noam Chomsky, Barbara Ehrenreich, Al Franken, Al Gore, John Gray, Naomi Klein, George Monbiot and Michael Moore, which all argued to a greater or lesser degree that (a) the world is a terrible place; (b) it's getting worse; (c) it's mostly the fault of commerce; and (d) a turning point has been reached. I did not see a single optimistic book.

Even the good news is presented as bad news. Reactionaries and radicals agree that 'excessive choice' is an acute and present danger – that it is corrupting, corroding and confusing to encounter ten thousand products in the supermarket, each reminding you of your limited budget and of the impossibility of ever satisfying your demands. Consumers are 'overwhelmed with relatively trivial choices' says a professor of psychology. This notion dates from Herbert Marcuse, who turned Marx's notion of the 'immiseration of the proletariat' by steadily declining living standards on its head and argued that capitalism forced excessive consumption on the working class instead. It resonates well in the academic seminar, causing heads to nod in agreement, but it is sheer garbage in the real world. When I go into the local superstore, I never see people driven to misery by the impossibility of choice. I see people choosing.

The problem is partly nostalgia. Even back in the golden age itself, in the eighth century BC, the poet Hesiod was nostalgic for a lost golden age when people 'dwelt in ease and peace upon their lands with many good things'. There has probably never been a generation since the Palaeolithic that did not deplore the fecklessness of the next and worship a golden memory of the past. The endless modern laments about how texting and emails are shortening the attention span go back to Plato, who deplored writing as a destroyer of memorising. The 'youth of today' are shallow, selfish, spoiled, feral good-for-nothings full of rampant narcissism and trained to have ephemeral attention spans, says one commentator. They spend too long in cyberspace, says another, where their grey matter is being 'scalded and defoliated by a kind of cognitive Agent Orange, depriving them of moral agency, imagination and awareness of consequences'. Balderdash. Of course, there are twerps and geeks in every generation, but today's young are volunteering for charities, starting companies, looking after their relatives, going to work – just like any other generation, maybe more so. Mostly when they are staring at screens it is to indulge in rampant social engagement. The *Sims 2* game, which sold more than a million copies in ten days when launched in 2004, is a game in which the players – often girls – get virtual people to live complex, realistic, highly social lives and then chat about it with their friends. Not much scalding and defoliating there. The psychoanalyst Adam Phillips believes that 'for increasing numbers of Britons and Americans, the "enterprise culture" means a life of overwork, anxiety and isolation. Competition reigns supreme, with even small children forced to compete against each other and falling ill as a result.' I have news for him: small children were more overworked, and fell a lot more ill, in the industrial, feudal, agrarian, Neolithic or hunter-gatherer past than in the free-market present.

Or how about the 'end of nature'? Bill McKibben's bestselling dirge of 1989 insisted that a turning point was at hand: 'I believe that without recognizing it we have already stepped over the threshold of such a change; that we are at the end of nature.'

Or the 'coming anarchy'? Robert Kaplan told the world in 1994, in a much discussed article in the *Atlantic Monthly* that became a bestselling book, how a turning point had been reached and 'scarcity, crime, overpopulation, tribalism, and disease are rapidly destroying the social fabric of our planet'. His evidence for this thesis was in essence that he had discovered urban west Africa to be a lawless, impoverished, unhealthy and rather dangerous place.

Or 'our stolen future'? In 1996 a book with this title claimed that sperm counts were falling, breast cancer was increasing, brains were becoming malformed and fish were changing sex, all because of synthetic chemicals that act as 'endocrine disruptors', which alter the hormonal balance of bodies. As usual, the scare proved greatly exaggerated: sperm counts are not falling, and no significant effect on human health from endocrine disruption has been detected.

In 1995 the otherwise excellent scientist and writer Jared Diamond fell under the spell of fashionable pessimism when he promised: 'By the time my young sons reach retirement age, half the world's species will be extinct, the air radioactive and the seas polluted with oil.' Let me reassure his sons that species extinction, though terrible, is so far under-shooting that promise by a wide margin. Even if you take E.O. Wilson's wildly pessimistic guess that 27,000 species are dying out every year, that equates to just 27 per cent a century (there are thought to be at least ten million species), a long way short of 50 per cent in sixty years. As for Diamond's other worries, the trends are getting better, not worse: the radioactive dose his sons receive today from weapons tests and nuclear accidents is 90 per cent down on what their father received in the early 1960s and is anyway less than 1 per cent of natural background radiation. The amount of oil spilled in the sea has been falling steadily since before the young Diamonds were born: it now is down by 90 per cent since 1980.

One ingenious argument for apocalypse relies on statistics. As related by Martin Rees in his book *Our Final Century*, Richard Gott's argument goes like this: given that I am roughly the sixty billionth person to live upon this planet, it is plausible to believe that I come roughly half way through my species' run on Broadway, rather than near the beginning of a million-year run. If you pull a number from an urn and it reads sixty, you would conclude that there are more likely to be 100 numbers in the urn than 1,000. Therefore, we are doomed. However, I do not intend to turn pessimist on the strength of a mathematical analogy. After all, the six billionth and six millionth person on the planet could have made exactly the same argument.

Pessimism has always been big box office. It plays into what Greg Easterbrook calls 'the collective refusal to believe that life is getting better'. People do not apply this to their own lives, interestingly: they tend to assume that they will live longer, stay married longer and travel more than they do. Some 19 per cent of Americans believe themselves to be in the top 1 per cent of income earners. Yet surveys consistently reveal individuals to be personally optimistic yet socially pessimistic. Dane Stangler calls this 'a non-burdensome form of cognitive dissonance we all walk around with'. About the future of society and the human race people are naturally gloomy. It goes with the fact that they are risk-averse: a large literature confirms that people much more viscerally dislike losing a sum of money than they like winning the same sum. And it seems that pessimism genes might quite literally be commoner than optimism genes: only about 20 per cent of people are homozygous for the long version of the serotonin transporter gene, which possibly endows them with a genetic tendency to look on the bright side. (Willingness to take risks, a possible correlate of optimism, is also partly heritable: the 7-repeat version of the DRD4 gene accounts for 20 per cent of financial risk taking in men – and is commoner in countries where most people are descended from immigrants.)

As the average age of a country's population rises, so people get more and more neophobic and gloomy. There is immense vested interest in pessimism, too. No charity ever raised money for its cause by saying things are getting better. No journalist ever got the front page by telling his editor that he wanted to write a story about how disaster was now less likely. Good news is no news, so the media megaphone is at the disposal of any politician, journalist or activist who can plausibly warn of a coming disaster. As a result, pressure groups and their customers in the media go to great lengths to search even the most cheerful of statistics for glimmers of doom. The day I was writing a first draft of this paragraph, the BBC reported on its morning news headlines a study that found the incidence of heart disease among young and middle-aged British women had 'stopped falling'. Note what was not news: the incidence of heart disease had until recently been falling steeply among all women, was still falling among men, and was not yet rising even among the female age group where it had just 'stopped falling'. Yet all the discussion was of this 'bad' news. Or note how the *New York Times* reported the reassuring news in 2009 that world temperature had not risen for a decade: 'Plateau in temperature adds difficulty to task of reaching a solution'.

Apocaholics (the word is Gary Alexander's – he calls himself a recovering apocaholic) exploit and profit from the natural pessimism of human nature, the innate reactionary in every person. For 200 years pessimists have had all the headlines, even though optimists have far more often been right. Archpessimists are feted, showered with honours and rarely challenged, let alone confronted with their past mistakes.

Should you ever listen to pessimists? Certainly. In the case of the ozone layer, a briefly fashionable scare of the early 1990s, the human race probably did itself and its environment a favour by banning chlorofluorocarbons, even though the excess ultraviolet light getting through the ozone layer in the polar regions never even approached one-five-hundredth of the level that is normally experienced by somebody living in the tropics – and even though a new theory suggests that cosmic rays are a bigger cause of the Antarctic ozone hole than chlorine is. Still, I should stop carping: in this case, getting chlorine out of the atmosphere was on balance the wise course of action and the costs to human welfare, though not negligible, were small.

And there are things that are getting worse, without doubt. Traffic congestion and obesity would be two big ones, yet both are the products of plenty, and your ancestors would have laughed at the idea that such abundance of food and transport was a bad thing. There are also many occasions on which pessimists have been ignored too much. Too few people listened to anxieties expressed about Hitler, Mao, Al-Qaeda and sub-prime mortgages – to name a handful of issues at random. But pessimism is not without its cost. If you teach children that things can only get worse, they will do less to make it untrue. I was a teenager in Britain in the 1970s, when every newspaper I read told me not just that oil was running out, a chemical cancer epidemic was on the way, food was growing scarce and an ice age was coming, but that my own country's relative economic decline was inevitable and its absolute decline probable. The sudden burst of prosperity and accelerating growth that Britain experienced in the 1980s and 1990s, not to mention the improvements in health, lifespan and the environment, came as quite a shock to me. I realised about the age of twenty-one that nobody had ever said anything optimistic to me about the future of the human race – not in a book, a film or even a pub. Yet in the decade that followed, employment increased, especially for women, health improved, otters and salmon returned to the local river, air quality improved, cheap flights to Italy began from the local airport, telephones became portable, supermarkets stocked more and more kinds of cheaper and better food. I feel angry that I was not taught and told that the world could get much better; I was somehow given a counsel of despair. As are my children today.

# Cancer

By now this generation of human beings was supposed to be dying like flies from cancer caused by chemicals. Starting in the late 1950s, posterity was warned that synthetic chemicals were about to create an epidemic of cancer. Wilhelm Hueper, chief of environmental cancer research at America's National Cancer Institute, so convinced himself that exposure to small traces of synthetic chemicals was a big cause of cancer that he even refused to believe that smoking caused cancer – lung cancer came from pollution, he believed. Rachel Carson, influenced by Hueper, set out in her book *Silent Spring* (1962) to terrify her readers as she had terrified herself about the threat to human health caused by synthetic chemicals and especially by the pesticide DDT. Whereas childhood cancer had once been a medical rarity, she wrote, 'today, more American school children die of cancer than from any other disease'. This was actually a statistical sleight of hand; the statement was true not because cancer was increasing among children (it was not), but because other causes of childhood death were declining faster. She expected DDT to cause a major cancer epidemic in human beings as well as other animals, and to shorten human lifespan as a result.

It is not much of an exaggeration to say that an entire generation of Westerners grew up expecting Carson's cancer epidemic to strike them down. I was one of them: it genuinely scared me at school to know that my life would be short and sick. Influenced by Carson and her apostles I set out to do a biological project. I would walk the countryside and pick up the dying birds I found, have their cancers diagnosed, and publish. It was not a great success: I found one corpse, of a swan that had hit a power line. 'Individuals born since 1945,' wrote the environmentalist Paul Ehrlich in 1971, 'and thus exposed to DDT since before birth may well have shorter life expectancies than they would if DDT had never existed. We won't know until the first of these reach their forties and fifties.' Later he was more specific: 'The U.S. life expectancy will drop to forty-two years by 1980, due to cancer epidemics.'

What actually happened is that – excepting lung cancer – both cancer incidence and death rate from cancer fell steadily, reducing by 16 per cent between 1950 and 1997, with the rate of the fall accelerating after that; even lung cancer then joined the party as smoking retreated. The life expectancy of those born after 1945 broke new records. The search for a widespread epidemic of cancer caused by synthetic chemicals, relentlessly and enthusiastically pursued by many scientists ever since the 1960s, has been entirely in vain. By the 1980s, a study by the epidemiologists Richard Doll and Richard Peto had concluded that age-adjusted cancer rates were falling, that cancer is caused chiefly by cigarette smoke, infection, hormonal imbalance and unbalanced diet – and that chemical pollution causes less than 2 per cent of all cases of cancer. The premise on which much of the environmental movement had grown up – that cleaning up pollution would prevent cancer – proved false. As Bruce Ames famously demonstrated in the late 1990s, cabbage has forty-nine natural pesticides in it, more than half of which are carcinogens. In drinking a single cup of coffee you encounter far more car-cinogenic chemicals than in a year's exposure to pesticide residues in food. This does not mean that coffee is dangerous, or contaminated: the carcinogens are nearly all natural chemicals found in the coffee plant and the dose is too low to cause disease, as it is in the pesticide residue. Ames says, 'We've put a hundred nails in the coffin of the cancer story and it keeps coming back out.'

DDT's miraculous ability to halt epidemics of malaria and typhus, saving perhaps 500 million lives in the 1950s and 1960s (according to the US National Academy of Sciences), far outweighed any negative effect it had on human health. Ceasing to use DDT caused a resurgence of malaria in Sri Lanka, Madagascar and many other countries. Of course, DDT should have been used more carefully than it was, for although it was far less toxic to birds than previous pesticides, many of which were arsenic-based, it did have the subversive ability to accumulate in the livers of animals and wipe out populations of predators at the top of long food chains, such as eagles, falcons and otters. Replacing it with less persistent chemicals has brought otters, bald eagles and peregrine falcons bouncing back to relative abundance after an absence of several decades. Fortunately, DDT's modern pyrethroid successors do not persist and accumulate. Moreover, sparing, targeted use of DDT against malarial mosquitoes can be done without any such threat to wildlife, for example by spraying the inside walls of houses.

# Nuclear Armageddon

There were very good reasons to be a nuclear pessimist in the Cold War: the build-up of weapons, the confrontations over Berlin and Cuba, the gung-ho rhetoric of some military commanders. Given how most arms races end, it seemed only a matter of time before the Cold War turned hot, very hot. If you had said at the time that you believed that mutually assured destruction would prevent large direct wars between the superpowers, that the Cold War would end, the Soviet empire would disintegrate, global arms spending would fall by 30 per cent and three-quarters of all nuclear missiles would be dismantled, you would have been dismissed as a fool. 'Historians will view nuclear arms reduction as such an incredible accomplishment,' says Gregg Easterbrook, 'that it will seem bizarre in retrospect that so little attention was paid while it was happening.' Perhaps this was just a stroke of luck, and admittedly the danger is far from over (especially for Koreans and Pakistanis), but nonetheless notice that things have got better, not worse.

## Famine

One of the hoariest causes for pessimism about the fate of humanity is the worry that food will run out. The prominent eco-pessimist Lester Brown predicted in 1974 that a turning point had been reached and farmers could 'no longer keep up with rising demand'. But they did. In 1981 he said that 'global food insecurity is increasing'. It was not. In 1984, he proclaimed that 'the slim margin between food production and population growth continues to narrow'. Wrong again. In 1989 'population growth is exceeding farmers' ability to keep up.' No. In 1994, 'Seldom has the world faced an unfolding emergency whose dimensions are as clear as the growing imbalance between food and people' and 'After forty years of record food production gains, output per person has reversed with unanticipated abruptness.' (A turning point had been reached.) A series of bumper harvests followed and the price of wheat fell to record lows, where it stayed for a decade. Then in 2007 the wheat price suddenly doubled because of a combination of Chinese prosperity, Australian drought, pressure from environmentalists to encourage the growing of biofuels and willingness of American pork-barrel politicians to oblige them by sluicing subsidies towards ethanol producers. Sure enough Lester Brown was once again the darling of the media, his pessimism as impregnable as it was thirty-three years before: 'cheap food may now be history,' he said. A turning point had been reached. Once again, a record harvest followed and the wheat price halved.

The prediction of global famine has a long history, but it probably reached its apocaholic shrillest in 1967 and 1968 with two bestselling books. The first was by William and Paul Paddock (*Famine, 1975!*). 'Population-food collision is inevitable; it is foredoomed' was the title of the first chapter. The Paddocks even went so far as to argue that countries such as Haiti, Egypt and India were beyond saving and should be left to starve; the world's efforts should, on the Verdun principle of triage, be focused on the less desperate cases. By 1975, with the world not yet starving, William Paddock was calling for a moratorium on research programmes designed to increase food production in countries with high population growth rates – almost as if he wanted to bring about his own prediction.

The following year saw the publication of an even bigger bestseller that was even more misanthropic in tone. *The Population Bomb* allowed Paul Ehrlich, an obscure butterfly ecologist, to metamorphose into a guru of the environmental movement complete with MacArthur 'genius' award. 'In the 1970s and 1980s,' he promised, declaring a turning point, 'hundreds of millions of people will starve to death in spite of any crash programs embarked upon now. At this late date nothing can prevent a substantial increase in the world death rate.' Ehrlich not only argued that mass death was inevitable and imminent, that human numbers would fall to two billion and that the poor would get poorer, but that those who saw that population growth was already beginning to slow were as foolish as those who greet a slightly less freezing day in December as a sign of approaching spring; in later editions, he added that the Green Revolution then transforming Asian agriculture would 'at the very best buy us only a decade or two'. Four decades later, Ehrlich had learnt his lesson – not to give dates: in his book *The Dominant Animal*, co-written with his wife and published in

2008, he again foresaw an 'unhappy increase in death rates' but this time mentioned no timescale. Without a word about why his previous predictions of mass starvation and mass cancer had never happened, he remains confident in calling the top of the human happiness market: 'The world in general seems to be gradually awakening to a realization,' he regretted to say, 'that our long evolutionary story is, through our actions but not our intentions, coming to a turning point.'

For reasons I explained in chapter 4, famine is largely history. Where it still occurs – Darfur, Zimbabwe – the fault lies with government policy, not population pressure.

## Resources

The history of the world is replete with examples of the extinction or near-exhaustion of resources: mammoths, whales, herrings, passenger pigeons, white pine forests, Lebanon cedars, guano. They are all, note, 'renewable'. By striking contrast, there is not a single non-renewable resource that has run out yet: not coal, oil, gas, copper, iron, uranium, silicon, or stone. As has been said – the remark has been attributed to many people – the Stone Age did not come to an end for lack of stone. 'It is one of the safest predictions,' wrote the economist Joseph Schumpeter in 1943, 'that in the calculable future we shall live in an *embarras de richesse* of both foodstuffs and raw materials, giving all the rein to expansion of total output that we shall know what to do with. This applies to mineral resources as well.' It is also one of the safest predictions that people will always be warning that natural resources are running out.

Consider the humiliating failure of the predictions made by a computer model called World3 in the early 1970s. World3 attempted to predict the carrying capacity of the planet's resources and concluded, in a report called *Limits to Growth*, authored by the grandiosely titled 'Club of Rome', that

302

exponential use could exhaust known world supplies of zinc, gold, tin, copper, oil and natural gas by 1992 and cause a collapse of civilisation and population in the subsequent century. *Limits to Growth* was enormously influential, with school textbooks soon parroting its predictions minus the caveats. 'Some scientists estimate that the world's known supplies of oil, tin, copper, and aluminium will be used up within your lifetime,' said one. 'Governments must help save our fossil fuel supply by passing laws limiting their use,' opined another. It was misleading chiefly because, like Malthus, it underestimated the speed and magnitude of technological change, the generation of new recipes for rearranging the world – as its godfather, the engineer Jay Forester, has acknowledged. In 1990 the economist Julian Simon won $576.07 in settlement of a wager from the environmentalist Paul Ehrlich. Simon had bet him that the prices of five metals (chosen by Ehrlich) would fall during the 1980s and Ehrlich had accepted 'Simon's astonishing offer before other greedy people jump in' (though later, while calling Simon an imbecile, he claimed he was 'goaded' into it).

The amount of oil left, the food-growing capacity of the world's farmland, even the regenerative capacity of the biosphere – these are not fixed numbers; they are dynamic variables produced by a constant negotiation between human ingenuity and natural constraints. Embracing dynamism means opening your mind to the possibility of posterity making a better world rather than preventing a worse one. We now know, as we did not in the 1960s, that more than six billion people can live upon the planet in improving health, food security and life expectancy and that this is compatible with cleaner air, increasing forest cover and some booming populations of elephants. The resources and technologies of 1960 could not have supported six billion – but the technologies changed and so the resources changed. Is six billion the turning point? Seven? Eight? At a time when glass fibre is replacing copper cable, electrons are replacing paper and most employment involves more software than hardware, only the most static of imaginations could think so.

## Clean air

In 1970, *Life* magazine promised its readers that scientists had 'solid experimental and theoretical evidence' that 'within a decade, urban dwellers will have to wear gas masks to survive air pollution … by 1985 air pollution will have reduced the amount of sunlight reaching earth by one half.' Urban smog and other forms of air pollution refused to follow the script, as technology and regulation rapidly improved air quality. So by the 1980s the script switched to acid rain. It is worth exploring the history of this episode because it was a dress rehearsal for global warming: atmospheric, international and with fossil fuels as the villains. The conventional story you will read in your children's textbooks is as follows: sulphuric and nitric acid, made mainly from smoke belched from coal-fired power stations, fell on lakes and forests in Canada, Germany and Sweden and devastated them. In the nick of time laws were passed limiting emissions and ecosystems slowly recovered.

Certainly, in the mid-1980s, a combination of scientists scenting grants and environmentalists scenting donations, led to some apocalyptic predictions. In 1984 the German magazine *Stern* reported that a third of Germany's forests were already dead or dying, that experts believed all its conifers would be gone by 1990 and that the Federal Ministry of the Interior predicted all forests would be gone by 2002. All! Professor Bernd Ulrich said it was already too late for Germany's forests: 'They cannot be saved.' Across the Atlantic, similar predictions of doom were made. Trees were said to be dying at an unnatural rate in 100 per cent of the forests on the eastern seaboard. 'The tops of the Blue Ridge Mountains are becoming tree graveyards,' said a plant pathology professor. Half of all lakes were becoming dangerously acidified. The New York Times declared 'a scientific consensus': it was time for action, not further research.

What actually happened? History shows that the biomass of European forests actually increased during the 1980s, during the time when unconstrained acid rain was supposed to be killing them and before any laws were passed to limit emissions. It continued to increase in the 1990s. Sweden's government eventually admitted that nitric acid – a fertiliser – had increased the overall growth rate of its trees. European forests not only did not die; they thrived. As for North America, the official, ten-year, half-a-billion-dollar, 700-scientist, government-sponsored study did a great rash of experiments and found that: 'there is no evidence of a general or unusual decline of forests in the United States or Canada due to acid rain' and 'there is no case of forest decline in which acidic deposition is known to be a predominant cause.' When asked if he had been pressured to be optimistic, one of the authors said the reverse was true. 'Yes, there were political pressures ... Acid rain had to be an environmental catastrophe, no matter what the facts revealed. Since we could not support this claim ... the [Environmental Protection Agency] worked to keep us from providing Congress with our findings.' The truth is that there were small pockets of damage to forests in the 1980s some of which were caused by pests, others by natural senescence or competition and a few by local pollution. There was no great forest die-off due to acid rain. At all.

It would be wrong to conclude that the anti-acid rain legislation did no good at all. The acidification of mountain lakes by distant power-station emissions was a real (though relatively rare) phenomenon, and this was indeed reversed by the legislation. But even this harm was vastly exaggerated during the debate: far from 50 per cent of lakes being affected, it was 4 per cent, said the official study. Some of these continue to be acid even after the clean-up, because of the chemistry of the surrounding rocks. The fact is, if you read the history of the episode carefully, acid rain was a minor and local nuisance that could be relatively cheaply dealt with, not a huge threat to large stretches of the planet. The ultra-pessimists were simply wrong.

## Genes

Every advance in human genetics and reproductive medicine is greeted with predictions of Frankenstein doom. The first attempts at genetic engineering of bacteria in the 1970s led to moratoria and bans. The activist Jeremy Rifkin said that biotechnology threatened 'a form of annihilation every bit as deadly as nuclear holocaust'. Yet the result was life-saving therapies for diabetics and haemophiliacs. Shortly after, the pioneers of in-vitro fertilisation, Robert Edwards and Patrick Steptoe, were vilified on all sides, even by their fellow doctors, for their supposedly dangerous experiments. When Louise Brown was born in 1978, the Vatican called it 'an event that can have very grave consequences for humanity'. Yet their invention has brought no eugenic abuse and heartfuls of individual happiness to millions of childless couples.

When the human genome was sequenced in 2000, pessimism soon dominated the commentary. People will fiddle with their children's genes, moaned some: yes, to avoid passing on terrible inherited diseases like Tay-Sachs or Huntingdon's. Predicting illness will make health insurance impossible, groaned others – yet health insurers' rates are so high precisely because they cannot predict who will get ill, so predicting and preventing will bring some costs down. Diagnosis will run far ahead of therapy, wailed others, so people will know their fate but not know how to cure themselves. In practice, there are very few diseases for which some kind of preventive intervention cannot be tried once a predisposition is known, and knowing is still and always should be voluntary. Then, to cap it all, within a few years the pessimists were complaining that genetic insights were coming disappointingly slowly.

## Plague

By the late 1990s, the modish cause for doom was the resurgence of infectious disease. The combination of an incurable and brand-new sexually transmitted disease, AIDS, with a growing resistance to antibiotics among hospital bacteria gave genuine cause for fear. But it also sparked a search for the next and still more lethal plague. Book after book trumpeted the alarm: *The Hot Zone, Outbreak, Virus X, The Coming Plague*. Hundreds of millions of people were going to die. Infectious disease was on its way back into human affairs as part of the planet's revenge for human despoliation of the environment. The human race was due a culling. Some of the more misanthropic authors, sounding like Puritan preachers, even expressed something approaching satisfaction at the thought. Yet once again, the auction of competitively pessimistic forecasts that surrounded Ebola virus, Lassa fever, hanta virus and SARS proved ridiculously overblown. Ebola outbreaks – which wreaked horrible disintegration upon their victims and wiped out whole villages in the Congo a handful of times in the 1990s before fading away each time – proved to be very local, easily controlled and partly man-made. That is to say, it soon emerged that what was turning this occasional bat-borne infection into a raging local epidemic was such things as quinine injections given by well-meaning nuns with re-usable syringes. Even AIDS, while terrible especially in Africa, has failed to live up to the dire predictions commonly made in the late 1980s for its global effects. The number of new cases of HIV/AIDS worldwide has been falling for nearly a decade, and the number of deaths from the disease has been falling since 2005. The proportion of the population infected with HIV is falling, even in southern Africa. The epidemic is far from over, and much more could be done, but the news is getting slowly better, not worse.

Remember mad-cow disease? Between 1980 and 1996 about 750,000 cattle infected with the brain-destroying prion called vCJD entered the human food chain in Britain. When it became clear in 1996 that some people were dying of the same agent, acquired from eating infected beef, there was perhaps understandably a competitive auction in doom-laden predictions. The winner, whose views were broadcast repeatedly as a result, was a bacteriology professor named Hugh Pennington who said things like 'we have to prepare for perhaps thousands, tens of thousands, hundreds of thousands, of cases of vCJD coming down the line.' Even the 'official' models warned that the true figure could be as high as 136,000 victims. In fact, as of this writing, the number of deaths has reached 166, of which just one was in 2008 and two in 2009. Only four people are now living with definite or probable vCJD. Each one is a tragedy, but a threat to humanity it is not.

(The numbers are surprisingly similar to those from Chernobyl. At least 500,000 people would die from cancers caused by the nuclear accident there in 1986, said the sobering early reports, and there will be many birth defects. The latest estimate is that less than 4,000 will die of Chernobyl cancer, compared with 100,000 natural cancer deaths among the exposed population, and that there were no extra birth defects at all. In addition, fifty-six died during the accident itself. The evacuation of the area has caused wildlife to flourish there to an extraordinary degree, without any unusual genetic changes at all in the rodents that have been studied.)

In the 2000s influenza, too, proved to be a paper tiger. H5N1 strains of the virus ('bird flu') jumped into human beings via free-range ducks on Chinese farms and, in 2005, the United Nations predicted five million – 150 million deaths from bird flu. Yet, contrary to what you have read, when H5N1 did infect human beings it proved neither especially virulent nor especially contagious. It has so far killed fewer than 300 people worldwide. As one commentator concluded: 'Hysteria over an avian flu pandemic has been very good for the Chicken Little media, authors, ambitious health officials, drug companies … But even as many of the panic-mongers have begun to lie low, the vestiges of hysteria remain – as do the misallocations of billions of dollars from more serious health problems. Too bad no one ever holds the doomsayers accountable for the damage they've done.'

I suspect this is too strong, and that flu may yet mount a serious epidemic in some form. But the H1N1 swine flu epidemic of 2009 that began in Mexico also followed the usual path of new flu strains, towards low virulence – about one death for every 1,000–10,000 infected people. This is no surprise. As the evolutionary biologist Paul Ewald has long argued, viruses undergo natural selection as well as mutation once established in a new species of host and casually transmitted viruses like flu replicate more successfully if they cause mild disease, so that the host keeps moving about and meeting new people. A victim lying in a darkened room alone is not as much use to the virus as somebody who feels just well enough to struggle into work coughing. The modern way of life, with lots of travel but also rather more personal space, tends to encourage mild, casual-contact viruses that need their victims to be healthy enough to meet fresh targets fleetingly. It is no accident that modern people suffer from more than 200 kinds of cold, the supreme viral exploiters of the modern world.

If this is so, why then did H1N1 flu kill perhaps fifty million people in 1918? Ewald and others think the explanation lies in the trenches of the First World War. So many wounded soldiers, in such crowded conditions, provided a habitat ideally suited to more virulent behaviour by the virus: people could pass on the virus while dying. Today you are far more likely to get the flu from a person who is well enough to go to work than one who is ill enough to stay at home. By contrast, it is no accident that water-borne and insect-borne diseases such as typhoid, cholera, yellow fever, typhus and malaria are so much more virulent, because they can spread from immobilised victims. Malaria spreads more easily if its victims are laid low in a darkened room – bait for mosquitoes. But in most of the modern world, people are increasingly protected from dirty water and from insects and therefore lethal diseases that debilitate their victims are in retreat.

On top of this, the weapons in the physician's armoury just keep on getting better. Diseases of my childhood, like measles, mumps and rubella, are now prevented by a single vaccine. Where it took more than ten years to understand HIV, it took just three weeks a couple of decades later to sequence the entire genome of the SARS virus and begin a search for its vulnerabilities. It took just months in 2009 to generate large doses of vaccines for swine flu.

The total eradication of many diseases is now a realistic prospect. Although it is now more than forty years since smallpox was exterminated and hopes of sending polio after it to the grave have been repeatedly dashed, none the less, the retreat of infectious killers from many parts of the world is little short of astounding. Polio is confined to a few parts of India and West Africa, malaria is gone from Europe, North America and nearly the whole of the Caribbean, measles is reduced to a tiny percentage of the numbers recorded even a few decades ago; sleeping sickness, filariasis and onchoceriasis are being steadily eliminated from country after country.

In the centuries to come there will certainly be new human diseases, but very few of them will be both lethal and contagious. Measures to cure and prevent them will come quicker and quicker.

# Sounding the retreat

Many of today's extreme environmentalists not only insist that the world has reached a 'turning point' – quite unaware that their predecessors have made the same claim for two hundred years about many different issues – but also insist that the only sustainable solution is to retreat, to cease economic growth and enter progressive economic recession. What else can they mean by demanding a campaign to 'de-develop the United States', in the words of President Obama's science adviser John Holdren; or 'isn't the only hope for the planet that the industrialised civilisations collapse? Isn't it our responsibility to bring that about?', in the words of Maurice Strong, first executive director of the United Nations Environment Programme (UNEP); or that what is needed is 'an ordered and structured down-sizing of the global economy' in the words of the journalist George Monbiot? This retreat must be achieved, says Monbiot, by 'political restraint'. This means not just that growing your company's sales would be a crime, but failing to shrink them; not just that travelling further than your ration of miles would be an offence, but failing to travel fewer miles each year; not just that inventing a new gadget would be illegal, but failing to abandon existing technologies; not just that growing more food per acre would be a felony, but failing to grow less – because these are the things that constitute growth.

Here's the rub: this future sounds awfully like the feudal past. The Ming and Maoist emperors had rules restricting the growth of businesses; forbidding unauthorised travel; punishing innovation; limiting family size. They do not say so, but that is the inevitable world the pessimists want to return to when they speak of retreat.

# CHAPTER 10

# The two great pessimisms of today: Africa and climate after 2010

It is possible to believe that all the past is but the beginning of a beginning, and that all that is and has been is but the twilight of the dawn.

H.G. WELLS
*The Discovery of the Future*

**GREENLAND ICE CAP TEMPERATURE FROM ICE CORES**

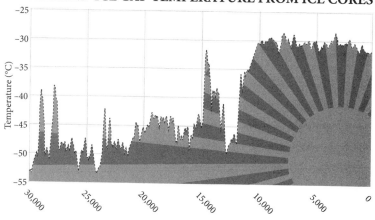

Years before present (data ends in 1909)

Sooner or later, the ubiquitous pessimist will confront the rational optimist with his two trump cards: Africa and climate. It is all very well Asia lifting itself out of poverty, and perhaps Latin America too, but surely, says the pessimist, it is hard to imagine Africa following suit. The continent is doomed by its population boom, its endemic diseases, its tribalism, its corruption, its lack of infrastructure, even – whisper some, more in sorrow than in prejudice – its genes. 'It's blindingly obvious,' says the environmentalist Jonathan Porritt: 'completely unsustainable population growth in most of Africa will keep it permanently, hopelessly, stuck in deepest, darkest poverty.'

And in any case, continues the pessimist, Africa cannot hope to boom because climate change will devastate the continent during the coming century before it can prosper. At the time of writing, global warming is by far the most fashionable reason for pessimism. The earth's atmosphere has warmed, and it seems that the great 100,000-year experiment of human progress is about to be tested against rising sea levels, melting ice caps, droughts, storms, famines, pandemics and floods. Human activity is causing much of this change, especially by the burning of fossil fuels, whose energy has been responsible for raising the living standards of many of the world's nearly seven billion people, so humankind faces a stark dilemma in the coming century between continuing a carbon-fuelled prosperity until global warming brings it to a calamitous halt, or restricting the use of carbon and risking a steep decline in living standards because of the lack of alternative sources of energy that are cheap enough. Either prospect might be catastrophic.

Africa and climate therefore confront the rational optimist with a challenge, to say the least. For somebody who has spent 300 pages looking on the bright side of human endeavour, arguing along the way that the population explosion is coming to a halt, that energy will not soon run out, that pollution, disease, hunger, war and poverty can all be expected to continue declining if human beings are not impeded from exchanging goods, services and ideas freely – for such a person as your author, African poverty and rapid global warming are indeed acute challenges.

Moreover, the two issues are connected, because the models that predict rapid global warming take as their assumption that the world will prosper mightily, and that the poorest countries on the planet – most of which are African – will by the end of this century be about nine times as rich as they are today. Unless they are, carbon dioxide emissions will be insufficient to cause such rapid warming. And at present there is no way to make Africans as rich as Asians except by them burning more fossil fuels per head. So Africa faces an especially stark dilemma: get rich by burning more carbon and then suffer the climate consequences; or join the rest of the world in taking action against climate change and continue to wallow in poverty.

That is the conventional wisdom. I think it is a false dilemma and that an honest appraisal of the facts leads to the conclusion that by far the most likely outcome of the next nine decades is both that Africa gets rich and that no catastrophic climate change happens.

## Africa's bottom billion

Of course, not all poverty is in Africa. I am well aware that there is terrible want in many other parts of the world, in Haiti and Afghanistan, in Bolivia and Cambodia, in Calcutta and São Paolo, even in parts of Glasgow and Detroit. But compared with a generation ago, thanks chiefly to progress elsewhere, poverty has come to be concentrated in that one continent as never before. Of the 'bottom billion' left behind by recent booms – Paul Collier's phrase – more than 600 million are Africans. The average African lives on just $1 a day. Saving Africa has become both the goal of idealists and the despair of pessimists. Not only has Africa failed to join Asia's boom since 1990, it has spent much of the time stagnant or going backwards. Between 1980 and 2000, the number of Africans living in poverty doubled. War in the west of the continent, genocide in the east, AIDS in the south, hunger in the north, dictators in the middle, population growth all over: no part of the continent has escaped the horror. Sudan, Ethiopia, Somalia, Kenya, Uganda, Rwanda, Congo, Zimbabwe, Angola, Liberia, Sierra Leone – the very names of countries have taken their turn as synonyms of chaos on the lips of newsreaders in the West.

Moreover, although Africa's demographic transition has begun, it has a long way to go before population growth decelerates. Nigeria's birth rate may have halved, but it is still twice as high as 'replacement rate'. Where will Africa's ghost acres, its emigration valve, or its industrial revolution come from?

There are hopeful exceptions, like Mali, Ghana, Mauritius and South Africa – countries that have achieved a measure of freedom, economic progress and peace. All across the continent, economic growth has picked up in recent years, and in Kenya, Uganda, Tanzania, Malawi, Zambia and Botswana even life expectancy is rising rapidly after falling while AIDS took its toll (South Africa and Mozambique have yet to follow suit). It is a false Western cliché that all African lives are spent dodging poverty, corruption, violence and disease. But far too many are, and the contrast with much of Asia grows more acute by the year. Whereas income per head stood still in Africa in the past twenty-five years, in Asia it trebled. Then tragically, Africa's promising economic boom in the 2000s was cut short by the credit crunch.

Some Westerners have been heard to say that growth is not what matters, that what Africa needs is an improvement in the human development index, towards the Millennium Development goals and to erase suffering without raising income, or that it needs a new kind of sustainable growth. Paul Collier and his colleagues at the World Bank encountered a storm of protest from non-governmental organisations when they published a study entitled *Growth Is Good for the Poor*. This suspicion of growth is a luxury that only wealthy Westerners can indulge. What Africans need is better living standards and these come chiefly from economic growth.

## Aid's test

Some of the most urgent needs of Africa can surely be met by increased aid from the rich world. Aid can save lives, reduce hunger, deliver a medicine, a mosquito net, a meal or a metalled road. Combating malaria has economic as well as medical benefits. But statistics, anecdotes and case histories all demonstrate that the one thing aid cannot reliably do is to start or accelerate economic growth. Aid to Africa doubled in the 1980s as a percentage of the continent's GDP; growth simultaneously collapsed from 2 per cent to zero. The aid that Zambia has received since 1960, if invested instead in assets giving a reasonable rate of return, would by now have given Zambians the income per head of the Portuguese – $20,000 instead of $500. Although in the early 2000s some studies managed to find evidence that certain kinds of aid sometimes trigger growth in countries with specific economic policies, even these conclusions were later dashed by Raghuram Rajan and Arvind Subramanian of the International Monetary Fund in 2005. They could find no evidence that aid resulted in growth in any countries. Ever.

It is worse than that. Most aid is delivered by governments to governments. It can therefore be a source of both corruption and discouragement to entrepreneurship. Some ends up in dictators' Swiss bank accounts; some goes to make billion-dollar steel mills that never work; some is given on condition of importing certain goods from a Western country; some is not independently evaluated for efficacy either by the donor or the recipient. Some African leaders are so disenchanted with government aid that they even embraced the recommendations of the Zambian economist Dambisa Moyo who concludes, bleakly, 'aid doesn't work, hasn't worked, and won't work ... no longer part of the potential solution, it's part of the problem – in fact, aid *is* the problem.'

Moreover, in recent years much aid has been granted on condition of free-market economic reform, which far from kick-starting economic growth, frequently proves damaging to local traditions, undermining the very mechanisms that get enrich-ment started. As William Easterly puts it while criticising the shock therapy that did such harm in both the Soviet bloc and Africa, 'you can't plan a market'. The top-down imposition of a bottom-up system is bound to fail.

Easterly cites the example of insecticide-treated mosquito bed nets, which are a cheap and proven way of preventing malaria. A bed net costs about $4. Encouraged by a flurry of publicity at the Davos World Economic Forum in 2005 from Gordon Brown, Bono and Sharon Stone, bed nets became a fashionable icon of the aid industry. Unfortunately, when given out free by donor agencies, they often become fashion items instead, being sold on the black market for wedding veils or used as fishing nets. They undercut local merchants supplying them for money. One American charity, Population Services International, came up with a better idea. It sold the nets for fifty cents to mothers attending antenatal clinics in Malawi and subsidised this price by selling the nets for $5 to richer urban Malawians. The poor mothers who bought these nets with half a day's wages made sure they were used properly. In four years, the proportion of children under five sleeping under such nets went up from 8 per cent to 55 per cent.

To do more good and less harm, says Easterly, the aid business could be transformed into a more transparent marketplace where donations compete to fund projects and projects compete to attract donations. Fortunately, the internet makes this possible for the first time. Globalgiving.com, for instance, allows projects to bid for donations from any donor. In the week I was writing this paragraph, projects that needed funding on the site ranged from feeding Ethiopian refugees, to building the fence around a retirement home for a pet cheetah used to inspire underprivileged children about conservation in South Africa.

In forums like this, aid could be democratised, taken out of the hands of inefficient international bureaucrats and corrupt African officials, taken away from idealistic free-market shock-therapists, separated from arms deals, removed from big industrial projects, distanced from patronising do-gooders and given person-to-person. A rich country could give each taxpayer a tax break for each suitable donation. To those who say that this would make an uncoordinated, unplanned business, I reply: exactly. Grandiose goals and centralised plans have just as long and just as disastrous a history in aid as they do in politics. Nobody planned the industrial revolution, or China's economic surge. The planners' role was to get out of the way of bottom-up evolutionary solutions.

## Bound to fail?

Most economists are agreed on a list of reasons for the failure of Africa to generate economic growth. Many African countries are more or less landlocked, which cuts them off from world trade. They have poor and deteriorating roads linking distant cities. They have exploding birth rates. They suffer from epidemic malaria, AIDS and other diseases such as sleeping sickness and guinea worm. Their institutions have never fully recovered from the disruptions caused by the slave trade. They were once colonies, which meant rule by minorities uninterested in allowing the development of an entrepreneurial class. Thanks to their imperial colonisers, their Marxist independence leaders and their monetarist aid donors, most African countries have lost many of their informal social traditions and institutions, so property rights and justice have become arbitrary and insecure. Their most promising industry – agriculture – is usually stifled by price controls and bureaucratic marketing agencies imposed by urban elites, and stymied by trade barriers and subsidies in Europe and America, not to mention devastated by a proliferation of over-grazing goats. Ethnic strife between the biggest tribe, which maintains one-party rule, and its hated rival usually poisons politics. Paradoxically, African countries are often also cursed by sudden windfalls of rich mineral wealth, such as oil or diamonds, which serve only to corrupt democratic politicians, strengthen the power of dictators, distract entrepreneurs, spoil the terms of trade of exporters and encourage reckless state borrowing.

Take, therefore, one such typical African country. It is land-locked, drought-prone and has a very high population growth rate. Its people belong to eight different tribes speaking different tongues. When freed from colonial rule in 1966 it had eight miles of paved road (for an area the size of Texas), twenty-two black university graduates, and only 100 secondary school graduates. It was later cursed by a huge diamond mine, crippled by AIDS, devastated by cattle disease, and ruled by one party with little effective opposition. Government spending has remained high; so has wealth inequality. This country, the fourth poorest in the entire world in 1950, has every one of Africa's curses. Its failure was inevitable and predictable.

But Botswana did not fail. It succeeded not just moderately well, but spectacularly. In the thirty years after independence it grew its per capita GDP faster on average (nearly 8 per cent) than any other country in the entire world – faster than Japan, China, South Korea and America during that period. It multiplied its per capita income thirteen times so that its average

320

citizens are now richer than Thais, Bulgarians or Peruvians. It has had no coups, civil wars or dictators. It has experienced no hyperinflation or debt default. It did not wipe out its elephants. It is consistently the most successful economy in the world in recent decades.

It is true that Botswana has a small and ethnically somewhat homogeneous population, unlike many other countries. But its biggest advantage is one that the rest of Africa could easily have shared: good institutions. In particular, Botswana turns out to have secure, enforceable property rights that are fairly widely distributed and fairly well respected. When Daron Acemoglu and his colleagues compared property rights with economic growth throughout the world, they found that the first explained an astonishing three quarters of the variation in the second and that Botswana was no outlier: the reason it had flourished was because its people owned property without fear of confiscation by chiefs or thieves to a much greater extent than in the rest of Africa. This is much the same explanation for why England had a good eighteenth century while China did not.

So give the rest of Africa good property rights and sit back and wait for enterprise to work its magic? If only it were that easy. Good institutions cannot usually be imposed from above: that way they are oxymorons. They must evolve from below. And it turns out that Botswana's institutions have deep evolutionary roots. The Tswana people who conquered the native Khoisan tribes in the eighteenth century (and still do not necessarily treat them well) had a political system that was remarkably, well, democratic. Cattle were privately owned, but land was owned collectively. The chiefs, who in theory allocated land and grazing rights, were under a strong obligation to consult an assembly, or *kgotla*. The Tswana were also inclusive, happy to bring other tribes into their system, which stood them in good stead when a collective army was needed to repel the Boers at the battle of Dimawe in 1852.

This was a good start, but Botswana then had a stroke of good fortune in its colonial experience. It was incorporated into the British empire in such a half-hearted and inattentive fashion that it barely experienced colonial rule. The British took it mainly to stop the Germans or Boers getting it. 'Doing as little in the way of administration or settlement as possible' was explicitly stated as government policy in 1885. Botswana was left alone, experiencing almost as little direct European imperialism as those later success stories of Asia – places like Thailand, Japan, Taiwan, Korea and China. In 1895, three Tswana chiefs went to Britain and successfully pleaded with Queen Victoria to be kept out of the clutches of Cecil Rhodes; in the 1930s, two chiefs went to court to prevent another attempt at more intrusive colonial rule and though they failed, the war then kept bossy commissioners at bay. Benign neglect continued.

After independence, Botswana's first president, Seretse Khama, one of the chiefs, behaved like most African leaders in setting out to build a strong state and disenfranchise the chiefs, as well as to win all future elections (so far so good for his party under two successors). This, together with the extreme poverty of the country and its dependence on foreign aid, foreign labour markets (in South Africa) and the sale of mineral rights to de Beers surely boded badly. Yet Botswana went from strength to strength by carefully investing its cattle export income and then its diamond windfall to develop other parts of the economy. Only a devastating AIDS epidemic, which lowered life expectancy between 1992 and 2002, mars the picture, and even that is now retreating.

## The world is your oyster

It is not as if Africa needs to invent enterprise: the streets of Africa's cities are teeming with entrepreneurs, adept at doing deals, but they cannot grow their businesses because of blockages in the system. The slums of Nairobi and Lagos are terrible places, but the chief fault lies with governments, which place bureaucratic barriers in the way of entrepreneurs trying to build affordable homes for people. Unable to negotiate the maze of regulations that govern planning, developers leave the poor to build their own slums, brick by brick as they can afford them, outside the law – and then await the official bulldozers. In Cairo it would take seventy-seven bureaucratic procedures involving thirty-one agencies and up to fourteen years to acquire and register a plot of state-owned land on which to build a house. No wonder nearly five million Egyptians have decided to build illegal dwellings instead. Typically, a Cairo house owner will build up to three illegal storeys on top of his house and rent them out to relatives.

Good for him. However, entrepreneurs who start businesses in the West usually finance them with mortgages, and you cannot get a mortgage on an illegal dwelling. The Peruvian economist Hernando de Soto estimates that Africans own an astonishing $1 trillion in 'dead capital' – savings that cannot be used as collateral because they are invested in ill-documented property. He draws an instructive parallel with the young United States in the early nineteenth century, where the formal codified law was fighting a rearguard action against an increasingly chaotic confusion of informal squatters' rights to property. More and more states were tolerating and even legalising pre-emption – ownership acquired by settling land and improving it. In the end it was the law that had to give, not the squatters – the law allowing itself to change by bottom-up evolution, not top-down planning. The retreat culminated in the Homestead Act of 1862, which formalised what had been happening for many years and signified 'the end of a long, exhausting and bitter struggle between elitist law and a new order brought about by massive migration and the needs of an open and sustainable society'. The result was a property-owning democracy in which almost everybody had 'live' capital, which could be used as collateral for starting a business. Enclosure had played a similar role in Britain earlier, though lack of unoccupied land made the result far less equitable. Revolution eventually achieved property rights for the French poor, too, rather more bloodily, and would probably have done the same for Russians, but for the Bolshevik coup.

The importance of property rights can even be demonstrated in the laboratory. Bart Wilson and his colleagues set up a land of three virtual villages inhabited by real undergraduates of two kinds – merchants and producers – making and needing three kinds of unit: red, blue and pink. Since no village can make all three units, the subjects had to start trading among themselves and did. Unlike in the previous, simpler experiment (see pages 89–90) they graduated to impersonal, market-like exchange. But when the players had a history of no property rights – i.e., they were able to steal units from each other's caches – the trading never flourished and the undergraduates went home poorer than if they had a history of property rights. It is exactly what de Soto and economists like Douglass North have been saying about the real world for some time.

(Incidentally, there is now overwhelming evidence that well crafted property rights are also the key to wildlife and nature conservation. Whether considering fish off Iceland, kudu in Namibia, jaguars in Mexico, trees in Niger, bees in Bolivia or water in Colorado, the same lesson applies. Give local people the power to own, exploit and profit from natural resources in a sustainable way and they will usually preserve and cherish those resources. Give them no share in a wildlife resource that is controlled – nay 'protected' – by a distant government and they will generally neglect, ruin and waste it. That is the real lesson of the tragedy of the commons.)

Property rights are not a silver bullet. In some countries, their formalisation simply creates a rentier class. And China experienced an explosion of enterprise after 1978 without ever giving its people truly secure property rights. But it did allow people to start businesses with relatively little bureaucratic fuss, so another of De Soto's recommendations is to free up the rules governing business. Whereas it takes a handful of steps to set up a company in America or Europe, De Soto's assistants found that to do the same in Tanzania would take 379 days and cost $5,506. Worse, to have a normal business career in Tanzania for fifty years, you would have to spend more than a thousand days in government offices petitioning for permits of one kind or another and spending $180,000 on them.

Little wonder that a staggering 98 per cent of Tanzanian businesses are extralegal. That does not mean they are governed by no rules: far from it. De Soto's study found thousands of examples of documents being used by people on the ground to attest ownership, record loans, embody contracts and settle disputes. Handwritten papers, sometimes signed with thumb-prints, are being drafted, witnessed, stamped, revised, filed and adjudicated all over the country. Just as Europeans did before the formal law gradually 'nationalised' their indigenous customs, Tanzanians are evolving a system of self-organised complexity to allow them to do business with strangers as well as neighbours. One handwritten, single-page document, for example, records a contract for a business loan between two individuals – the amount of the loan, the interest rate, the payment period and the collateral (the debtor's house) – and is signed, witnessed and stamped by the local elder.

But these customs, these laws of the people, are a fragmented jigsaw. They work well for sole traders in small communities, but being dependent on local people and local rules they cannot help the ambitious entrepreneur who tries to expand beyond his local community. What Tanzania needs to do, as Europe and America did hundreds of years ago, is not to enforce its unaffordable official legal system, but gradually to encourage this bottom-up, informal law to broaden and standardise itself.

De Soto's team identified sixty-seven bottlenecks that prevent the poor using the legal system to generate wealth.

It is this kind of institutional reform that will in the end do far more for African living standards than dams, factories, aid or population control. In the 1930s, Nashville, Tennessee, was rescued from poverty by its music entrepreneurs, using good local copyright laws to start recording indigenous music, not by the giant dams of the Tennessee Valley Authority. Likewise, Bamako in Mali could build upon its strong musical traditions given the right copyright laws and entrepreneurial spirit.

In a neat example of bottom-up change, the poor have taken to mobile telephones with unexpected gusto all across Africa, to the surprise of those who thought this a luxury technology for a later stage of development. In Kenya, despairing of state-controlled landlines, one-quarter of the population acquired a mobile phone after 2000. Kenyan farmers call different markets to find the best prices before setting out with their produce, and are better off for it. Studies of rural villages in Botswana find that the ones that have mobile reception have more non-farm jobs than the ones that do not. Mobile phones not only enable people to get work, but also to pay for and be paid for services – mobile phone credits having become in effect a form of informal banking and payments system. In Ghana, manufacturers of T-shirts can be paid directly by American buyers using phone credits. Micro-finance banking, mobile telephony and the internet are now merging to produce systems that allow individuals in the West to make small loans to entrepreneurs in Africa (through websites like Kiva), who can then use their mobile phone credits to deposit receipts and pay bills without waiting for banks to open and without handling vulnerable cash. These developments offer opportunities to the poor of Africa that were not available to the poor of Asia a generation ago. They are one reason that Africa saw economic growth rise to Asian-tiger levels in the late 2000s.

The role of the mobile phone in enriching the poor was

especially well illustrated by a study of the sardine fishermen of Kerala in southern India (though similar stories can now be told about Africa). As documented by the economist Robert Jensen, on 14 January 1997, a typical day, eleven fishermen landed good catches at the village of Badagara only to find that there were no buyers left: the local market was sated and the price of the perishable sardines was zero. Just ten miles away in both directions along the coast, at Chombala and Quilandi, that morning there were twenty-seven willing buyers getting ready to leave the markets empty-handed because they could find no sardines to buy, even at the inflated price of nearly ten rupees per kilogram they were offering. Had the Badagara fishermen known, they could have diverted to the other markets and pocketed on average 3,400 rupees of profit each, after fuel costs. Later that year, using mobile phones on the newly installed cellular network (whose signals could be picked up twelve miles out to sea), the Kerala fishermen started doing just that: they called ahead to find out where best to land their catch. The result was that fishermen's profits increased by 8 per cent, sardine prices to consumers fell by 4 per cent and sardine wastage fell from more than 5 per cent to virtually nil. Everybody gained (except the sardines). As Robert Jensen commented: 'Overall the fisheries sector was transformed from a series of essentially autarkic fishing markets to a state of nearly perfect spatial arbitrage.'

Using such technologies, Africa can follow the same route to prosperity that the rest of the world is following: to specialise and exchange. Once two individuals find ways to divide labour between them, both are better off. The future for Africa lies in trade – in selling tea, coffee, sugar, rice, beef, cashews, cotton, oil, bauxite, chrome, gold, diamonds, cut flowers, green beans, mangoes and more – but it is almost impossible for poor Africans in the informal economy to be entrepreneurs in such international trade. A handwritten contract between two people in Tanzania may be affordable and enforceable, but it is little help if the debtor wishes to start an export business supplying cut flowers to a London-based supermarket.

Of course, it will not all be easy or smooth, but I refuse to be pessimistic about Africa when such an opportunity is available at a few strokes of a pen and when the evidence of entrepreneurial vitality in the extralegal sector is so strong. Besides, as its population growth rates fall, Africa is about to reap a 'demographic dividend' when its working-age population is large relative to both the dependent elderly and the dependent young: such a demographic bonanza gave Asia perhaps one third of its miracle of growth. The key policies for Africa are to abolish Europe's and America's farm subsidies, quotas and import tariffs, formalise and simplify the laws that govern business, undermine tyrants and above all encourage the growth of free-trading cities. In 1978 China was about as poor and despotic as Africa is now. It changed because it deliberately allowed free-trading zones to develop in emulation of Hong Kong. So, says the economist Paul Romer, why not repeat the formula? Use Western aid to create a new 'charter city' in Africa on uninhabited land, free to trade with the rest of the world, and allow it to draw in people from the surrounding nations. It worked for Tyre 3,000 years ago, for Amsterdam 300 years ago and for Hong Kong thirty years ago. It can work for Africa today.

If, that is, the climate does not lurch into chaos.

## Climate

In the mid-1970s it was briefly fashionable for journalists to write scare stories about the recent cooling of the globe, which was presented as undiluted bad news. Now it is fashionable for them to write scare stories about the recent warming of the globe, which is presented as undiluted bad news. Here are two quotes from the same magazine three decades apart. Can you tell which is about cooling and which about warming?

The weather is always capricious, but last year gave new meaning to the term. Floods, hurricanes, droughts – the only plague missing was frogs. The pattern of extremes fit scientists' forecasts of what a —— world would be like.

Meteorologists disagree about the cause and extent of the —— trend, as well as over its specific impact on local weather conditions. But they are almost unanimous in the view that the trend will reduce agricultural productivity for the rest of the century … The longer the planners delay, the more difficult will they find it to cope with climatic change once the results become grim reality.

The point I am making is not that one prediction proved wrong, but that the glass was half empty in both cases. Cooling and warming were both predicted to be disastrous, which implies that only the existing temperature is perfect. Yet climate has always varied; it is a special sort of narcissism to believe that only the recent climate is perfect. (The answer by the way is that the first one was a recent warning about warming; the second an old warning about cooling – both are from *Newsweek*.)

I could plunge into the scientific debate and try to persuade you and myself that the competitive clamour of alarm is as exaggerated as it proved to be on eugenics, acid rain, sperm counts and cancer – that the warming the globe faces in the next century is more likely to be mild than catastrophic; that the last three decades of relatively slow average temperature changes are more compatible with a low-sensitivity than a high-sensitivity model of greenhouse warming; that clouds may slow the warming as much as water vapour may amplify it; that the increase in methane has been (erratically) decelerating for twenty years; that there were warmer periods in earth's history in medieval times and about 6,000 years ago yet no accelerations or 'tipping points' were reached; and that humanity and nature survived much faster warming lurches in climate during the ice ages than anything predicted for this century. There are respectable scientific arguments to support all these arguments – and in some cases respectable scientific ripostes to them, too. But this is not a book about the climate; it is about the human race and its capacity for change. Besides, even if the current alarm does prove exaggerated, there is now no doubt that the climate of this planet has been subject to natural lurches in the past, and that though luckily there has been no huge lurch for 8,200 years, there have been some civilisation-killing perturbations – as the ruins at both Angkor Wat and Chichen Itza probably testify. So if only hypothetically, it is worth asking whether civilisation could survive climate change at the rate assumed by the consensus of scientists who comprise the Intergovernmental Panel on Climate Change (IPCC) – that is, that the earth will warm during this century by around 3°C.

However, that is just a mid-range figure. In 2007 the IPCC used six 'emissions scenarios', ranging from a fossil-fuel-intensive, centennial global boom to something that sounds more like a sustainable, groovy fireside sing-along, to calculate how much temperature will increase during the century. The average temperature increases predicted for the end of this century ranged from 1.8°C to 4°C above 1990 levels. Include the 95 per cent confidence intervals and the range is 1–6°C. In some cities the warming will be – has already been – even more, thanks to the 'urban heat island' effect. On the other hand, all experts agree that the warming will happen disproportionately at night, in winter and in cold regions, so cold times and places will get less cold more than hot ones will get hotter.

As for what might happen after 2100, in 2006 the British government appointed a civil servant, Nicholas Stern, to count the potential cost of extreme climate change far into the future. He came up with the answer that the cost was so high that almost any price to mitigate it now would be worth paying. But he only managed this by first cherry-picking high estimates of harm; and second using an unusually low discount rate to measure the present value of future loss. Where the Dutch economist Richard Tol had estimated costs as 'likely to be substantially smaller' than $14 per tonne of carbon dioxide, Stern simply doubled the figure to $29 per tonne. Tol – no sceptic – called the Stern report alarmist, incompetent and preposterous. As for discount rates, Stern used 2.1 per cent for the twenty-first century, 1.9 per cent for the twenty-second, and 1.4 per cent for subsequent centuries. Compared with a typical discount rate of about 6 per cent, this multiplies the apparent cost of harm in the twenty-second century one hundredfold. In other words, he said that a life saved from coastal flooding in 2200 should have almost the same spending priority *now* as a life saved from AIDS or malaria today. Hordes of economists, including notable names like William Nordhaus, quickly pointed out how this made no sense. It implies that your impoverished great great great grandfather, whose standard of living was roughly that of a modern Zambian, should have put aside most of his income to pay your bills today. With a higher discount rate, Stern's argument collapses because, even in the worst case, harm done by climate change in the twenty-second century is far less costly than harm done by climate-mitigation measures today. Nigel Lawson asks, reasonably enough: 'How great a sacrifice is it either reasonable or realistic to ask the present generation, particularly the present generation in the developing world, to make, in the hope of avoiding the prospect that the people of the developing world in a hundred years time may not be 9.5 times as well off as they are today, but only 8.5 times?'

Your grandchildren will be that rich. Do not take my word for it: all six of the IPCC's scenarios assume that the world will experience so much economic growth that the people alive in 2100 will be on average four to eighteen times as wealthy as we are today. The scenarios assume that the entire world will have a mean standard of living somewhere between today's Portugal and Luxembourg, and even the citizens of developing countries will have incomes between those of today's Malaysians and Norwegians. In the hottest scenario, income rises from $1,000 per head in poor countries today to more than $66,000 in 2100 (adjusted for inflation). Posterity in these futures is staggeringly wealthier than today, even in Africa – an interesting starting assumption for an attempt to warn us of a terrible future. Note that this is true even if climate change itself cuts wealth by Stern's 20 per cent by 2100: that would mean the world becoming 'only' two to ten times as rich. The paradox was stark when the Prince of Wales said in 2009 that humanity had '100 months left to take the necessary steps to avert irretrievable climate and ecosystem collapse', then went on in the same speech to say that, by 2050, there will be nine billion people on the planet, mostly consuming at Western levels.

The reason for these rosy assumptions about wealth is that the only way the world can get that hot is by getting very rich through emitting lots of carbon dioxide. Many economists think these futures, wonderful as they sound, are unrealistic. In one IPCC future, world population reaches fifteen billion by 2100, nearly double what demographers expect. In another, the poorest countries grow their per capita income four times as fast as Japan grew in the twentieth century. All the futures use market exchange rates instead of purchasing power parities for GDP, further exaggerating warming. In other words, the high-end projections have pretty wild assumptions, so the 4°C warming, let alone the unlikely 6°C, will only happen if it is also accompanied by truly astonishing increases in human prosperity. And if it is possible to get so prosperous, then the warming cannot have been doing much economic harm along the way.

To this some economists such as Martin Weitzman reply that even if the risk of catastrophe is vanishingly small, the cost would be so great that the normal rules of economics do not apply: so long as there is some possibility of a huge disaster, the world should take all steps to avoid it. The trouble with this reasoning is that it applies to all risks, not just climate change. The annual risk of collision with a very large asteroid, such as wiped out the dinosaurs, is put at about one in 100 billion. Given that such an event would greatly reduce human prosperity, it seems to be rather cheap of humankind to be spending as little as $4m a year to track such asteroids. Why are we not spending large sums stockpiling food caches in cities so that people can survive the risks from North Korean missiles, rogue robots, alien invaders, nuclear war, pandemics, super-volcanoes? Each risk may be very unlikely, but with the potential harm so very great, almost infinite resources deserve to be spent on them, and almost nothing on present causes of distress, under Weitzman's argument.

In short, the extreme climate outcomes are so unlikely, and depend on such wild assumptions, that they do not dent my optimism one jot. If there is a 99 per cent chance that the world's poor can grow much richer for a century while still emitting carbon dioxide, then who am I to deny them that chance? After all, the richer they get the less weather dependent their economies will be and the more affordable they will find adaptation to climate change.

## Warmer and richer or cooler and poorer?

So much for the outlying risks. Now consider the IPCC's much more probable central case: a 3°C rise by 2100. (I say more probable, but note that the rate of increase of temperature will have to be double that experienced in the 1980s and 1990s to hit this level – and the rate has been decelerating, not accelerating.) Count the cost – and benefit – of the extra warmth in terms of sea level, water, storms, health, food, species and ecosystems.

Sea level is by far the most worrisome issue, because the

current sea level is indeed the best of all possible sea levels: any change – up or down – will leave ports unusable. The IPCC forecasts that average sea level will rise by about 2–6 millimetres a year, compared with a recent rate of about 3.2 millimetres a year (or about a foot per century). At such rates, although coastal flooding will increase slightly in some places (local rising of the land causes sea level to fall in many areas), some countries will continue to gain more land from siltation than they lose to erosion. The Greenland land-based ice cap will melt a bit around the edge – many Greenland glaciers retreated in the last few decades of the twentieth century – but even the highest estimates of Greenland's melting are that it is currently losing mass at the rate of less than 1 per cent *per century*. It will be gone by AD 12,000. Of course, there is a temperature at which the Greenland and west Antarctic ice caps would disintegrate, but according to the IPCC scenarios if it is reached at all it is certainly not going to be reached in the twenty-first century.

As for fresh water, the evidence suggests, remarkably, that, other things being equal, warming will itself reduce the total population at risk from water shortage. Say again? Yes, reduce. On average rainfall will increase in a warmer world because of greater evaporation from the oceans, as it did in previous warm episodes such as the Holocene (when the Arctic ocean may have been almost ice-free in summer), the Egyptian, Roman and medieval warm periods. The great droughts that changed history in western Asia happened, as theory predicts, in times of cooling: 8,200 years ago and 4,200 years ago especially. If you take the IPCC's assumptions and count the people living in zones that will have more water versus zones that will have less water, it is clear that the net population at risk of water shortage by 2100 falls under all their scenarios. Although water will continue to be fought over, polluted and exhausted, while rivers and boreholes may dry up because of over-use, that will happen in a cool world too. As climate zones shift, southern Australia and northern Spain may get drier, but the Sahel and northern Australia will probably continue their recent wetter trend. Nor is there any evidence for the oft-repeated assertion that climate will be more volatile when wetter. Ice cores confirm that volatility of climate from year to year decreases markedly when the earth warms from an ice age. There will probably be some increase in the amount of rain that falls in the most extreme downpours, and perhaps more flooding as a result, but it is a sad truth that the richer people are, the less likely they are to drown, so the warmer and richer the world, the better the outcome.

The same is true for storms. During the warming of the twentieth century there was no increase in either the number or the maximum wind speed of Atlantic hurricanes making landfall. Globally, tropical cyclone intensity hit a thirty-year low in 2008. The cost of the damage done by hurricanes has increased greatly, but that is because of the building and insuring of expensive coastal properties, not because of storm intensity or frequency. The global annual death rate from weather-related natural disasters has declined by a remarkable 99 per cent since the 1920s – from 242 per million in the 1920s to three per million in the 2000s. The killing power of hurricanes depends far more on wealth and weather forecasts than on wind speed. Category 5 Hurricane Dean struck the well-prepared Yucatan in 2007 and killed nobody. A similar storm struck impoverished and ill-prepared Burma the next year and killed 200,000. If they are freed to prosper, the future citizens of Burma will be able to afford protection, rescue and insurance by 2100.

In measuring health, note that globally the number of excess deaths during cold weather continues to exceed the number of excess deaths during heat waves by a large margin – by about five to one in most of Europe. Even the notorious one-off death rate in the European summer heat wave of 2003 failed to match the number of excess cold deaths recorded in Europe during most winters. Besides, once again, people will adapt, as they do today. People move happily from London to Hong Kong or Boston to Miami and do not die from heat, so why should they die if their home city gradually warms by a few degrees? (It already has, because of the urban heat island effect.)

What about malaria? Even distinguished scientists have been heard to claim that malaria will spread northwards and uphill in a warming world. But malaria was rampant in Europe, North America and even arctic Russia in the nineteenth and early twentieth century, when the world was nearly a degree cooler than now. It disappeared, while the world was warming, because people kept their cattle in barns (providing mosquitoes with an alternative dining option), moved indoors at night behind closed windows, and to a lesser extent because swamps were drained and pesticides used. Today malaria is not limited by climate: there are lots of areas where it could rampage but does not. The same is true of malaria's mountain limitations. Just 2 per cent of Africa is too high for malarial mosquitoes now, and where highland areas have become malarial in the past century, such as in Kenya and New Guinea, the cause is human migration and habitat change, not climate change. 'There is no evidence that climate has played any role in the burgeoning tragedy of this disease at any altitude,' says Paul Reiter, a malaria expert. Should we not do something to prevent a million people dying of preventable malaria each year now, before worrying about the possibility that global warming might increase that number by 30,000 – at the very most? Likewise, a jump in tick-borne disease in eastern Europe around 1990, initially blamed on climate change, turned out to be caused by the fact that people who lost their jobs after the collapse of communism spent more time foraging for mushrooms in the forests.

Many commentators seized on the World Health Organisation's 2002 estimate that 150,000 people were dying each year as a result of climate change. The calculation assumed that an arbitrary 2.4 per cent of diarrhoea deaths were due to extra warmth breeding extra pathogenic bacteria; that some proportion of malaria deaths were due to extra rainfall breeding extra mosquitoes, and so on. But even if you accept these guesses, the WHO's own figures showed that climate change was dwarfed as a cause of death by iron deficiency, cholesterol, unsafe sex, tobacco, traffic accidents and other things, not to mention 'ordinary' diarrhoea and malaria. Even obesity, according to the same report, was killing more than twice as many people as climate change. Nor was any attempt made to estimate the number of lives saved by carbon emissions – by the provision of electric power to a village where people suffer from ill health due to indoor air pollution from cooking over open fires, say, or the deaths from malnutrition prevented by the higher productivity of agriculture using fertiliser made from natural gas. In 2009 Kofi Annan's Global Humanitarian Forum doubled the number of climate deaths to 315,000 a year, but only by ignoring these points, arbitrarily doubling the diarrhoea deaths caused by climate, and adding in ludicrous assumptions about how climate change was responsible for 'inter-clan fighting in Somalia', Hurricane Katrina and other disasters. Remember that every year fifty to sixty million people die: even going by the GHF figures less than 1 per cent of those die from climate change.

The global food supply will probably increase if temperature rises by up to 3°C. Not only will the warmth improve yields from cold lands and the rainfall improve yields from some dry lands, but the increased carbon dioxide will itself enhance yields, especially in dry areas. Wheat, for example, grows 15–40 per cent faster in 600 parts per million of carbon dioxide than it does in 295 ppm. (Glasshouses often use air enriched in carbon dioxide to 1,000 ppm to enhance plant growth rates.) This effect, together with greater rainfall and new techniques, means that less habitat will probably be lost to farming in a warmer world.

Indeed under the warmest scenario, much land could revert to wilderness, leaving only 5 per cent of the world under the plough in 2100, compared with 11.6 per cent today, allowing more space for wilderness. The richest and warmest version of the future will have the least hunger, and will have ploughed the least extra land to feed itself. These calculations come not from barmy sceptics, but from the IPCC's lead authors. And this is before taking into account the capacity of human societies to adapt to a changing climate.

The four horsemen of the human apocalypse, which cause the most premature and avoidable death in poor countries, are and will be for many years the same: hunger, dirty water, indoor smoke and malaria, which kill respectively about seven, three, three and two people per minute. If you want to do your fellow human beings good, spend your effort on combating those so that people can prosper, ready to meet climate challenges as they arrive. Economists estimate that a dollar spent on mitigating climate change brings ninety cents of benefits compared with $20 benefits per dollar spent on healthcare and $16 per dollar spent on hunger. Keeping climate at 1990 levels, assuming it could be done, would leave more than 90 per cent of human mortality causes untouched.

## Saving ecosystems

Ah, but that is the human race. What about other species? Will the warmth cause a wave of extinctions? Perhaps, but not necessarily. So far, despite two bursts of twentieth-century warming, not a single species has unambiguously been shown to succumb to global climate trends. The golden toad of Costa Rica, sometimes cited as the first casualty, died out either from a fungal disease or because of the drying of its cloud forest, probably caused by deforestation on the lower slopes of its mountain home: a local, not a global cause. The polar bear, still thriving today (eleven of thirteen populations are growing or steady) but threatened by the loss of Arctic sea ice in high summer, may contract its range further north, but it already adapts to ice-free summer months in Hudson's Bay by fasting on land till the sea re-freezes; and there is good evidence from northern Greenland of a briefly almost ice-free summer sea in the Arctic about 5,500 years ago, during a period that was markedly warmer than today. Arguably, the orang-utan, being devastated by the loss of forest to palm oil bio-fuel plantations in Borneo, is under greater threat from renewable energy than the polar bear is from global warming.

Do not get me wrong, I am not denying that species extinctions are occurring. I passionately believe in saving threatened species from extinction and I have twice worked on projects attempting to rescue endangered species – the cheer pheasant and the lesser florican. But the threats to species are all too prosaic: habitat loss, pollution, invasive competitors and hunting being the same four horsemen of the ecological apocalypse as always. Suddenly many of the big environmental organisations have lost interest in these threats as they chase the illusion of stabilising a climate that was never stable in the past. It is as if the recent emphasis on climate change has sucked the oxygen from the conservation movement. Conservationists, who have done tremendous good over the past half-century protecting and restoring a few wild ecosystems, and encouraging local people to support and value them, risk being betrayed by the new politicised climate campaigners, whose passion for renewable energy is eating into those very ecosystems and drawing funds away from their efforts.

Take coral reefs, which are suffering horribly from pollution, silt, nutrient runoff and fishing – especially the harvesting of herbivorous fishes that otherwise keep reefs clean of algae. Yet environmentalists commonly talk as if climate change is a far greater threat than these, and they are cranking up the apocalyptic statements just as they did wrongly about forests and acid rain. Charlie Veron, an Australian marine biologist: 'There is no hope of reefs surviving to even mid-century in any form that we now recognise.' Alex Rogers of the Zoological Society of London pledges 'an absolute guarantee of their annihilation'. No wiggle room there. It is true that rapidly heating the water by a few degrees can devastate reefs by 'bleaching' out the corals' symbiotic algae, as happened to many reefs in the especially warm El Niño year of 1998. But bleaching depends more on rate of change than absolute temperature. This must be true because nowhere on the planet, not even in the Persian Gulf where water temperatures reach 35°C, is there a sea too warm for coral reefs. Lots of places are too cold for coral reefs – the Galapagos, for example. It is now clear that corals rebound quickly from bleaching episodes, repopulating dead reefs in just a few years, which is presumably how they survived the warming lurches at the end of the last ice age. It is also apparent from recent research that corals become more resilient the more they experience sudden warmings. Some reefs may yet die if the world warms rapidly in the twenty-first century, but others in cooler regions may expand. Local threats are far more immediate than climate change.

Ocean acidification looks suspiciously like a back-up plan by the environmental pressure groups in case the climate fails to warm: another try at condemning fossil fuels. The oceans are alkaline, with an average pH of about 8.1, well above neutral (7). They are also extremely well buffered. Very high carbon dioxide levels could push that number down, perhaps to about 7.95 by 2050 – still highly alkaline and still much higher than it was for most of the last 100 million years. Some argue that this tiny downward shift in average alkalinity could make it harder for animals and plants that deposit calcium carbonate in their skeletons to do so. But this flies in the face of chemistry: the reason the acidity is increasing is that the dissolved bicarbonate is increasing too – and increasing the bicarbonate concentration increases the ease with which carbonate can be precipitated out with calcium by creatures that seek to do so. Even with tripled bicarbonate concentrations, corals show a continuing increase in both photosynthesis and calcification. This is confirmed by a rash of empirical studies showing that increased carbonic acid either has no effect or actually increases the growth of calcareous plankton, cuttlefish larvae and coccolithophores.

My general optimism is therefore not dented by the undoubted challenge of global warming by carbon dioxide. Even if the world warms as much as the consensus expects, the net harm still looks small alongside the real harm now being done by preventable causes; and if it does warm this much, it will be because more people are rich enough to afford to do something about it. As usual, optimism gets a bad press in this debate. Optimists are dismissed as fools, pessimists as sages, by a media that likes to be spoon-fed on scary press releases. That does not make the optimists right, but the poor track record of pessimists should at least give one pause. After all, we have been here before. 'I want to stress the urgency of the challenge,' said Bill Clinton once: 'This is not one of the summer movies where you can close your eyes during the scary parts.' He was talking not about climate change but about Y2K: the possibility that all computers would crash at midnight on 31 December 1999.

## Decarbonising the economy

In short, a warmer and richer world will be more likely to improve the well-being of both human beings and ecosystems than a cooler but poorer one. As Indur Goklany puts it, 'neither on grounds of public health nor on ecological factors is climate change likely to be the most important problem facing the globe this century.' The results of thirteen economic analyses of climate change, assuming consensus amounts of warming, conclude that it will either add or subtract about one year of global economic growth in the second half of the twenty-first century. Critics of this view often argue that development and carbon reduction need not be alternatives, and that it is the poor who are hit hardest by climate change. True, but it is a point that cuts both ways – it is the poor who are hit hardest by high energy costs, too. If mismanaged, climate mitigation could prove just as damaging to human welfare as climate change. A child that dies from indoor smoke in a village denied fossil-fuel electricity is just as great a tragedy as a child that dies in a flood caused by climate change. A forest that is cut down by people deprived of fossil fuels is just as felled as one lost to climate change. If climate change proves to be mild but cutting carbon causes real pain, we may find we have stopped a nose bleed by putting a tourniquet round our neck.

And cutting carbon will mean costly energy: so says the IPCC. If I am to accept the IPCC's estimate of temperature rise for the sake of this argument, then I should also accept its estimate of the cost of carbon rationing – which it puts at 5.5 per cent of GDP after about 2050, and that is after making highly unlikely assumptions of (quoting from the IPCC's 2007 report) 'transparent markets, no transaction costs, and thus perfect implementation of policy measures throughout the twenty-first century, leading to the universal adoption of cost-effective mitigations measures, such as carbon taxes or universal cap-and-trade programmes'.

The world economy needs plentiful joules of energy if it is not to run on slaves, and at the moment by far the cheapest source of those joules is the burning of hydrocarbons. About 600 kilograms of carbon dioxide are emitted per thousand dollars of economic activity. No country 'is remotely on a path' towards cutting that number substantially, says the physicist David MacKay. It could be done, but only at vast cost. The cost would be environmental as well as financial. Take Britain, an 'average rich' country. Burning hydrocarbon still provides 106 of the 125 kilowatt-hours per day per person of work that give Britons their standard of living. How could Britain power itself without fossil fuels? Suppose that an aggressive and expensive plan of pumped heat, waste incineration and loft insulation knocks twenty-five of that demand off, leaving 100kWh per day to find. Divide that 100 in four and ask for twenty-five from nuclear, twenty-five from wind, twenty-five from solar and five each from biofuel, wood, wave, tide and hydro. What would the country look like?

There would be sixty nuclear power stations around the coasts, wind farms would cover 10 per cent of the entire land (or a big part of the sea), there would be solar panels covering an area the size of Lincolnshire, eighteen Greater Londons growing bio-fuels, forty-seven New Forests growing fast-rotation harvested timber, hundreds of miles of wave machines off the coast, huge tidal barrages in the Severn estuary and Strangford Lough, and twenty-five times as many hydro dams on rivers as there are today. The prospect is unappetising: the entire country would look like a power station, pylons would march across the uplands and convoys of trucks would cart timber along the roads. Power cuts would be frequent – imagine a still, cold foggy day in January when the slack tide in the Severn estuary coincides with peak demand, when the solar panels are dead and the wind turbines still. Wildlife would suffer from the loss of estuaries, free-flowing rivers and open country. Powering the world with such renewables now is the surest way to spoil the environment. (Of course, coal mining and oil drilling can and do spoil the environment, too, but compared with most renewables their footprints are surprisingly small for the energy they yield.)

Besides, there is just no sign of most renewables getting cheaper. The cost of wind power has been stuck at three times the cost of coal power for many years. To get a toehold in the electricity market at all, wind power requires a regressive transfer from ordinary working people to rent-seeking rich landowners and businesses: as a rule of thumb, a wind turbine generates more value in subsidy than it does in electricity. Even in 6,000-turbine Denmark, not a single emission has been saved because intermittent wind requires fossil-fuel back-up (Denmark's wind power is exported to Sweden and Norway, which can turn their hydro plants back on quickly when the Danish wind drops). Meanwhile a Spanish study confirms that wind power subsidies destroy jobs: for each worker who moves from conventional electricity generation to renewable electricity generation, 'two jobs at a similar rate of pay must be forgone elsewhere in the economy, otherwise the funds to pay for the excess costs of renewable generation cannot be provided.' Although green campaigners are wont to argue that raising the cost of energy is a good thing, by definition it destroys jobs by reducing investment in other sectors. 'The suggestion that we can lift ourselves out of the economic doldrums by spending lavishly on exceptionally expensive new sources of energy is absurd,' writes Peter Huber.

But that's today. Tomorrow, there may well be carbon-free energy sources that do not have these disadvantages. It is possible, though unlikely, that these will include hot, dry geothermal power, offshore wind, wave and tide, or even ocean thermal energy conversion, using the temperature difference between the deep sea and the surface. They may include better biofuels from algal lagoons, though personally I would rather see a nuclear power plant so the lagoons can be used for fish farming or nature reserves. It is also possible that quite soon engineers will be able to use sunlight to make hydrogen directly from water with ruthenium dye as a catalyst – replicating photosynthesis, in effect. Clean-coal, with its carbon dioxide reinjected into the rocks, may play a part if its cost can be brought down (a mighty big 'if').

A big contribution will surely come from solar power, the least land-hungry of the renewables. Once solar panels can be mass-produced at $200 per square metre and with an efficiency of 12 per cent, they could generate the equivalent of a barrel of oil for about $30. Then, instead of drilling for $40 oil, everybody will be rushing to cover their roofs, and large parts of Algeria and Arizona with cheap solar panels. Most of Arizona gets about six kilowatt-hours of sunlight per square metre per day so, assuming 12 per cent efficiency, it would take about one-third of Arizona to supply Americans with all their energy: a lot of land, but not unimaginable. Apart from cost, solar's big problem, like wind's, is its intermittent nature: it does not work at night, for instance.

But the obvious way to go low-carbon is nuclear. Nuclear power plants already produce more power from a smaller footprint, with fewer fatal accidents and less pollution than any other energy technology. The waste they produce is not an insoluble issue. It is tiny in volume (a Coke can per person per lifetime), easily stored and unlike every other toxin gets safer with time – its radioactivity falls to one-billionth of the starting level in two centuries. These advantages are growing all the time. Better kinds of nuclear power will include small, disposable, limited-life nuclear batteries for powering individual towns for limited periods and fast-breeder, pebble-bed, inherent-safe atomic reactors capable of extracting 99 per cent of uranium's energy, instead of 1 per cent as at present, and generating even smaller quantities of short-lived waste while doing so. Modern nuclear reactors are already as different from the inherently unstable, uncontained Chernobyl ones as a jetliner is from a biplane. Perhaps one day fusion will contribute, too, but do not hold your breath.

The Italian engineer Cesare Marchetti once drew a graph of human energy use over the past 150 years as it migrated from wood to coal to oil to gas. In each case, the ratio of carbon atoms to hydrogen atoms fell, from ten in wood to one in coal to a half in oil to a quarter in methane. In 1800 carbon atoms did 90 per cent of combustion, but by 1935 it was 50:50 carbon and hydrogen, and by 2100, 90 per cent of combustion may come from hydrogen – made with nuclear electricity, most probably. Jesse Ausubel predicts that 'if the energy system is left to its own devices, most of the carbon will be out of it by 2060 or 2070.'

The future will feature ideas that are barely glints in engineers' eyes right now – devices in space to harness the solar wind, say, or the rotational energy of the earth; or devices to shade the planet with mirrors placed at the Lagrange Point between the sun and the earth. How do I know? Because ingenuity is rampant as never before in this massively net-worked world and the rate of innovation is accelerating, through serendipitous searching, not deliberate planning. When asked at the Chicago World Fair in 1893 which invention would most likely have a big impact in the twentieth century, nobody mentioned the automobile, let alone the mobile phone. So even more today you cannot begin to imagine the technologies that will be portentous and commonplace in 2100.

They may not even tackle man-made carbon, but may go for the natural cycle instead. Each year more than 200 billion tonnes of carbon are removed from the atmosphere by growing plants and plankton, and 200 billion tonnes returned to it by rotting, digestion and respiration. Human activity adds less than ten billion tonnes to that cycle, or 5 per cent. It cannot be beyond the wit of twenty-first century humankind to nudge the natural carbon cycle into taking up 5 per cent more than it releases by fertilising desert stretches of the ocean with iron or phosphorus; by encouraging the growth of carbon-rich oceanic organisms called salps, which sink to the bottom of the ocean; or by burying 'biochar' – powdered charcoal made from crops.

The way to choose which of these technologies to adopt is probably to enact a heavy carbon tax, and cut payroll taxes

(National Insurance in Britain) to the same extent. That would encourage employment and discourage carbon emissions. The way not to get there is to pick losers, like wind and biofuel, to reward speculators in carbon credits and to load the economy with rules, restrictions, subsidies, distortions and corruption. When I look at the politics of emissions reduction, my optimism wobbles. The Copenhagen conference of December 2009 came worryingly close to imposing a corruptible and futile system of carbon rationing, which would have hurt the poor, damaged ecosystems and rewarded bootleggers and dictators.

Remember I am not here attempting to resolve the climate debate, nor saying that catastrophe is impossible. I am testing my optimism against the facts, and what I find is that the probability of rapid and severe climate change is small; the probability of net harm from the most likely climate change is small; the probability that no adaptation will occur is small; and the probability of no new low-carbon energy technologies emerging in the long run is small. Multiply those small probabilities together and the probability of a prosperous twenty-first century is therefore by definition large. You can argue about just how large, and therefore about how much needs to be spent on precaution; but you cannot on the IPCC's figures make it anything other than very probable that the world will be a better place in 2100 than it is today.

And there is every reason to think that Africa can share in that prosperity. Despite continuing war, disease and dictators, inch by inch its population will stabilise; its cities will flourish; its exports will grow; its farms will prosper; its wildernesses will survive and its people will experience peace. In the mega-droughts of the ice ages, Africa could support very few early hunter-gatherers; in a warm and moist interglacial, it can support a billion mostly urban exchanger-specialisers.

# CHAPTER 11

# The catallaxy:
# rational optimism about 2100

I hear babies cry, I watch them grow,
They'll learn much more than I'll ever know,
And I think to myself, what a wonderful world.
BOB THIELE and GEORGE DAVID WEISS
*What a Wonderful World*

## IPCC PROJECTIONS FOR WORLD GDP PER CAPITA

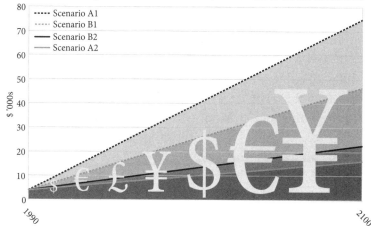

In this book I have tried to build on both Adam Smith and Charles Darwin: to interpret human society as the product of a long history of what the philosopher Dan Dennett calls 'bubble-up' evolution through natural selection among cultural rather than genetic variations, and as an emergent order generated by an invisible hand of individual transactions, not the product of a top-down determinism. I have tried to show that, just as sex made biological evolution cumulative, so exchange made cultural evolution cumulative and intelligence collective, and that there is therefore an inexorable tide in the affairs of men and women discernible beneath the chaos of their actions. A flood tide, not an ebb tide.

Somewhere in Africa more than 100,000 years ago, a phenomenon new to the planet was born. A Species began to add to its habits, generation by generation, without (much) changing its genes. What made this possible was exchange, the swapping of things and services between individuals. This gave the Species an external, collective intelligence far greater than anything it could hold in its admittedly capacious brain. Two individuals could each have two tools or two ideas while each knowing how to make only one. Ten individuals could know between them ten things, while each understanding one. In this way exchange encouraged specialisation, which further increased the number of different habits the Species could have, while shrinking the number of things that each individual knew how to make. Consumption could grow more diversified, while production grew more specialised. At first, the progressive expansion of the Species' culture was slow, because it was limited by the size of each connected population. Isolation on an island or devastation by a famine could reduce the population and so diminish its collective intelligence. Bit by bit, however, the Species expanded both in numbers and in prosperity. The more habits it acquired, the more niches it could occupy and the more individuals it could support. The more individuals it could support, the more habits it could acquire. The more habits it acquired, the more niches it could create.

The cultural progress of the Species encountered impediments along the way. Overpopulation was a constant problem: as soon as the capacity of the local environment to support the population began to suffer, so individuals began to retreat from specialisation and exchange into defensive self-sufficiency, broadening their production and narrowing their consumption. This reduced the collective intelligence they could draw upon, which reduced the size of the niche they occupied, putting further pressure on population. So there were crashes, even local extinctions. Or the Species found itself expanding in numbers but not in living standards. Yet, again and again the Species found ways to recover through new kinds of exchange and specialisation. Growth resumed.

Other impediments were of the Species' own making. Equipped by their animal ancestry with an ambitious and jealous nature, individuals were often tempted to predate upon and parasitise their fellows' productivity – to take and not to give. They killed, they enslaved, they extorted. For millennium after millennium this problem remained unsolved and the expansion of the Species, both its living standards and its population, was sporadically slowed, set back and reversed by the enervating greed of the parasites. Not all of the hangers-on were bad: there were rulers and public servants who lived off the traders and producers but dispensed justice and defence, or built roads and canals and schools and hospitals, making the lives of the specialise-and-exchange folk easier, not harder. These behaved like symbionts, rather than parasites (government can do good, after all). Yet still the Species grew, both in numbers and in habits, because the parasites never quite killed the system off which they fed.

Around 10,000 years ago, the pace of the Species' progress leapt suddenly ahead thanks to the suddenly greater stability of the climate, which allowed the Species to co-opt other species and enable them to evolve into exchange-and-specialise partners, generating services for the Species in exchange for their needs. Now, thanks to farming, each individual had not only other members of the Species working for her (and vice versa), but members of other species as well, such as cows and corn. Around 200 years ago, the pace of change quickened again thanks to the Species' new ability to recruit extinct species to its service as well, through the mining of fossil fuels and the releasing of their energy in ways that generated still more services. By now the Species was the dominant large animal on its planet and was suddenly experiencing rapidly rising living standards because of falling birth rates. Parasites plagued it still – starting wars, demanding obedience, building bureaucracies, committing frauds, preaching schisms – but the exchange and specialisation continued, and the collective intelligence of the Species reached unprecedented levels. By now almost the entire world was connected by a web so that ideas from everywhere could meet and mate. The pace of progress picked up once more. The future of the Species was bright, though it did not know it.

## Onward and upward

I have presented the case for sunny optimism. I have argued that now the world is networked, and ideas are having sex with each other more promiscuously than ever, the pace of innovation will redouble and economic evolution will raise the living standards of the twenty-first century to unimagined heights, helping even the poorest people of the world to afford to meet their desires as well as their needs. I have argued that although such optimism is distinctly unfashionable, history suggests it is actually a more realistic attitude than apocalyptic pessimism. 'It is the long ascent of the past that gives the lie to our despair,' said H.G. Wells.

These are great sins against conventional wisdom. Worse, they may even leave the impression of callous indifference to the fact that a billion people have not enough to eat, that a billion lack access to clean water, that a billion are illiterate. The opposite is true. It is precisely because there is still far more suffering and scarcity in the world than I or anybody else with a heart would wish that ambitious optimism is morally mandatory. Even after the best half-century for poverty reduction, there are still hundreds of millions going blind for lack of vitamin A in their monotonous diet, or watching their children's bellies swell from protein deficiency, or awash with preventable dysentery caused by contaminated water, or coughing with avoidable pneumonia caused by the smoke of indoor fires, or wasting from treatable AIDS, or shivering with unnecessary malaria. There are people living in hovels of dried mud, slums of corrugated iron, or towers of soulless concrete (including the 'Africas within' the West), people who never get a chance to read a book or see a doctor. There are young boys who carry machine guns and young girls who sell their bodies. If my great granddaughter reads this book in 2100 I want her to know that I am acutely aware of the inequality of the world I inhabit, a world where I can worry about my weight and a restaurant owner can moan about the iniquity of importing green beans by air from Kenya in winter, while in Darfur a child's shrunken face is covered in flies, in Somalia a woman is stoned to death and in Afghanistan a lone American entrepreneur builds schools while his government drops bombs.

It is precisely this 'evitable' misery that is the reason for pressing on urgently with economic progress, innovation and change, the only known way of bringing the benefits of a rising living standard to many more people. It is precisely because there is so much poverty, hunger and illness that the world must be very careful not to get in the way of the things that have bettered so many lives already – the tools of trade, technology and trust, of specialisation and exchange. It is precisely because there is still so much further to go that those who offer counsels of despair or calls to slow down in the face of looming environmental disaster may be not only factually but morally wrong.

It is a common trick to forecast the future on the assumption of no technological change, and find it dire. This is not wrong. The future would indeed be dire if invention and discovery ceased. As Paul Romer puts it: 'Every generation has perceived the limits to growth that finite resources and undesirable side effects would pose if no new recipes or ideas were discovered. And every generation has underestimated the potential for finding new recipes and ideas. We consistently fail to grasp how many ideas remain to be discovered.' By far the most dangerous, and indeed unsustainable thing the human race could do to itself would be to turn off the innovation tap. Not inventing, and not adopting new ideas, can itself be both dangerous and immoral.

## How good could it get?

Futurology always ends up telling you more about your own time than about the future. H.G. Wells made the future look like Edwardian England with machines; Aldous Huxley made it feel like 1920s New Mexico on drugs; George Orwell made it sound like 1940s Russia with television. Even Arthur C. Clarke and Isaac Asimov, more visionary than most, were steeped in the transport-obsessed 1950s rather than the communication-obsessed 2000s. So in describing the world of 2100, I am bound to sound like somebody stuck in the world of the early twenty-first century, and make laughable errors of extrapolation. 'It's tough to make predictions,' joked somebody, perhaps Yogi Berra: 'especially about the future.' Technologies I cannot even conceive will be commonplace and habits I never knew human beings needed will be routine. Machines may have become sufficiently intelligent to design themselves, in which case the rate of economic growth may by then have changed as much as it did at the start of the industrial revolution – so that the world economy will be doubling in months or even weeks, and accelerating towards a technological 'singularity' where the rate of change is almost infinite.

But here goes, none the less. I forecast that the twenty-first century will show a continuing expansion of catallaxy – Hayek's word for spontaneous order created by exchange and specialisation. Intelligence will become more and more collective; innovation and order will become more and more bottom-up; work will become more and more specialised, leisure more and more diversified. Large corporations, political parties and government bureaucracies will crumble and fragment as central planning agencies did before them. The *Bankerdämmerung* of 2008 swept away a few leviathans but fragmented and short-lived hedge funds and boutiques will spring up in their place. The collapse of Detroit's big car makers in 2009 leaves a flock of entrepreneurial startups in charge of the next generation of cars and engines. Monolithic behemoths, whether private or nationalised, are vulnerable as never before to this Lilliputian assault. They are steadily being driven extinct not just by small firms, but by ephemeral aggregations of people that form and reform continuously. The big firms that survive will do so by turning themselves into bottom-up evolvers. Google, dependent on millions of instantaneous auctions to raise revenue from its AdWords, is 'an economy unto itself, a seething laboratory', says Stephen Levy. But Google will seem monolithic compared with what comes next.

The bottom-up world is to be the great theme of this century. Doctors are having to get used to well-informed patients who have researched their own illnesses. Journalists are adjusting to readers and viewers who select and assemble their news on demand. Broadcasters are learning to let their audiences choose the talent that will entertain them. Engineers are sharing problems to find solutions. Manufacturers are responding to consumers who order their products *à la carte*. Genetic engineering is going to become open-source, where people, not corporations, decide what combinations of genes they want. Politicians are increasingly corks tossed on the waves of public opinion. Dictators are learning that their citizens can organise riots by text message. 'Here comes everybody' says the author Clay Shirky.

People will more and more freely find ways to exchange their specialised production for diversified consumption. This world can already be glimpsed on the web, in what John Barlow calls 'dot-communism': a workforce of free agents bartering their ideas and efforts barely interested in whether the barter yields 'real' money. The explosion of interest in the free sharing of ideas that the internet has spawned has taken everybody by surprise. 'The online masses have an incredible willingness to share' says Kevin Kelly. Instead of money, 'peer producers who create the stuff gain credit, status, reputation, enjoyment, satisfaction and experience'. People are willing to share their photographs on Flickr, their thoughts on Twitter, their friends on Facebook, their knowledge on Wikipedia, their software patches on Linux, their donations on GlobalGiving, their community news on Craigslist, their pedigrees on Ancestry.com, their genomes on 23andMe, even their medical records on PatientsLikeMe. Thanks to the internet, each is giving according to his ability to each according to his needs, to a degree that never happened in Marxism.

This catallaxy will not go smoothly, or without resistance. Natural and unnatural disasters will still happen. Governments will bail out big corporations and big bureaucracies, hand them special favours such as subsidies or carbon rations and regulate them in such a way as to create barriers to entry, slowing down creative destruction. Chiefs, priests, thieves, financiers, consultants and others will appear on all sides, feeding off the surplus generated by exchange and specialisation, diverting the life-blood of the catallaxy into their own reactionary lives. It happened in the past. Empires bought stability at the price of creating a parasitic court; monotheistic religions bought social cohesion at the price of a parasitic priestly class; nationalism bought power at the expense of a parasitic military; socialism bought equality at the price of a parasitic bureaucracy; capitalism bought efficiency at the price of parasitic financiers. The online world will attract parasites too: from regulators and cyber-criminals to hackers and plagiarists. Some of them may temporarily throttle their generous hosts.

It is just possible that the predators and parasites will actually win altogether, or rather that ambitious ideological busybodies will succeed in shutting down the catallaxy and crashing the world back into pre-industrial poverty some time during the coming century. There is even a new reason for such pessimism: the integrated nature of the world means that it may soon be possible to capture the entire world on behalf of a foolish idea, where before you could only capture a country, or perhaps if you were lucky an empire. (The great religions all needed empires within which to flourish and become powerful: Buddhism within the Mauryan and Chinese, Christianity within the Roman, Islam within the Arab.)

Take the twelfth century as an example of how close the world once came to turning its back on the catallaxy. In one fifty-year period, between 1100 and 1150, three great nations shut down innovation, enterprise and freedom all at once. In Baghdad, the religious teacher Al-Ghazali almost single-handedly destroyed the tradition of rational enquiry in the Arab world and led a return to mysticism intolerant of new thinking. In Peking, Su-Sung's astronomical clock, the 'cosmic engine', probably the most sophisticated mechanical device ever built at that date, was destroyed by a politician suspicious of novelty and (t)reason, setting the tone for the retreat to autarky and tradition that would be China's fate for centuries to come. In Paris, St Bernard of Clairvaux persecuted the scholar Peter Abelard, criticised the rational renaissance centred on the University of Paris and supported the disastrous fanaticism of the second crusade. Fortunately, the flames of free thought and reason and catallaxy were kept burning – in Italy and North Africa, especially. But imagine if they had not been. Imagine if the entire world had turned its back on the catallaxy then. Imagine if the globalised world of the twenty-first century allows a globalised retreat from reason. It is a worrying thought. The wrong kind of chiefs, priests and thieves could yet snuff out future prosperity on earth. Already lords don boiler suits to destroy genetically modified crops, presidents scheme to prevent stem-cell research, prime ministers trample on habeas corpus using the excuse of terrorism, metastasising bureaucracies interfere with innovation on behalf of reactionary pressure groups, superstitious creationists stop the teaching of good science, air-headed celebrities rail against free trade, mullahs inveigh against the empowerment of women, earnest princes lament the loss of old ways and pious bishops regret the coarsening effects of commerce. So far they are all sufficiently localised in their effects to achieve no more than limited pauses in the happy progress of the species, but could one of them go global?

I doubt it. It will be hard to snuff out the flame of innovation, because it is such an evolutionary, bottom-up phenomenon in such a networked world. However reactionary and cautious Europe and the Islamic world and perhaps even America become, China will surely now keep the torch of catallaxy alight, and India, and maybe Brazil, not to mention a host of smaller free cities and states. By 2050, China's economy may well be double the size of America's. The experiment will go on. So long as human exchange and specialisation are allowed to thrive somewhere, then culture evolves whether leaders help it or hinder it, and the result is that prosperity spreads, technology progresses, poverty declines, disease retreats, fecundity falls, happiness increases, violence atrophies, freedom grows, knowledge flourishes, the environment improves and wilderness expands. Said Lord Macaulay, 'We see in almost every part of the annals of mankind how the industry of individuals, struggling up against wars, taxes, famines, conflagrations, mischievous prohibitions, and more mischievous protections, creates faster than governments can squander, and repairs whatever invaders can destroy.'

Human nature will not change. The same old dramas of aggression and addiction, of infatuation and indoctrination, of charm and harm, will play out, but in an ever more prosperous world. In Thornton Wilder's play *The Skin of Our Teeth*, the Antrobus family (representing humankind) just manages to survive the ice age, the flood and a world war, but their natures do not change. History repeats itself as a spiral not a circle, Wilder implied, with an ever-growing capacity for both good and bad, played out through unchanging individual character. So the human race will continue to expand and enrich its culture, despite setbacks and despite individual people having much the same evolved, unchanging nature. The twenty-first century will be a magnificent time to be alive.

Dare to be an optimist.

# POSTSCRIPT

In 1755, Lisbon was struck by a fearful earthquake. Much of the city was destroyed, massive fires broke out and a tsunami engulfed the remains. Perhaps as many as 60,000 people died. But as well as the human tragedy, the disaster at Lisbon struck a blow against the philosophy of 'optimism'.

The word was fairly new, having been coined in 1737, and it did not then mean what it means now: a hopeful view of the future. It meant almost the opposite, namely that the world was at its 'optimum', that this was the best of all possible worlds and that nothing could be better. The philosopher Gottfried Leibniz had come up with this bright idea as part of his philosophy of theodicy, which argued that God is a benevolent deity so even bad things must have a good purpose. Nothing could be better than this world.

The deaths at Lisbon must therefore be a good thing, not a bad thing. Lisbon was being punished for its sinfulness, and the world would be cleansed by its devastation. This was too much for the French playwright and philosopher Voltaire. In his *Poème sur le désastre de Lisbonne*, he wondered what God had against the Portuguese:

Was then more vice in fallen Lisbon found,
Than Paris, where voluptuous joys abound?

Was less debauchery to London known,
Where opulence luxurious holds the throne?

Lisbon, and Jean-Jacques Rousseau's critique of the poem,
led Voltaire to write his famous novel *Candide*, in which he
ridiculed the philosophy of optimism. Candide, the protagon-
ist, naively and disingenuously accepts all the misfortunes that
befall him, taught by his mentor, Dr Pangloss, that all is always
for the best in the best of all possible worlds. When they are
shipwrecked in Lisbon harbour just before the earthquake, and
their companion Jacques drowns, Pangloss comforts Candide
by explaining that Lisbon harbour was created in order to
drown Jacques. Pangloss's conviction that this is the best of all
possible worlds persists – despite the fact that he experiences
syphilis, shipwreck, earthquake, fire, torture by the inquisition
as a heretic, hanging and slavery.

I am not an optimist of the Panglossian variety. I do not
think this is the best of all possible worlds. Indeed, a large part
of my motivation to write this book was the realisation that not
only are huge improvements in living standards still possible,
but that too many of my fellow human beings think they are
not. I think this is a vale of tears compared with what we can
achieve next. As I wrote in Chapter 1, 'It is precisely because
so much human betterment has been shown to be possible in
recent centuries that the continuing imperfection of the world
places a moral duty on humanity to allow economic evolution
to continue. To prevent change, innovation and growth is to
stand in the way of potential compassion.'

The true Panglossians of today are those who urge retreat,
caution and fear of where change may lead. They think this is
the least worst of all possible worlds, that change can only
make it worse. If I have done anything in this book, I hope I
have opened your eyes, as I have my own, to the immense
potential the human race has to improve its lot much, much

Postscript

further. We can continue to raise the per capita income of the people of the planet while continuing to reduce our ecological footprint.

I am writing this at the end of a decade that saw many natural disasters similar to the earthquake at Lisbon. There was the Indian Ocean tsunami of 2004, Hurricane Katrina in 2005, Burma's cyclone in 2008, China's earthquake in 2008, Australia's fires of 2009, Haiti's earthquake in 2010, Pakistan's floods in 2010. Yet the natural disasters of recent years have strongly vindicated optimism – not of the Leibnizian, but of the modern, hopeful kind. The difference between Haiti's death toll of up to 300,000 and Chile's of less than 500 a month later can be attributed in large part to the difference in their wealth. Likewise, Category 5 Hurricane Dean struck the well-prepared Yucatán in 2007 and killed no one, but when a similar storm struck impoverished and ill-prepared Burma the next year, it killed more than 200,000. Pakistan's floods killed 1,800; Poland's less than fifty. As I write this, Java's Mount Merapi has killed 130 people; Iceland's Eyjafjallajökull killed no one.

In short, prosperity buys survival. (The shocking thing about Hurricane Katrina was not that it killed so many, but that it did so in such a prosperous country.) The scholar Indur Goklany has calculated that, as the world has grown richer in the past ninety years, the death rate per 100,000 people from extreme weather events – floods, droughts, fires, storms and freezes – has fallen by 98 per cent since it peaked in the 1920s. Even the actual number of deaths due to extreme weather has fallen by 93 per cent, despite a quadrupling of the population and an increase in the number of (recorded) extreme weather events.

In my adult lifetime, the near complete defeat of famine at a time when human population has doubled is extraordinary proof that the path we have been treading is on the whole heading in a good direction. Suppose world per capita income

363

were to octuple in the next ninety years, as it did, roughly, in the last ninety. So long as all get their share of this prosperity, we can expect most of the world in 2100 to become as nearly disaster-proof as the rich west is today: through building standards, warning systems, emergency services, health services, trade networks and technology. A mega-volcano or a big asteroid would still test any society, but much less pain will come from nature the richer we become.

This is not, and will never be, the best of all possible worlds. But it can get much, much better.

Matt Ridley, 2011

# ACKNOWLEDGEMENTS

It is one of the central arguments of this book that the special feature of human intelligence is that it is collective, not individual – thanks to the invention of exchange and specialisation. The same is true of the ideas in this book. I have done no more in writing it than try to open my mind to the free flow and exchange of ideas of others and hope that those ideas will mate furiously within my own cortex. The writing of the book has been a sort of continuous conversation, therefore, with friends, experts, mentors and strangers, carried on in person, by email, by exchange of papers and references, in person and by telephone. The internet is truly a great gift to authors, providing boundless access to sources of knowledge on which to draw, a virtual library of unlimited size and speed (and of course variable quality).

I am immensely grateful to all those who have allowed me to converse with them in this way and I have met nothing but freely given help and advice from all. I am especially grateful to Jan Witkowski, Gerry Ohrstrom and Julian Morris, who helped me organise a meeting on 'The Retreat from Reason' at Cold Spring Harbor to begin to explore my ideas; and then to Terry Anderson and Monika Cheney who two years later arranged a seminar in Napa, California, for me to bounce a first draft of my book off some remarkable people for two days.

Here, in alphabetical order, are some of those whose ideas and thoughts I sampled most fruitfully. Their collective generosity and perspicacity have been astounding. The mistakes, of course, are mine. They include: Bruce Ames, Terry Anderson, June Arunga, Ron Bailey, Nick Barton, Roger Bate, Eric Beinhocker, Alex Bentley, Carl Bergstrom, Roger Bingham, Doug Bird, Rebecca Bliege Bird, the late Norman Borlaug, Rob Boyd, Kent Bradford, Stewart Brand, Sarah Brosnan, John Browning, Erwin Bulte, Bruce Charlton, Monika Cheney, Patricia Churchland, Greg Clark, John Clippinger, Daniel Cole, Greg Conko, Jack Crawford, the late Michael Crichton, Helena Cronin, Clive Crook, Tony Curzon Price, Richard Dawkins, Tracey Day, Dan Dennett, Hernando de Soto, Frans de Waal, John Dickhaut, Anna Dreber, Susan Dudley, Emma Duncan, Martin Durkin, David Eagleman, Niall Ferguson, Alvaro Fischer, Tim Fitzgerald, David Fletcher, Rob Foley, Richard Gardner, Katya Georgieva, Gordon Getty, Jeanne Giaccia, Urs Glasser, Indur Goklany, Allen Good, Oliver Goodenough, Johnny Grimond, Monica Guenther, Robin Hanson, Joe Henrich, Dominic Hobson, Jack Horner, Sarah Hrdy, Nick Humphrey, Anya Hurlbert, Anula Jayasuriya, Elliot Justin, Anne Kandler, Ximena Katz, Terence Kealey, Eric Kimbrough, Kari Kohn, Meir Kohn, Steve Kuhn, Marta Lahr, Nigel Lawson, Don Leal, Gary Libecap, Brink Lindsey, Robert Litan, Bjørn Lomborg, Marcus Lovell-Smith, Qing Lu, Barnaby Marsh, Richard Maudslay, Sally McBrearty, Kevin McCabe, Bobby McCormick, Ian McEwan, Al McHughen, Warren Meyer, Henry Miller, Alberto Mingardi, Graeme Mitchison, Julian Morris, Oliver Morton, Richard Moxon, Daniel Nettle, Johann Norberg, Jesse Norman, Haim Ofek, Gerry Ohrstrom, Kendra Okonski, Svante Paabo, Mark Pagel, Richard Peto, Ryan Phelan, Steven Pinker, Kenneth Pomeranz, David Porter, Virginia Postrel, C.S. Prakash, Chris Pywell, Sarah Randolph, Trey Ratcliff, Paul Reiter, Eric Rey, Pete Richerson, Luke Ridley,

Russell Roberts, Paul Romer, David Sands, Rashid Shaikh, Stephen Shennan, Michael Shermer, Lee Silver, Vernon Smith, Dane Stangler, James Steele, Chris Stringer, Ashley Summerfield, Ray Tallis, Dick Taverne, Janice Taverne, John Tooby, Nigel Vinson, Nicholas Wade, Ian Wallace, Jim Watson, Troy Wear, Franz Weissing, David Wengrow, Tim White, David Willetts, Bart Wilson, Jan Witkowski, Richard Wrangham, Bob Wright and last, but certainly not least, Paul Zak, who employed me as white-coated lab assistant for a day.

My agent, Felicity Bryan, is, as ever, a godmother of this book – encouraging and reassuring at all the right moments. She and Peter Ginsberg have been champions of the project throughout, as have my editors Terry Karten, Mitzi Angel and Louise Haines and other supportive friends at 4th Estate and HarperCollins, especially Elizabeth Woabank. Huge thanks too to Kendra Okonski for invaluable help in making rationaloptimist.com a reality, and to Luke Ridley for help with research. Thanks also to Roger Harmar, Sarah Hyndman and MacGuru Ltd for the charts at the start of each chapter.

My greatest debt is to my family, not least for helping me to find the space and time to write. Anya's inspiration, insight and support are immeasurably valuable. It has been a great joy to have for the first time the unflinchingly sharp mind of my son to discuss ideas with, and check facts, as I write. He helped prepare most of the charts. And my daughter led me to a bridge in Paris one evening, to listen to Dick Miller and his group singing 'What a Wonderful World'.

# NOTES AND REFERENCES

These notes will be continuously corrected and expanded on the website www.rationaloptimist.com.

## Prologue

p. 1 'In other classes of animals, the individual advances from infancy to age or maturity'. Ferguson, A. 1767. *An Essay on the History of Civil Society.*

pp. 1–2 'On my desk as I write sit two artefacts of roughly the same size'. Photographs of the hand axe and computer mouse reproduced by permission of John Watson.

p. 3 'from perhaps 3 million to nearly 7 billion people'. Kremer, M. 1993. Population growth and technical change, one million B.C. to 1990. *Quarterly Journal of Economics* 108:681–716.

p. 4 'The human being is the only animal that ...' Gilbert, D. 2007. *Stumbling on Happiness.* Harper Press.

p. 4 'with the possible exception of language'. Pagel, M. 2008. Rise of the digital machine. *Nature* 452:699.

p. 4 'compared with even chimpanzees humans are almost obsessively interested in faithful imitation'. Horner, V. and Whiten, A. 2005. Causal knowledge and imitation/emulation switching in chimpanzees (*Pan troglodytes*) and children (*Homo sapiens*). *Animal Cognition* 8:164–81.

p. 5 'We may call it social evolution when an invention quietly spreads through imitation.' Tarde, G. 1969/1888. *On Communication and Social Influence.* Chicago University Press.

p. 5 'selection by imitation of successful institutions and habits'. Hayek, F.A. 1960. *The Constitution of Liberty.* Chicago University Press.

p. 5 'Richard Dawkins in 1976 coined the term "meme" for a unit of cultural imitation'. Dawkins, R. 1976. *The Selfish Gene.* Oxford University Press.

p. 5 'Richard Nelson in the 1980s proposed that whole economies evolve by natural selection'. Nelson, R.R. and Winter, S.G. 1982. *An Evolutionary Theory of Economic Change.* Harvard University Press.

p. 6 'a culture or a camera'. Richerson, P. and Boyd, R. 2005. *Not by Genes Alone*. Chicago University Press: 'adding one innovation after another to a tradition until the results resemble organs of extreme perfection'.

p. 7 '"To create is to recombine" said the molecular biologist François Jacob'. Jacob, F. 1977. Evolution and tinkering. *Science* 196:1163.

p. 8 'what Adam Smith said in 1776'. Smith, A. 1776. *The Wealth of Nations*.

p. 9 'sluiced artificially cheap money towards bad risks'. For a good account of this see Norberg, J. 2009. *Financial Fiasco*. Cato Institute.

p. 9 'The crisis has at least as much political as economic causation'. Friedman, J. 2009. A crisis of politics, not economics: complexity, ignorance and policy failure. *Critical Review* 23 (introduction to special issue).

## Chapter 1

p. 11 'On what principle is it, that when we see nothing but improvement behind us, we are to expect nothing but deterioration before us?' Macaulay, T.B. 1830. Review of Southey's Colloquies on Society. *Edinburgh Review*, January 1830.

p. 11 World GDP graph. Maddison, A. 2006. *The World Economy*. OECD Publishing.

p. 12 'But the vast majority of people are much better fed, much better sheltered, much better entertained, much better protected against disease and much more likely to live to old age than their ancestors have ever been'. Kremer, M. 1993. Population growth and technical change, one million BC to 1990. *Quarterly Journal of Economics* 108:681–716. See Brad De Long's estimates at http://econ161.berkeley.edu/TCEH/1998_Draft/World_GDP/Estimating_World_GDP.html.

p. 12 'the number of different products that you can buy in New York or London tops ten billion'. Beinhocker, E. 2006. *The Origin of Wealth*. Harvard Business School Press.

p. 13 'As for the bird outside the window, tomorrow it will be trapped and eaten by the boy'. See McCloskey, D. 2006. *The Bourgeois Virtues*. Chicago University Press: 'Let us then be rich. Remember smoky crofters' cabins. Remember being tied in Japan by law and cost to one locale. Remember American outhouses and iced-over rain barrels and cold and wet and dirt. Remember in Denmark ten people living in one room, the cows and chickens in the other room. Remember in Nebraska sod houses and isolation.'

p. 14 'income has risen more than nine times'. Maddison, A. 2006. *The World Economy*. OECD Publishing.

p. 15 'The proportion of Vietnamese living on less than $2 a day'. Norberg, J. 2006. *When Man Created the World*. Published in Swedish as *När människan skapade världen*. Timbro.

p. 15 'The poor in the developing world grew their consumption twice as fast as the world as a whole between 1980 and 2000'. Lal, D. 2006.

*Reviving the Invisible Hand.* Princeton University Press. See also Bhalla, S. 2002. *Imagine There's No Country.* Institute of International Economics.

p. 15 'The percentage living in such absolute poverty has dropped by more than half – to less than 18 per cent'. Chen, S. and Ravallion, M. 2007. Absolute poverty measures for the developing world, 1981–2004. *Proceedings of the National Academy of Sciences* USA (*PNAS*). 104: 16757–62.

p. 15 'The United Nations estimates that poverty was reduced more in the last fifty years than in the previous 500.' Lomborg, B. 2001. *The Sceptical Environmentalist.* Cambridge University Press.

p. 16 'In 1958 J.K. Galbraith declared'. Galbraith, J.K. 1958. *The Affluent Society.* Houghton Mifflin.

p. 16 'This would have been unthinkable at mid-century'. Statistics from Lindsey, B. 2007. *The Age of Abundance: How Prosperity Transformed America's Politics and Culture.* Collins.

p. 17 'Today, a car emits less pollution travelling at full speed than a parked car did in 1970 from leaks.' Pollution facts from Norberg, J. 2006. *When Man Created the World.* Published in Swedish as *När människan skapade världen.* Timbro.

p. 17 'Within just five years both predictions were proved wrong in at least one country.' Oeppen, J. and Vaupel, J.W. 2002. Demography. Broken limits to life expectancy. *Science* 296:1029–31.

p. 18 'People are not only spending a longer time living, but a shorter time dying.' Tallis, R. 2006. 'Sense about Science' annual lecture. http://www.senseaboutscience.org.uk/pdf/Lecture2007Transcript.pdf.

p. 18 'The same is true of cancer, heart disease and respiratory disease: they all still increase with age, but they do so later and later, by about ten years since the 1950s.' Fogel, R.W. 2003. *Changes in the Process of Aging during the Twentieth Century: Findings and Procedures of the Early Indicators Project.* NBER Working Papers 9941, National Bureau of Economic Research.

p. 19 'Yet the global effect of the growth of China and India has been to reduce the difference between rich and poor worldwide.' This is especially clear in Hans Rosling's animated graphs of global income distribution at www.gapminder.com. Incidentally, the individualisation of life that brought personal freedom after the 1960s also brought less loyalty towards the group, a process that surely reached crisis point in the bonus rows of 2009: see Lindsey, B. 2009. *Paul Krugman's Nostalgianomics: Economic Policy, Social Norms and Income Inequality.* Cato Institute.

p. 19 'As Hayek put it'. Hayek, F.A. 1960. *The Constitution of Liberty.* Chicago University Press.

p. 19 'Known as the Flynn effect, after James Flynn who first drew attention to it'. Flynn, J.R. 2007. *What Is Intelligence? Beyond the Flynn Effect.* Cambridge University Press.

pp. 19–20 'To date 234 innocent Americans have been freed'.http://www. innocenceproject.org/know.

p. 20 'the average family house probably costs slightly less today than it did in 1900 or even 1700'. Comparing house prices over long periods of time is fraught with difficulty, because houses vary so much, but Piet Eichholtz has tried to index house prices by comparing the same area of Amsterdam, the Herengracht, over nearly 400 years: Eichholtz, P.M.A. 2003. A long run house price index: The Herengracht Index, 1628–1973. *Real Estate Economics* 25:175–92.

p. 20 'the same amount of artificial lighting'. Pearson, P.J.G. 2003. *Energy History, Energy Services, Innovation and Sustainability.* Report and Proceedings of the International Conference on Science and Technology for Sustainability 2003: Energy and Sustainability Science, Science Council of Japan, Tokyo.

pp. 20–1 'an hour of work in 1800 earned you ten minutes of reading light'. Nordhaus, W. 1997. *Do Real-Output and Real Wage Measures Capture Reality? The History of Lighting Suggests Not.* Cowles Foundation Paper no. 957, Yale. A modern check using British figures of £479 average weekly income and £0.09 per kilowatt-hour electricity cost produces a similar result: ¼ second of work for 18 watt-hours, plus a little more for the cost of the bulb.

p. 21 'using the currency that counts, your time'. Nordhaus, W. 1997. *Do Real-Output and Real Wage Measures Capture Reality? The History of Lighting Suggests Not.* Cowles Foundation Paper no. 957, Yale.

p. 21 'The economist Don Boudreaux'. http://cafehayek.typepad.com/hayek /2006/08/were_much_wealt.html.

p. 22 'The average Briton today consumes roughly 40,000 times as much artificial light as he did in 1750.' Fouquet, R., Pearson, P.J.G., Long run trends in energy services 1300–2000. Environmental and Resource Economists 3rd World Congress, via web, Kyoto.

p. 23 'Healthcare and education are among the few things that cost more in terms of hours worked now than they did in the 1950s.' Cox, W.M. and Alm, R. 1999. *Myths of Rich and Poor – Why We Are Better Off Than We Think.* Basic Books. See also Easterbrook, G. 2003. *The Progress Paradox.* Random House.

p. 23 'observe what *Harper's Weekly* had to say'. Gordon, J.S. 2004. *An Empire of Wealth: the Epic History of American Power.* Harper Collins.

p. 23 'They were enricher-barons, too'. McCloskey, D. 2006. *The Bourgeois Virtues.* Chicago University Press.

p. 24 'Henry Ford got rich by making cars cheap'. Moore, S. and Simon, J. 2000. *It's Getting Better All the Time.* Cato Institute.

p. 24 'The price of aluminium fell from $545 a pound in the 1880s to twenty cents a pound in the 1930s'. Shermer, M. 2007. *The Mind of the Market.* Times Books.

p. 24 'When Juan Trippe sold cheap tourist class seats on his Pan Am airline in 1945'. Norberg, J. 2006. *When Man Created the World*. Published in Swedish as *När människan skapade världen*. Timbro.

p. 25 'Where it took sixteen weeks to earn the price of 100 square feet of housing in 1956, now it takes fourteen weeks and the housing is of better quality.' Cox, W.M. and Alm, R. 1999. *Myths of Rich and Poor – Why We Are Better Off Than We Think*. Basic Books.

p. 25 'To remedy this, governments then have to enforce the building of more affordable housing, or subsidise mortgage lending to the poor'. Woods, T.E. 2009. *Meltdown*. Regnery Press.

p. 25 'according to Richard Layard'. Layard, R. 2005. *Happiness: Lessons from a New Science*. Penguin.

p. 26 'The hippies were right all along'. Oswald, Andrew. 2006. The hippies were right all along about happiness. *Financial Times*, 19 January 2006.

p. 26 'a study by Richard Easterlin in 1974'. Easterlin, R.A. 1974. Does economic growth improve the human lot? in Paul A. David and Melvin W. Reder (eds). *Nations and Households in Economic Growth: Essays in Honor of Moses Abramovitz*. Academic Press.

p. 26 'the Easterlin paradox does not exist'. Stevenson, B. and Wolfers, J. 2008. *Economic Growth and Subjective Well-Being: Reassessing the Easterlin Paradox*. NBER Working Papers 14282, National Bureau of Economic Research; Ingleheart, R., Foa, R., Peterson, C. and Welzel, C. 2008. Development, freedom and rising happiness: a global perspective, 1981–2007. *Perspectives on Psychological Science* 3:264–86.

p. 26 'In the words of one of the studies'. Stevenson, B. and Justin Wolfers, J. 2008. *Economic Growth and Subjective Well-Being: Reassessing the Easterlin Paradox*. NBER Working Papers 14282, National Bureau of Economic Research.

p. 27 'a tax on consumption to encourage saving for investment instead'. Frank, R.H. 1999. *Luxury Fever: Why Money Fails to Satisfy in an Era of Excess*. The Free Press.

p. 27 'to be well off and unhappy is surely better than to be poor and unhappy.' The journalist Gregg Easterbrook's prayer goes: 'thank you that I and five hundred million others are well-housed, well-supplied, overfed, free, and not content; because we might be starving, wretched, locked under tyranny and still not content.' Easterbrook, G. 2003. *The Progress Paradox*. Basic Books.

p. 27 'psychologists find people to have fairly constant levels of happiness'. Gilbert, D. 2007. *Stumbling on Happiness*. Harper Press.

p. 27 'political scientist Ronald Ingleheart'. Ingleheart, R., Foa, R., Peterson, C. and Welzel, C. 2008. Development, freedom and rising happiness: a global perspective, 1981–2007. *Perspectives on Psychological Science* 3:264–86.

p. 28 'Ruut Veenhoven finds'. Veenhoven, R. 1999. Quality-of-life in individualistic society: A comparison of 43 nations in the early 1990's. *Social Indicators Research* 48:157–86.

p. 28 'some pressure groups may have exacerbated real hunger in Zambia'. Paarlberg, R. 2008. *Starved for Science*. Harvard University Press.

p. 28 'The precautionary principle'. Ron Bailey points out that most renditions of the precautionary principle boil down to the injunction: 'Never do anything for the first time.' http://reason.com/archives/2003/07/02/making-the-future-safe.

p. 29 'By the same age, human hunter-gatherers have consumed about 20 per cent of their lifetime calories, but produced just 4 per cent.' Kaplan, H.E. and Robson, A.J. 2002. The emergence of humans: the co-evolution of intelligence and longevity with intergenerational transfers. *PNAS* 99:10221–6; see also Kaplan, H. and Gurven, M. 2005. The natural history of human food sharing and cooperation: a review and a new multi-individual approach to the negotiation of norms. In *Moral Sentiments and Material Interests* (eds H. Gintis, S. Bowles, R. Boyd and E.Fehr). MIT Press.

p. 31 'curse of resources'. Ferguson, N. 2008. *The Ascent of Money*. Allen Lane.

p. 31 'the Great Depression of the 1930s is just a dip in the slope'. Findlay, R. and O'Rourke, K.H. 2007. *Power and Plenty: Trade, War and the World Economy*. Princeton University Press.

p. 31 'All sorts of new products and industries were born during the Depression'. Nicholas, T. 2008. Innovation lessons from the 1930s. *McKinsey Quarterly*, December 2008.

p. 31 'Arcadia Biosciences in northern California'. http://www. arcadiabio. com/pr_0032.php.

p. 33 'Henry David Thoreau asked'. Thoreau, H.D. 1854. *Walden: Or Life in the Woods*. Ticknor and Fields.

p. 34 'In 1900, the average American spent $76 of every $100 on food, clothing and shelter. Today he spends $37'. Cox, W.M. and Alm, R. 1999. *Myths of Rich and Poor – Why We Are Better Off Than We Think*. Basic Books.

p. 34 'To produce implies that the producer desires to consume' said John Stuart Mill; 'why else should he give himself useless labour?'. Mill, J.S. 1848. *Principles of Political Economy.*

p. 34 'Thomas Thwaites set out to make his own toaster'. http://www.the toasterproject.org. 'Kelly Cobb of Drexel University set out to make a man's suit'. http://www.wired.com/print/culture/design/news/2007/03/100milesuit0330. See also http://www.thebigquestions.com/2009/10/30/the-10000-suit.

p. 37 'In civilized society,' wrote Adam Smith'. Smith, A. 1776. *The Wealth of Nations.*

p. 38 'Leonard Read's classic 1958 essay "I, Pencil"'. Read, L.E. 1958. I, Pencil. *The Freeman*, December 1958. For a fine modern rerun of the same subject see the novel by Roberts, R. 2008. *The Price of Everything*. Princeton University Press.

p. 38 'As Friedrich Hayek first clearly saw'. Hayek, F.A. 1945. The use of knowledge in society. *American Economic Review* 35:519–30.

p. 39 'a smaller quantity of labour produce a greater quantity of work'. Smith, A. 1776. *The Wealth of Nations*.

p. 39 'you would have spent your after-tax income in roughly the following way'. Data from the Bureau of Labour Statistics: www.bls.org.

p. 40 'An English farm labourer in the 1790s spent his wages roughly as follows'. Clark, G. 2007. *A Farewell to Alms*. Princeton University Press.

p. 40 'A rural peasant woman in modern Malawi spends her time roughly as follows'. Blackden, C.M. and Wodon, Q. 2006. *Gender, Time Use and Poverty in SubSaharan Africa*. World Bank.

p. 40 'the Shire River in Machinga province'. http://allafrica.com/stories/200712260420.html.

p. 41 'not just the services you need but also those you crave.' The distinction between needs and wants, as expressed by Abraham Maslow's hierarchy of needs, is a mischievous one: people evolved to be ambitious, to start exaggerating their social status or sexual worth, long before they have satisfied their basic needs. See Miller, G. 2009. *Spent*. Heinemann.

p. 41 'the entire concept of food miles is "a profoundly flawed sustainability indicator"'. Bailey, R. 2008. The food miles mistake. *Reason*, 4 November 2008. http://www.reason.com/news/show/129855. html.

p. 41 'Ten times as much carbon'. See https://statistics.defra.gov.uk/esg/reports/foodmiles/final.pdf.

p. 42 'six times the carbon footprint of a Kenyan rose'. Specter, M. 2008. Big foot. *The New Yorker,* 25 February 2008. http://www.new yorker.com/reporting/2008/02/25/080225fa_fact_specter. See also http://grown underthesun.com.

p. 42 'just as it did in Europe in 1315–18'. Jordan, W.C. 1996. *The Great Famine: Northern Europe in the Early Fourteenth Century*. Princeton University Press.

p. 43 'Today, 1 per cent works in agriculture and 24 per cent in industry'. Statistics in this paragraph from Angus Maddison (*Phases of Capitalist Development*), cited in Kealey, T. 2008. *Sex, Science and Profits*. Heinemann.

p. 43 'the original affluent society'. Sahlins, M. 1968. Notes on the original affluent society. In *Man the Hunter* (eds R.B. Lee and I. DeVore). Aldine. Pages 85–9.

p. 43 'They lived into old age far more frequently than their ancestors had done.' Caspari, R. and Lee, S.-H. 2006. Is human longevity a consequence of cultural change or modern biology? *American Journal of Physical Anthropology* 129:512–17.

p. 43 'they had largely wiped out the lions and hyenas'. Ofek, H. 2001. *Second Nature: Economic Origins of Human Evolution*. Cambridge University Press.

p. 44 'Geoffrey Miller, for example, in his excellent book *Spent*'. Miller, G. 2009. *Spent*. Heinemann.

p. 44 'The warfare death rate of 0.5'. Keeley, L. 1996. *War Before Civilization*. Oxford University Press.

p. 44 'a cemetery uncovered at Jebel Sahaba'. Otterbein, K.F. 2004. *How War Began*. Texas A & M Press.

p. 45 'asks Geoffrey Miller'. Miller, G. 2009. *Spent*. Heinemann.

## Chapter 2

p. 47 'He steps under the shower, a forceful cascade pumped down from the third floor.' McEwan, I 2005. *Saturday.* Jonathan Cape. The person taking the shower is Perowne, the surgeon at the centre of the plot.

p. 47 Life expectancy graph. World Bank Development Indicators.

p. 48 'One day a little less than 500,000 years ago, near what is now the village of Boxgrove'. Potts, M. and Roberts, M. 1998. *Fairweather Eden.* Arrow Books.

p. 49 'a single twitch of progress in biface hand-axe history'. Klein R.G. and Edgar B. 2002. *The Dawn of Human Culture*. Wiley.

p. 49 'Its brain was almost as big as a modern person's'. Rightmire, G.P. 2003. Brain size and encephalization in early to Mid-Pleistocene *Homo*. *American Journal of Physical Anthropology* 124: 109–23.

p. 51 'the erectus hominid species'. For simplicity, I am going to call all the species of hominid that lived between about 1.5 million and 300,000 years ago 'erectus hominid' after the longest-established and most comprehensive name used for hominids of this period. The current fashion is to include four species within this group: *H. ergaster* earliest in Africa, *H. erectus* a little later in Asia, *H. heidelbergensis* coming out of Africa later into Europe and its descendant, *H. neanderthalensis*. See Foley, R.A. and Lahr, M.M. 2003. On stony ground: Lithic technology, human evolution, and the emergence of culture. *Evolutionary Anthropology* 12:109–22.

p. 51 'it was a natural expression of human development'. See Richerson, P. and Boyd, R. 2005. *Not by Genes Alone*. Chicago University Press: 'Perhaps we need to entertain the hypothesis that Acheulean bifaces were innately constrained rather than wholly cultural and that their temporal stability stemmed from some component of genetically transmitted psychology.'

p. 51 'Meat enabled them to cut down on the huge gut'. Aiello, L.C. and Wheeler, P. 1995. The expensive tissue hypothesis: the brain and the digestive system in human and primate evolution. *Current Anthropology* 36:199–221.

p. 52 'the toolkit was showing signs of change as early as 285,000 years ago'. McBrearty, S. and Brooks, A. 2000. The revolution that wasn't: a new interpretation of the origin of modern human behavior. *Journal of Human Evolution* 39:453–563. Morgan, L.E. and Renne, P.R. 2008. Diachronous dawn of Africa's Middle Stone Age: New 40Ar/39Ar ages from the Ethiopian Rift. *Geology* 36:967–70.

p. 52 'by at least 160,000 years ago'. White T.D. et al. 2003. Pleistocene Homo sapiens from Middle Awash, Ethiopia. *Nature* 423:742–7; Willoughby, P. R. 2007. *The Evolution of Modern Humans in Africa: a Comprehensive Guide.* Rowman AltaMira.

p. 52 'Pinnacle Point in South Africa'. Marean, C.W. et al. 2007. Early human use of marine resources and pigment in South Africa during the Middle Pleistocene. *Nature* 449:905–8.

p. 53 'a few slender-headed Africans did begin to colonise the Middle East'. Stringer, C. and McKie, R. 1996. *African Exodus.* Jonathan Cape.

p. 53 'at Grottes des Pigeons near Taforalt in Morocco'. Bouzouggar, A. et al. 2007. 82,000-year-old shell beads from North Africa and implications for the origins of modern human behavior. *PNAS* 2007 104:9964–9; Barton R.N.E., et al. 2009. OSL dating of the Aterian levels at Dar es-Soltan I (Rabat, Morocco) and implications for the dispersal of modern Homo sapiens. *Quaternary Science Reviews.* doi:10.1016/j.quascirev.2009.03.010.

p. 53 'obsidian may have begun to move over long distances'. Negash, A., Shackley, M.S. and Alene, M. 2006. Source provenance of obsidian artefacts from the Early Stone Age (ESA) site of Melka Konture, Ethiopia. *Journal of Archeological Science* 33:1647–50; and Negash, A. and Shackley, M.S. 2006. Geochemical provenance of obsidian artefacts from the MSA site of Porc Epic, Ethiopia. *Archaeometry* 48:1–12.

p. 54 'Lake Malawi, whose level dropped 600 metres'. Cohen, A.S. et al. 2007. Ecological consequences of early Late Pleistocene megadroughts in tropical Africa. *PNAS* 104:16422–7.

p. 54 'Their genes, marked by the L3 mitochondrial type, suddenly expanded and displaced most others in Africa'. Atkinson, Q.D., Gray, R.D. and Drummond, A.J. 2009. Bayesian coalescent inference of major human mitochondrial DNA haplogroup expansions in Africa. *Proceedings of the Royal Society B* 276:367–73.

p. 55 'living in large social groups on a plentiful diet both encourages and allows brain growth'. Dunbar, R. 2004. *The Human Story.* Faber and Faber.

p. 55 'a fortuitous genetic mutation triggered a change in human behaviour'. Klein, R.G. and Edgar, B. 2002. *The Dawn of Human Culture.* John Wiley.

p. 55 'FOXP2, which is essential to speech and language in both people and songbirds'. Fisher, S.E. and Scharff, C. 2009. FOXP2 as a molecular window into speech and language. *Trends in Genetics* 25:166–77.doi: 10.1016/j.tig.2009.03.002 A.

p. 55 'the mutations even change the way mice pups squeak'. Enard, W. et al. 2009. A humanized version of FOXP2 affects cortico-basal ganglia circuits in mice. *Cell* 137:961–71.

p. 55 'Neanderthals share the very same two mutations'. Krause, J. et al. 2007. The derived FOXP2 variant of modern humans was shared with Neandertals. *Current Biology* 17:1908–12.

p. 57 'as Leda Cosmides and John Tooby put it'. Cosmides, L. and Tooby, J. 1992. Cognitive adaptations for social exchange. In *The Adapted Mind* (eds J.H. Barkow, L. Cosmides and J. Tooby). Oxford University Press.

p. 57 'In Adam Smith's words'. Both Adam Smith quotes are from book 1, part 2, of *The Wealth of Nations* (1776).

p. 57 'In the grasslands of Cameroon'. Rowland and Warnier, quoted in Shennan, S. 2002. *Genes, Memes and Human History*. Thames & Hudson.

p. 59 'The primatologist Sarah Brosnan tried to teach two different groups of chimpanzees about barter'. Brosnan, S.F., Grady, M.F., Lambeth, S.P., Schapiro, S.J. and Beran, M.J. 2008. Chimpanzee autarky. PLOS ONE 3(1):e1518. doi:10.1371/journal.pone.0001518.

p. 59 'Chimpanzees and monkeys can be taught to exchange tokens for food'. Chen, M.K. and Hauser, M. 2006. How basic are behavioral biases? Evidence from capuchin monkey trading behavior. *Journal of Political Economy* 114:517–37.

p. 59 'not even a hint of this complementarity is found among nonhuman primates.' Wrangham, R. 2009. *Catching Fire: How Cooking Made Us Human*. Perseus Books.

p. 60 'Birute Galdikas reared a young orang utan'. Galdikas, B. 1995. *Reflections of Eden*. Little, Brown.

p. 60 'fire itself is hard to start, but easy to share'. Ofek, H. 2001. *Second Nature: Economic Origins of Human Evolution*. Cambridge University Press.

p. 61 'males and females specialise and then share food'. Low, B. 2000. *Why Sex Matters: a Darwinian Look at Human Behavior*. Princeton University Press.

p. 61 'men hunt, women and children gather'. Kuhn, S.L. and Stiner, M.C. 2006. What's a mother to do? A hypothesis about the division of labour and modern human origins. *Current Anthropology* 47:953–80.

p. 61 'making strikingly different decisions about how to obtain resources within that habitat'. Kaplan, H. and Gurven, M. 2005. The natural history of human food sharing and cooperation: a review and a new multi-individual approach to the negotiation of norms. In *Moral Sentiments and Material Interests* (eds H. Gintis, S. Bowles, R. Boyd and E. Fehr). MIT Press.

p. 62 'Martu women in western Australia hunt goanna lizards'. Bliege Bird, R. 1999. Cooperation and conflict: the behavioural ecology of the sexual division of labour. *Evolutionary Anthropology* 8:65–75.

p. 62 'Women demand meat as their social right, and they get it – otherwise they leave their husbands, marry elsewhere or make love to other men'. Biesele, M. 1993. *Women Like Meat*. Indiana University Press.

p. 62 'In the Mersey estuary near Liverpool'. Stringer, C. 2006. *Homo Britannicus*. Penguin.

p. 63 'In the Alyawarre aborigines of Australia'. Bliege Bird, R. and Bird, D.

2008. Why women hunt: risk and contemporary foraging in a Western Desert Aboriginal community. *Current Anthropology* 49:655–93.

p. 63 'A sexual division of labour would exist even without childcare constraints.' It is reasonable to wonder if a hundred thousand years of doing different things have not left their mark on at least some of the modern leisure pursuits of the two sexes. Shopping for shoes is a bit like gathering – picking out the perfect item in a crowd of possibilities. Playing golf is a bit like hunting – aiming a ballistic projectile at a target in the great outdoors. It is also noticeable how much more carnivorous most men are than most women. In the West, female vegetarians outnumber male ones by more than two to one, but even among non-vegetarians it is common to find men who take only a token nibble at the vegetables on their plate, and women who do the same with meat. Of course, it is part of my case that in the Stone Age men supplied gathering women with meat and women supplied hunting men with veg, so both sexes were omnivores, but perhaps when it came to 'stopping for lunch', the women would eat the nuts they had gathered while elsewhere the men cooked up a tortoise or cut a steak off their first kill. Such speculation is not, I admit, very scientific.

p. 63 'It is as if the species now has two brains'. Joe Henrich first made this point to me late at night in a bar in Indiana.

p. 63 'men seem to strive to catch big game to feed the whole band'. Bliege Bird, R. and Bird, D. 2008. Why women hunt: risk and contemporary foraging in a Western Desert Aboriginal community. *Current Anthropology* 49:655–93.

p. 63 'Hadza men spend weeks trying to catch a huge eland antelope'. Hawkes, K. 1996. Foraging differences between men and women. In *The Archaeology of Human Ancestry* (eds James Steele and Stephen Shennan). Routledge.

p. 63 'men on the island of Mer in the Torres Strait'. Bliege Bird, R. 1999. Cooperation and conflict: the behavioural ecology of the sexual division of labour. *Evolutionary Anthropology* 8:65–75.

p. 64 'Steven Kuhn and Mary Stiner think that modern, African-origin *Homo sapiens* had a sexual division of labour and Neanderthals did not'. Kuhn, S.L. and Stiner, M.C. 2006. What's a mother to do? A hypothesis about the division of labour and modern human origins. *Current Anthropology* 47:953–80.

p. 64 'first advocated by Glyn Isaac in 1978'. Isaac, G.L. and Isaac, B. 1989. *The Archaeology of Human Origins: Papers by Glyn Isaac.* Cambridge University Press.

p. 65 'To paraphrase H.G. Wells'. Wells, H.G. 1902. 'The Discovery of the Future'. Lecture at the Royal Institution, 24 January 1902, published in *Nature* 65:326–31. Reproduced with the permission of AP Watt Ltd on behalf of the Literary Executors of the Estate of H.G. Wells.

p. 66 'to land, probably around 45,000 years ago, on the continent of Sahul'.

O'Connell, J.F. and Allen, J. 2007. Pre-LGM Sahul (Pleistocene Australia-New Guinea) and the archaeology of Early Modern Humans. In Mellars, P., Boyle, K., Bar-Yosef, O. et al., *Rethinking the Human Revolution*, Cambridge: McDonald Institute for Archaeological Research, pp. 395–410.

p. 66 'genetics tell an unambiguous story of almost complete isolation since the first migration'. Thangaraj, K. et al. 2005. Reconstructing the origin of Andaman Islanders. *Science* 308: 996; Macaulay, V. et al. 2005. Single, rapid coastal settlement of Asia revealed by analysis of complete mitochondrial genomes. *Science* 308:1034–6; Hudjashov et al. 2007. Revealing the prehistoric settlement of Australia by Y chromosome and mtDNA analysis. *PNAS*. 104: 8726–30.

p. 67 'Jonathan Kingdon first suggested'. Kingdon, J. 1996. *Self-Made Man: Human Evolution from Eden to Extinction.* John Wiley.

p. 67 'All along the coast of Asia, the beachcombers would have found fresh water'. Faure, H., Walter, R.C. and Grant, D.E. 2002. The coastal oasis: Ice Age springs on emerged continental shelves. *Global and Planetary Change* 33:47–56.

p. 68 'so louse genes suggest'. Pennisi, E. 2004. Louse DNA suggests close contact between Early Humans. *Science* 306:210.

p. 68 'conceivably even close enough to acquire a smattering of their cousins' genes'. Svante Paabo, personal communication. See also Evans, P.D. et al. 2006. Evidence that the adaptive allele of the brain size gene microcephalin introgressed into Homo sapiens from an archaic Homo lineage. *PNAS* 103:18178–83.

p. 69 'driven to the brink of extinction by human predation'. Stiner, M. C. and Kuhn, S. L. 2006. Changes in the 'connectedness' and resilience of palaeolithic societies in Mediterranean ecosystems. *Human Ecology* 34:693–712.

p. 69 'in the Mojave desert of California, ravens occasionally kill tortoises for food'. http://www.scienceblog.com/community/older/archives/E/usgs 398.html.

p. 70 'shells, fossil coral, steatite, jet, lignite, hematite, and pyrite were used to make ornaments and objects'. Stringer, C. and McKie, R. 1996. *African Exodus.* Jonathan Cape.

p. 70 'A flute made from the bone of a vulture'. Conard, N.J., Maline, M. and Munzel, S.C. 2009. New flutes document the earliest musical tradition in southwestern Germany. *Nature* 46:737–740.

p. 71 'jewellery made of shells from the Black Sea and amber from the Baltic'. Ofek, H. 2001. *Second Nature: Economic Origins of Human Evolution.* Cambridge University Press.

p. 71 'This is in striking contrast to the Neanderthals, whose stone tools were virtually always made from raw material available within an hour's walk of where the tool was used'. Stringer. C. 2006. *Homo Britannicus.* Penguin: 'Whereas virtually all Neanderthal stone tools were made from raw materials sourced within an hour's walk from their sites, Cro-Magnons were either much more mobile or had exchange networks for their resources covering hundreds of miles'.

p. 73 'say the evolutionary biologists Mark Pagel and Ruth Mace'. Pagel, M. and Mace, R. 2004. The cultural wealth of nations. *Nature* 428:275–8.

p. 73 'Ian Tattersall remarks'. Tattersall, I. 1997. *Becoming Human.* Harcourt.

p. 73 'It is such a human a thing to do, and so obvious an explanation of the thing that needs explaining: the capacity for innovation'. See for example Horan, R.D., Bulte, E.H. and Shogren, J.F. 2005. How trade saved humanity from biological exclusion: the Neanderthal enigma revisited and revised. *Journal of Economic Behavior and Organization* 58:1–29.

p. 75 'defined by the stockbroker David Ricardo in 1817'. Ricardo, D. 1817. *The Principles of Political Economy and Taxation.* John Murray.

p. 75 'It is such an elegant idea that it is hard to believe that Palaeolithic people took so long to stumble upon it (or economists to define it)'. It is also surprising how hard it is for many intellectuals to grasp its essentials. For a catalogue of its misrepresentations, see Paul Krugman's essay 'Ricardo's Difficult Idea': http://web.mit.edu/krugman/www/ricardo.htm.

pp. 75–6 'Insect social life is built not on increases in the complexity of individual behaviour, "but instead on specialization among individuals".' Holldobbler, B. and Wilson, E.O. 2008. *The Superorganism.* Norton.

p. 77 'Even Charles Darwin reckoned'. Darwin, C. R. 1871. *The Descent of Man.* Quoted in Ofek, H. 2001. *Second Nature: Economic Origins of Human Evolution.* Cambridge University Press.

p. 77 'According to the anthropologist Joe Henrich'. Henrich, J. 2004. Demography and cultural evolution: how adaptive cultural processes can produce maladaptive losses – the Tasmanian case. *American Antiquity* 69:197–214.

p. 78 'The most striking case of technological regress is Tasmania'. Henrich, J. 2004. Demography and cultural evolution: how adaptive cultural processes can produce maladaptive losses – the Tasmanian case. *American Antiquity* 69:197–214.

p. 79 'it was not that there was no innovation; it was that regress overwhelmed progress'. Diamond, J. 1993. Ten thousand years of solitude. *Discover*, March 1993.

p. 80 'The Tasmanian market was too small to sustain many specialised skills'. Heinrich, J. 2004. Demography and cultural evolution: how adaptive cultural processes can produce maladaptive losses – the Tasmanian case. *American Antiquity* 69:197–214.

p. 81 'On Kangaroo Island and Flinders Island, human occupation petered out, probably by extinction, a few thousand years after isolation'. Bowdler, S. 1995. Offshore island and maritime explorations in Australian prehistory. *Antiquity* 69:945–58.

p. 81 'causing the anthropologist W.H.R. Rivers to puzzle'. Shennan, S. 2002.

*Genes, Memes and Human History*. Thames & Hudson.

p. 81 'Shell beads had been moving long distances across Australia since at least 30,000 years ago.' Balme, J. and Morse, K. 2006. Shell beads and social behaviour in Pleistocene Australia. *Antiquity* 80: 799–811.

p. 81 'The best stone axes travelled up to 500 miles from where they were mined.' Flood, J. 2006. *The Original Australians: the Story of the Aboriginal People*. Allen & Unwin.

p. 81 'In contrast to Tasmania, Tierra del Fuego'. Heinrich, J. 2004. Demography and cultural evolution: how adaptive cultural processes can produce maladaptive losses – the Tasmanian case. *American Antiquity* 69:197–214.

p. 82 'The success of human beings depends crucially, but precariously, on numbers and connections.' Incidentally, the story of the Greenland Norse, or of the inhabitants of Easter Island, told so eloquently as tales of ecological exhaustion in Jared Diamond's book *Collapse*, probably say as much about isolation as ecology. Isolated from Scandinavia by a combination of the Black Death and the worsening climate, the Greenland Norse could not sustain their lifestyles; like the Tasmanians, they forgot how to fish. Easter Island Diamond may have partly misread: some argue that its society was possibly still flourishing, despite deforestation, when a holocaust of slave traders arrived in the 1860s – see Peiser, B. 2005. From genocide to ecocide: the rape of Rapa Nui. *Energy & Environment* 16:513–39.

p. 82 'This may explain why Australian aboriginal technology, although it developed and elaborated steadily over the ensuing millennia, was lacking in so many features of the Old World'. O'Connell, J.F. and Allen, J. 2007. Pre-LGM Sahul (Pleistocene Australia-New Guinea) and the archaeology of Early Modern Humans. In Mellars, P., Boyle, K., Bar-Yosef, O. et al. *Rethinking the Human Revolution*. Cambridge: McDonald Institute for Archaeological Research, pp. 395–410.

pp.82–3 'The "Tasmanian effect" may also explain why technological progress had been so slow and erratic in Africa after 160,000 years ago'. Richerson, P.J., Boyd, R. and Bettinger, R.L. 2009. Cultural innovations and demographic change. *Human Biology* 81:211–35; Powell, A., Shennan, S. and Thomas, M.G. 2009. Late Pleistocene demography and the appearance of modern human behaviour. *Science* 324:1298–1301.

p. 83 'As the economist Julian Simon put it'. Simon, J. 1996. *The Ultimate Resource 2*. Princeton University Press.

p. 84 'Tasmanians sold women to the sealers as concubines'. Flood, J. 2006. *The Original Australians: the Story of the Aboriginal People*. Allen & Unwin.

## Chapter 3

p. 85 'Money is not metal. It is trust inscribed'. Ferguson, N. 2008. *The Ascent of Money*. Allen Lane.

p. 85 Homicide rate graph. Spierenburg, P. 2008. *A History of Murder*. Polity

Press. See also Eisner, M. 2001. Modernization, Self-Control and Lethal Violence. The Long-term Dynamics of European Homicide Rates in Theoretical Perspective *The British Journal of Criminology* 41:618-638.

p. 85 'Greenstreet whispers to Bogart'. Siegfried, T. 2006. *A Beautiful Math: John Nash, Game Theory and the Modern Quest for a Code of Nature.* Joseph Henry Press.

p. 86 'As the economist Herb Gintis puts it'. http://www.reason.com/news/show/34772.html.

p. 86 'people in fifteen mostly small-scale tribal societies were enticed to play the Ultimatum Game'. Henrich, J. et al. 2005. 'Economic man' in cross-cultural perspective: Behavioral experiments in 15 small-scale societies. *Behavioral and Brain Sciences* 28:795–815.

p. 87 'costly punishment of selfishness may be necessary'. Fehr, E. and Gachter, S. 2000. Cooperation and punishment in public goods experiments. *American Economic Review*, Journal of the American Economic Association 90: 980–94; Henrich, J. et al. 2006. Costly punishment across human societies. *Science* 312:1767–70.

p. 88 'in other group-living species, such as ants or chimpanzees, the interactions between members of different groups are almost always violent'. Brosnan, S. 2008. Fairness and other-regarding preferences in nonhuman primates. In Zak, P. (ed.) 2008. *Moral Markets.* Princeton University Press.

p. 88 'human beings can treat strangers as honorary friends'. Seabright, P. 2004. *The Company of Strangers.* Princeton University Press.

p. 88 'primatologists such as Sarah Hrdy and Frans de Waal'. Hrdy, S. 2009. *Mothers and Others.* Belknap. De Waal, F. 2006. *Our Inner Ape.* Granta Books.

p. 89 'The traders of Malaysia, Indonesia and the Philippines were often women, who were taught to calculate and to account from an early age.' Pomeranz, K. and Topik, S. 2006. *The World That Trade Created.* M.E. Sharpe.

p. 89 'the British government trusted a Jewish lender named Nathan Rothschild'. Ferguson, N. 2008. *The Ascent of Money.* Allen Lane.

p. 90 'the experiment, run by Bart Wilson, Vernon Smith and their colleagues'. Crockett, S., Wilson, B. and Smith, V. 2009. Exchange and specialization as a discovery process. *Economic Journal* 119: 1162–88.

p. 90 'the Yir Yoront aborigines, in northern Australia'. Sharp, L. 1974. Steel axes for stone age Australians. In Cohen, Y. (ed.) 1974. *Man in Adaptation.* Aldine de Gruyter.

pp. 91–2 'a young naturalist named Charles Darwin came face to face with some hunter-gatherers'. Darwin, C.R. 1839. *The Voyage of the* Beagle. John Murray.

p. 92 'New Guinea highlanders, when first contacted by Michael Leahy and his fellow prospectors in 1933'. Connolly, R. and Anderson, R. 1987. *First Contact.* Viking.

p. 92 'The people of the Pacific coast of North America were sending seashells hundreds of miles inland, and importing obsidian from even farther afield.' Baugh, T.E. and Ericson, J.E. 1994. *Prehistoric Exchange Systems in North America.* Springer.

pp. 92–3 'The Chumash of the Californian channel islands'. Arnold, J.E. 2001. *The Origins of a Pacific Coast Chiefdom: The Chumash of the Channel Islands.* University of Utah Press.

p. 93 '*Das Adam Smith Problem*'. Coase, R. H. 1995. Adam Smith's view of man. In *Essays on Economics and Economists.* University of Chicago Press.

p. 93 'How selfish soever man may be supposed'. Smith, A. 1759. *The Theory of Moral Sentiments.*

p. 93 'Man has almost constant occasion for the help of his brethren'. Smith, A. 1776. *The Wealth of Nations.*

p. 93 'honorary friends'. Seabright, P. 2004. *The Company of Strangers.* Princeton University Press.

p. 94 'As the philosopher Robert Solomon put it'. Solomon, R.C. 2008. Free enterprise, sympathy and virtue. In Zak, P. (ed.). 2008. *Moral Markets.* Princeton University Press.

p. 94 'a baby smiling causes particular circuits in its mother's brain to fire'. Noriuchi, M., Kikuchi, Y. and Senoo, A. 2008. The functional neuroanatomy of maternal love: mother's response to infant's attachment behaviors. *Biological Psychiatry* 63:415–23.

p. 94 'the neuro-economist Paul Zak'. Zak, P. 2008. Values and value. In Zak, P. (ed.). 2008. *Moral Markets.* Princeton University Press.

p. 94 'Zak, together with Ernst Fehr and other colleagues, conducted one of the most revealing experiments in the history of economics'. Kosfeld, M., Henrichs, M., Zak, P.J., Fischbacher, U. and Fehr, E. 2005. Oxytocin increases trust in humans. *Nature* 435: 673–6.

p. 95 'by suppressing the activity of the amygdala, the organ that expresses fear'. Rilling, J.K., et al. 2007. Neural correlates of social cooperation and non-cooperation as a function of psychopathy. *Biological Psychiatry* 61:1260–71.

p. 96 'says the economist Robert Frank'. Frank, R. 2008. The status of moral emotions in consequentialist moral reasoning. In Zak, P. (ed.) 2008. *Moral Markets.* Princeton University Press.

p. 96 'people acutely remember the faces of those who cheat them'. Mealey, L., Daood, C. and Krage, M. 1996. Enhanced memory for faces of cheaters. *Ethology and Sociobiology* 17:119–28.

pp. 96–7 'Capuchin monkeys and chimpanzees are just as resentful of unfair treatment'. Brosnan, S. 2008. Fairness and other-regarding preferences in nonhuman primates. In Zak, P. (ed.) 2008. *Moral Markets.* Princeton University Press.

p. 97 'the more people trust each other in a society, the more prosperous that society is'. Zak, P. and Knack, S. 2001. Trust and growth. *Economic Journal* 111:295–321.

p. 99 'John Clippinger draws an optimistic conclusion'. Clippinger, J.H. 2007. *A Crowd of One*. Public Affairs Books.

p. 99 'as Robert Wright has argued'. Wright, R. 2000. *Non Zero: the Logic of Human Destiny*. Pantheon.

p. 101 'Michael Shermer thinks that is because in most of the Stone Age it was true'. Shermer, M. 2007. *The Mind of the Market*. Times Books.

p. 101 'incredible augmentation of the pots and pans of the country'. Quoted in O'Rourke, P.J. 2007. *On The Wealth of Nations*. Atlantic Monthly Press.

p. 102 'said the Archbishop of Canterbury in 2008'. *Spectator*, 24 September. 2008.

p. 102 'As the Australian economist Peter Saunders argues'. Saunders, P. 2007. Why capitalism is good for the soul. *Policy Magazine* 23:3–9.

p. 102 'Brink Lindsey writes'. Lindsey, B. 2007. *The Age of Abundance: How Prosperity Transformed America's Politics and Culture*. Collins.

p. 102 'Arnold Toynbee, lecturing working men on the English industrial revolution which had so enriched them'. Quoted in Phillips, A. and Taylor, B. 2009. *On Kindness*. Hamish Hamilton.

p. 103 'In 2009 Adam Phillips and Barbara Taylor argued'. Phillips, A. and Taylor, B. 2009. *On Kindness*. Hamish Hamilton.

p. 103 'As the British politician Lord Taverne puts it'. Lord Taverne, personal communication.

p. 103 'John Padgett at the University of Chicago compiled data on the commercial revolution in fourteenth-century Florence'. Described in Clippinger, J.H. 2007. *A Crowd of One*. Public Affairs Books.

p. 103 'observed Charles, Baron de Montesquieu'. Quoted in Hirschman, A. 1977. *The Passions and the Interests*. Princeton University Press.

p. 103 'David Hume thought commerce "rather favourable to liberty"'. McFarlane, A. 2002. David Hume and the political economy of agrarian civilization. *History of European Ideas* 27:79–91.

p. 104 'The rapid commercialisation of lives since 1800 has coincided with an extraordinary improvement in human sensibility'. Pinker, S. 2007. A history of violence. *The New Republic*, 19 March 2007.

p. 105 'it was the nouveau-riche merchants, with names like Wedgwood and Wilberforce, who financed and led the anti-slavery movement'. Desmond, A. and Moore, J. 2009. *Darwin's Sacred Cause*. Allen Lane.

p. 105 'Far from being a vice,' says Eamonn Butler'. Butler, E. 2008. *The Best Book on the Market*. Capstone.

p. 105 'When shown a photograph of an attractive man'. Miller, G. 2009. *Spent*. Heinemann.

p. 106 'As Michael Shermer comments'. Shermer, M. 2007. *The Mind of the Market*. Times Books.

p. 106 'your chances of being murdered have fallen steadily since the seventeenth century in every European country'. Eisner, M. 2001. Modernization, self-control and lethal violence. The long-term dynamics of European homicide rates in theoretical perspective. *British Journal of Criminology* 41:618–38.

p. 106 'Murder was ten times as common before the industrial revolution in Europe, per head of population, as it is today.' See also Spierenburg, P. 2009. *A History of Murder*. Polity Press.

p. 106 'the environmental Kuznets curve'. Yandle, B., Bhattarai, M. and Vijayaraghavan, M. 2004. *Environmental Kuznets Curves*. PERC.

p. 106 'when per capita income reaches about $4,000, people demand a clean-up of their local streams and air'. Goklany, I. 2008. *The Improving State of the World*. Cato Institute.

p. 107 'because people were enriching themselves and demanding higher standards'. Moore, S. and Simon, J. 2000. *It's Getting Better All the Time*. Cato Institute.

p. 107 'The "long tail" of the distribution'. Anderson, C. 2006. *The Long Tail: Why the Future of Business Is Selling Less of More*. Hyperion.

p. 108 'now-unfashionable philosopher Herbert Spencer who insisted that freedom would increase along with commerce'. Quotes are from 1842 essay for *The Nonconformist* and 1853 essay for *The Westminster Review*. Both quoted in Nisbet, R. 1980. *History of the Idea of Progress*. Basic Books.

p. 108 'The American civil rights movement drew its strength partly from a great economic migration'. Lindsey, B. 2007. *The Age of Abundance: How Prosperity Transformed America's Politics and Culture*. Collins.

p. 109 'much argument about whether democracy is necessary for growth'. Friedman, B. 2005. *The Moral Consequences of Economic Growth*. Knopf.

p. 109 'I am happy to cheer, with Deirdre McCloskey'. McCloskey, D. 2006. *The Bourgeois Virtues*. Chicago University Press.

p. 110 'One side denounced capitalism but gobbled up its fruits; the other cursed the fruits while defending the system that bore them.' Lindsey, B. 2007. *The Age of Abundance: How Prosperity Transformed America's Politics and Culture*. Collins.

p. 111 'Like Milton Friedman'. Quoted in Norberg, J. 2008. *The Klein Doctrine*. Cato Institute briefing paper no. 102. 14 May 2008.

p. 111 'serfs under feudal brandlords'. Klein, N. 2001. *No Logo*. Flamingo.

p. 111 'Shell may have tried to dump an oil-storage device'. Greenpeace claimed that the *Brent Spar* had 5,500 tonnes of oil in it, then later admitted the true figure was nearer 100 tonnes.

p. 111 'Enron funded climate alarmism'. Ken Lay had ambitions for Enron to 'become the world's leading renewable energy company' and it lobbied hard for renewable energy subsidies and mandates. See http://masterresource.org/?p=3302#more-3302.

p. 111 'half of today's biggest companies did not even exist in 1980'. Micklethwait, J. and Wooldridge, A. 2003. *The Company*. Weidenfeld.

p. 112 'According to Eric Beinhocker of McKinsey'. Beinhocker, E. 2006. *The Origin of Wealth*. Random House.

p. 113 'Like corrugated iron and container shipping'. The development of containerisation in the 1950s made the loading and unloading of ships roughly twenty times as fast and thereby dramatically lowered the cost of trade, helping to start the boom in Asian exports. Today, despite the advent of the weightless information age, the world's merchant fleet – at over 550 million gross registered tonnes – is twice the size it was in 1970 and ten times the size it was in 1920. See Edgerton, D. 2006. *The Shock of the Old: Technology and Global History since 1900*. Profile Books.

p. 113 'A single, routine, minuscule Wal-Mart decision in the 1990s'. Fishman, C. 2006. *The Wal-Mart Effect*. Penguin.

p. 114 'As Kodak and Fuji slugged it out for dominance in the 35mm film industry'. The remarkable thing about the death of film cameras is how blind the film companies were to it. As late as 2003, they were insisting that digital would only take some of the market and film would endure.

p. 114 'In America, roughly 15 per cent of jobs are destroyed every year'. Kauffman Foundation estimates: cited in *The Economist* survey of business in America, by Robert Guest, 30 May 2009.

p. 114 '"This isn't about auctions," said Meg Whitman, the chief executive of eBay'. 'ebay, inc'. Harvard Business School case study 9-700-007.

p. 117 'In a sample of 127 countries'. Carden, A. and Hall, J. 2009. Why are some places rich while others are poor? The institutional necessity of economic freedom (29 July 2009). Available at SSRN: http://ssrn.com/abstract=1440786.

p. 117 'the World Bank published a study of "intangible wealth"'. Bailey, R. 2007. The secrets of intangible wealth. *Reason*, 5 October 2007. http://reason.com/news/show/122854.html.

p. 118 '*lex mercatoria*'. I discuss this in more detail in *The Origins of Virtue* (1996).

p. 118 'When Michael Shermer and three friends started a bicycle race across America'. In Shermer, M. 2007. *The Mind of the Market*. Times Books.

## Chapter 4

p. 121 'Whoever could make two ears of corn'. Swift, J. 1726. *Gulliver's Travels*.

p. 121 Global cereal harvest graph. See FAOSTAT: http://faostat.fao.org.

p. 122 'Oetzi, the mummified "iceman"'. See http://www.mummytombs.com/otzi/scientific.htm for sources on Oetzi.

p. 122 'The biologist Lee Silver'. Lee Silver, personal communication.

p. 123 'For Adam Smith capital is "as it were, a certain quantity of labour stocked and stored up to be employed, if necessary, upon some other occasion".' Smith, A. 1776. *The Wealth of Nations*.

p. 124 'At one remarkable site, Ohalo II'. Piperno, D.R., Weiss, E., Holst, I. and Nadel, D. 2004. Processing of wild cereal grains in the Upper Palaeolithic revealed by starch grain analysis. *Nature* 430:670–3.

p. 124 'One study notes an "extreme reluctance to shift to domestic foods"'.

Johnson, A.W. and Earle, T.K. 2000. *The Evolution of Human Societies: from Foraging Group to Agrarian State.* Stanford University Press.

p. 125 'The probable cause of this hiatus was a cold snap'. Rosen, A.M. 2007. *Civilizing Climate: Social Responses to Climate Change in the Ancient Near East.* Rowman AltaMira.

p. 126 'the survivors took to nomadic hunter-gathering again'. Shennan, S. 2002. *Genes, Memes and Human History.* Thames & Hudson.

p. 126 'Peru by 9,200 years ago'. Dillehay, T.D. et al. 2007. Preceramic adoption of peanut, squash, and cotton in northern Peru. *Science* 316:1890–3.

p. 126 'millet and rice in China by 8,400 years ago'. Richerson, P.J., Boyd, R. and Bettinger, R.L. 2001. Was agriculture impossible during the Pleistocene but mandatory during the Holocene? A climate change hypothesis. *American Antiquity* 66:387–411.

p. 126 'maize in Mexico by 7,300 years ago'. Pohl, M.E.D. et al. 2007. Microfossil evidence for pre-Columbian maize dispersals in the neotropics from San Andrés, Tabasco, Mexico. *PNAS* 104: 11874–81.

p. 126 'taro and bananas in New Guinea by 6,900 years ago'. Denham, T.P., et al. 2003. Origins of agriculture at Kuk Swamp in the Highlands of New Guinea. *Science* 301: 189–93.

p. 126 'This phenomenal coincidence'. Recent scholarship has made the coincidence much more striking. Until recently, agriculture in Peru, Mexico and New Guinea was believed to have started much later.

p. 127 'agriculture was impossible during the last glacial, but compulsory in the Holocene.' Richerson, P.J., Boyd, R. and Bettinger, R.L. 2001. Was agriculture impossible during the Pleistocene but mandatory during the Holocene? A climate change hypothesis. *American Antiquity* 66(3): 387–411. Incidentally, there is a fascinating parallel between the sudden appearance of farming at the end of the last ice age and the sudden appearance of multicellular life after the mother of all ice ages, the snowball-earth period between 790 and 630 million years ago, when from time to time even the tropics lay under thick ice sheets. The isolated pockets of shivering bacterial refugees upon snowball earth found themselves so inbred, goes one ingenious argument, that individuals clubbed together as a 'body' and delegated breeding to specialised reproductive cells. See Boyle, R.A., Lenton, T.M., Williams, H.T.P. 2007. Neoproterozoic 'snowball Earth' glaciations and the evolution of altruism. *Geobiology* 5:337–49.

p. 127 'It is no accident that modern Australia, with its unpredictable years of drought followed by years of wet, still looks a bit like that volatile glacial world'. Lourandos, H. 1997. *Continent of Hunter-Gatherers.* Cambridge University Press.

p. 127 'One of the intriguing things about the first farming settlements is that they also seem to be trading towns'. Sherratt, A. 2005. The origins of farming in South-West Asia. ArchAtlas, January 2008, edition 3, http://

www.archatlas.org/OriginsFarming/Farming.php, accessed 30 January 2008.

p. 128 'Jane Jacobs suggested in her book *The Economy of Cities*'. Jacobs, J. 1969. *The Economy of Cities*. Random House.

p. 128 'In Greece, farmers arrived suddenly and dramatically around 9,000 years ago.' Perles, C. 2001. *The Early Neolithic in Greece*. Cambridge University Press.

p. 128 'so the genetic evidence suggests'. Cavalli-Sforza, L.L. and Cavalli-Sforza, E. C. 1995. *The Great Human Diasporas: the History of Diversity*. Addison-Wesley.

p. 129 'Other descendants of the Black Sea refugees took to the plains of what is now Ukraine'. Fagan, B. 2004. *The Long Summer*. Granta.

p. 129 'a genetic mutation, substituting G for A in a control sequence upstream of a pigment gene called OCA2'. Eiberg H. et al. 2008. Blue eye color in humans may be caused by a perfectly associated founder mutation in a regulatory element located within the HERC2 gene inhibiting OCA2 expression. *Human Genetics* 123:177–87.

p. 130 'The carbon dioxide released by the fires may even have helped to warm the climate to its 6,000-years-ago balmy maximum'. Ruddiman, W.F. and Ellis, E.C. 2009. Effect of per-capita land use changes on Holocene forest clearance and $CO_2$ emissions. *Quaternary Science Reviews*. (doi:10.1016/j.quascirev.2009.05.022).

p. 130 'the stamp seals of the Halaf people, 8,000 years ago'. http://www.tell-halaf-projekt.de/de/tellhalaf/tellhalaf.htm.

p. 131 'Haim Ofek writes'. Ofek, H. 2001. *Second Nature: Economic Origins of Human Evolution*. Cambridge University Press.

p. 131 'in the words of two theorists'. Richerson, P.J. and Boyd, R. 2007. The evolution of free-enterprise values. In Zak, P. (ed.) 2008. *Moral Markets* Princeton University Press.

p. 131 'very early mining of pure copper-metal deposits around Lake Superior'. Pledger, T. 2003. A brief introduction to the Old Copper Complex of the Western Great Lakes: 4000–1000 BC. In *Proceedings of the Twenty-seventh Annual Meeting of the Forest History Association of Wisconsin, Inc.* Oconto, Wisconsin, 5 October 2002, pp. 10–18. See also http://en.wikipedia.org/wiki/Old_Copper_Complex

p. 132 'the Mitterberg copper miners'. Shennan, S.J. 1999. Cost, benefit and value in the organization of early European copper production. *Antiquity* 73:352–63.

p. 132–3 'typical modern non-industrial people, living in traditional societies, directly consume between one-third and two-thirds of what they produce, and exchange the rest for other goods'. Davis, J. 1992. *Exchange*. Open University Press.

p. 133 'Up to about 300 kilograms of food per head per year, people eat what they grow'. Clark, C. 1970. *Starvation or Plenty?* Secker and Warburg.

p. 133 'Stephen Shennan satirises the attitude thus'. Shennan, S.J. 1999. Cost, benefit and value in the organization of early European copper production. *Antiquity* 73:352–63.

p. 134 'The 'kula' system of the south Pacific'. Davis, J. 1992. *Exchange*. Open University Press.

p. 135 'the worst mistake in the history of the human race'. Diamond. J. 1987. The worst mistake in the history of the human race? *Discover*, May: 64–6.

p. 136 'polygamy enables poor women to share in prosperity more than poor men'. Shennan, S. 2002. *Genes, Memes and Human History*. Thames & Hudson.

p. 137 'Fuegian men, who could not swim, left their wives to anchor canoes in kelp beds and swim ashore in snow storms'. Bridges, E.L. 1951. *The Uttermost Part of the Earth*. Hodder & Stoughton.

p. 137 'One commentator writes'. Wood, J.W. et al. 1998. A theory of preindustrial population dynamics: demography, economy, and well-being in Malthusian systems. *Current Anthropology* 39:99–135.

p. 137 'The archaeologist Steven LeBlanc says that the evidence of constant violence in the ancient past'. LeBlanc, S.A. and Register, K. 2003. *Constant Battles: Why We Fight*. St Martin's Griffin.

p. 138 'In the Merzbach valley in Germany'. Shennan, S. 2002. *Genes, Memes and Human History*. Thames & Hudson.

p. 138 'At Talheim around 4900 BC'. Bentley, R.A., Wahl, J., Price T.D. and Atkinson, T.C. 2008. Isotopic signatures and hereditary traits: snapshot of a Neolithic community in Germany. *Antiquity* 82:290–304.

p. 138 'As Paul Seabright has written'. Seabright, P. 2008. *Warfare and the Multiple Adoption of Agriculture after the Last Ice Age*, IDEI Working Paper no. 522, April 2008.

p. 138 'When Samuel Champlain accompanied (and assisted with his arquebus) a successful Huron raid upon the Mohawks in 1609'. Brook, T. 2008. *Vermeer's Hat*. Profile Books.

p. 139 'Robert Malthus'. Yes, Robert: to call Thomas Robert Malthus by his first name which he did not use, is like calling the first director of the FBI John Hoover.

p. 140 'the eminent British chemist Sir William Crookes gave a similar jeremiad'. Crookes, W. 1898. *The Wheat Problem*. Reissued by Ayers 1976.

p. 140 'Fritz Haber and Carl Bosch'. Smil, V. 2001. *Enriching the Earth*. MIT Press.

pp. 140–1 'As late as 1920, over three million acres of good agricultural land in the American Midwest lay uncultivated'. Clark, C. 1970. *Starvation or Plenty?* Secker and Warburg.

p. 142 'a scientist working in Mexico called Norman Borlaug'. Easterbrook, G. 1997. Forgotten benefactor of humanity. *The Atlantic Monthly*.

p. 143 'In 1968, after huge shipments of Mexican seed, the wheat harvest was extraordinary in both countries.' Hesser, L. 2006. *The Man Who Fed the*

*World.* Durban House. See Borlaug, N.E. 2000. Ending world hunger: the promise of biotechnology and the threat of antiscience zealotry. *Plant Physiology* 124:487–90. Also author's interview with N. Borlaug 2004.

p. 144 'Intensification has saved 44 per cent of this planet for wilderness.' Goklany I. 2001. Agriculture and the environment: the pros and cons of modern farming. *PERC Reports* 19:12–14.

p. 144 'Some argue that the human race already appropriates for itself an unsustainable fraction of the planet's primary production'. The World Wildlife Fund estimates that humankind is already overdrawn in its use of earth's resources, but it reaches this conclusion only by including a vast acreage of new forest planting needed to balance each person's carbon emissions.

p. 144 'HANPP – the "human appropriation of net primary productivity"'. Haberl, H. et al. 2007. Quantifying and mapping the human appropriation of net primary production in earth's terrestrial ecosystems. Proceedings of the National Academy of Sciences 104:12942–7.

p. 145 'These findings suggest that, on a global scale, there may be a considerable potential to raise agricultural output without necessarily increasing HANPP'. Haberl, H. et al. 2007. Quantifying and mapping the human appropriation of net primary production in earth's terrestrial ecosystems. Proceedings of the National Academy of Sciences 104:12942–7.

p. 145 'Even the confinement of chickens, pigs and cattle to indoor barns and batteries'. Dennis Avery of the Hudson Institute has written on this. See http://www.hudson.org/index.cfm?fuseaction=publication_details &id=3988.

p. 146 'Colin Clark calculated that human beings could in theory sustain themselves on just twenty-seven square metres of land each'. Clark, C. 1963. Agricultural productivity in relation to population. In *Man and His Future*, CIBA Foundation; also Clark, C. 1970. *Starvation or Plenty?* Secker and Warburg.

p. 146 'the world grew about two billion tonnes of rice, wheat and maize on about half a billion hectares of land'. Statistics taken from the FAO: www.faostat.fao.org.

p. 147 'would mean an extra seven billion cattle grazing an extra thirty billion acres of pasture'. Smil, V. 2001. *Enriching the Earth*. MIT Press. See also http://www.heartland.org/policybot/results/22792/Greenpeace_ Farming_Plan_Would_Reap_Environmental_Havoc_around_the_ World.html: Dennis Avery asked Vaclav Smil to make this calculation.

p. 147 'Lester Brown points out that India depends heavily on a rapidly depleting aquifer and a slowly drying Ganges'. Brown, L. 2008. *Plan B 3.0: Mobilizing to Save Civilisation*. Earth Policy Institute.

p. 148 'Once it is properly priced by markets, water is not only used more frugally'. Morriss, A.P. 2006. Real people, real resources and real choices:

the case for market valuation of water. *Texas Tech Law Review* 38.

p. 149 'as a professor and a chef have both suggested on my radio recently'. The professor and the chef I refer to are Tim Lang and Gordon Ramsay. 'Why are we buying food from other people which should be feeding developing countries?' asked Tim Lang, member of the Sustainable Development Commission, on the BBC *Today* programme, 4 March 2008. 'I don't want to see asparagus on menus in the middle of December. I don't want to see strawberries from Kenya in the middle of March. I want to see it home grown,' said Gordon Ramsay on 9 May 2008. (See 'Ramsay orders seasonal-only menu', http://news.bbc.co.uk/1/hi/uk/7390959.stm.) Apart from the increased emissions, imagine the terrible monotony of the British diet under these proposals. There would be no coffee or tea, no bananas or mangoes, no rice or curry powder, there would be strawberries only in June and July and no lettuce in winter. You would eat an awful lot of potatoes. The rich would heat their greenhouses and plant orange trees in them, or travel to foreign parts and smuggle papayas in their luggage. Meat would become a luxury only available to the professor and his fellow rich folk – for to grow a lamb chop requires ten times as much land as to grow a piece of bread of equivalent calories. There are no combine harvester factories in Britain, so unless the prof wants us hypocritically to import combines but not flour, we would all have to take our turn in the fields with sickles in August. These are no doubt mere inconveniences that the professor would sort out with some laws and some food police. The real problem lies elsewhere, conveniently out of sight in the developing world. The growers of coffee, tea, bananas, mangoes, rice and turmeric would all suffer. They would have to stop growing cash crops and start being more self-sufficient. Sounds charming, but self-sufficiency is the very definition of poverty. Unable to sell their cash crops, they would have to eat what they grow. As we in the north munched our potatoes and bread, so they in the tropics would be growing heartily sick of an endless diet of mangoes and turmeric. The cash economy enables me to eat mangoes and them to eat bread, thank goodness.

p. 149 'again the acreage under the plough will have to balloon'. Or, to put the point in academic-ese: 'The additional harvest of 4–7 Pg C/yr needed to achieve this level of bioenergy use would almost double the present biomass harvest and generate substantial additional pressure on ecosystems.' Haberl, H. et al. 2007. Quantifying and mapping the human appropriation of net primary production in earth's terrestrial ecosystems. *Proceedings of the National Academy of Sciences* 104:12942–7.

p. 149 'each needs little more than a thousand square metres, a tenth of a hectare'. Smil, V. 2000. *Feeding the World*. MIT Press.

p. 150 'Organic farming is low-yield, whether you like it or not'. Avery, A. 2006. *The Truth about Organic Foods*. Henderson Communications. See also Goulding, K.W.T. and Trewavas, A.J. 2009. Can organic feed the

world? AgBioview Special Paper 23 June 2009. http://www.agbioworld.
org/newsletter_wm/index.php?caseid=archive&newsid=2894.

p. 150 'With such help a particular organic plot can match non-organic
yields, but only by using extra land elsewhere to grow the legumes and
feed the cattle'. A recent study claimed that organic yields can be higher
than those of conventional farming (http://www.ns.umich.edu/htdocs/
releases/story.php?id=5936), but only by an extremely selective and
biased misuse of the statistics (see http://www.cgfi.org/2007/09/06/
organic-abundance-report-fatally-flawed/).

p. 150 'a pound of organic lettuce, grown without synthetic fertilisers or
pesticides in California, and containing eighty calories, requires 4,600
fossil-fuel calories to get it to a customer's plate'. Pollan, M. 2006 *The
Omnivore's Dilemma: the Search for the Perfect Meal in a Fast Food
World.* Bloomsbury.

p. 150 'when a technology came along that promised to make organic
farming both competitive and efficient, the organic movement promptly
rejected it'. Ronald, P. and Adamchak, R.W. 2008. *Tomorrow's Table:
Organic Farming, Genetics and the Future of Food.* Oxford University
Press.

p. 152 'a near-doubling of yield and a halving of insecticide use'. ISAAA
2009. *The Dawn of a New Era: Biotech Crops in India.* ISAAA Brief 39,
2009: http://www.isaaa.org/resources/publications/downloads/The-
Dawn-of-a-New-Era.pdf.

p. 152 'the use of insecticides is down by as much as 80 per cent'. Marvier
M., McCreedy, C., Regetz, J. and Kareiva, P. 2007. A meta-analysis of
effects of Bt cotton and maize on nontarget invertebrates. *Science*
316:1475–7; also Wu, K.-M. et al. 2008. Suppression of cotton bollworm
in multiple crops in China in areas with Bt Toxin-containing cotton.
*Science* 321:1676–8 (doi: 10.1126/science.1160550).

p. 152 'the leaders of the organic movement locked themselves out of a new
technology'. Ronald, P.C. and Adamchak, R.W. 2008. *Tomorrow's Table:
Genetics, and the Future of Food.* Oxford University Press.

p. 152 'the amount of pesticide *not* used because of genetic modification at
over 200 million kilograms of active ingredients'. Miller, J.K. and
Bradford, K.J. 2009. The pipeline of transgenic traits in specialty crops.
Unpublished paper, Kent Bradford.

p. 152 'writes the Missouri farmer Blake Hurst'. Hurst, B. 2009. The
omnivore's delusion: against the agri-intellectuals. *The American*, journal
of the American Enterprise Institute. 30 July 2009. http://www.american.
com/archive/2009/july/the-omnivore2019s-delusion-against-the-agri-
intellectuals.

p. 152 '*Silent Spring*'. Carson, R. 1962. *Silent Spring.* Houghton Mifflin.

p. 153 'These mutations were selected, albeit inadvertently'. Doebley, J. 2006.
Unfallen grains: how ancient farmers turned weeds into crops. *Science*
312:1318–19.

p. 153 'DNA sequences borrowed from mosses and algae'. Richardson, A.O. and Palmer, J.D. 2006. Horizontal gene transfer in plants. *Journal of Experimental Botany* 58:1–9.

p. 153 'DNA has even been caught jumping naturally from snakes to gerbils with the help of a virus.' Piskurek, O. and Okada, N. 2007. Poxviruses as possible vectors for horizontal transfer of retroposons from reptiles to mammals. *PNAS* 29:12046–51.

p. 154 'Only in parts of Europe and Africa were these crops denied to farmers and consumers'. Brookes, G. and Barfoot, P. 2007. Global impact of GM crops: socio-economic and environmental effects in the first ten years of commercial use. *AgBioForum* 9:139–51.

p. 154 'what Stewart Brand calls their "customary indifference to starvation"'. Brand, S. 2009. *Whole Earth Discipline*. Penguin.

p. 154 'Robert Paarlberg writes'. Paarlberg, R. 2008. *Starved for Science*. Harvard University Press.

p. 154 'Ingo Potrykus thinks'. Potrykus, I. 2006. *Economic Times of India*, 26 December 2005. Reprinted at http://www.fighting diseases.org/main/articles.php?articles_id=568.

p. 154 'Or as the Kenyan scientist Florence Wambugu puts it'. Quoted in Brand, S. 2009. *Whole Earth Discipline*. Penguin.

pp. 154–5 'Per capita food production in Africa has fallen 20 per cent in thirty-five years'. Collier, P. 2008. The politics of hunger: how illusion and greed fan the food crisis. *Foreign Affairs* November/December 2008.

p. 155 'Field trials begin in Kenya in 2010 of drought-resistant and insect-resistant maize'. Muthaka, B. 2009. GM maize for local trials. *Daily Nation* (Nairobi), 17 June 2009.

p. 156 'For example, modern plant oils and plentiful red meat make for a diet low in omega-3 fatty acids'. Morris, C.E. and Sands, D. 2006. The breeder's dilemma: resolving the natural conflict between crop production and human nutrition. *Nature Biotechnology* 24: 1078-80.

p. 156 'The Indian activist Vandana Shiva'. Quoted in Avery, D.T. 2000. What do environmentalists have against golden rice? Center for Global Food Issues, http://www.cgfi.org/materials/articles/2000/mar_7_00.htm. See also www.goldenrice.org for more of the shocking story of opposition to this humanitarian project.

## Chapter 5

p. 157 'Imports are Christmas morning; exports are January's MasterCard bill.' O'Rourke, P.J. 2007. *On The Wealth of Nations*. Atlantic Monthly Press.

p. 157 Death rates from water-related disease graph. Goklany, I. 2009. *Electronic Journal of Sustainable Development*. www.ejsd.org.

p. 158 'A modern combine harvester, driven by a single man, can reap enough wheat in a single day to make half a million loaves.' Half a kilogram of flour per loaf, 3,500 kg per acre, eighty acres per day =

560,000 loaves per day. These are numbers my colleagues achieve on my own farm.

p. 158 'a distinctive 'Ubaid' style of pottery, clay sickles and house design'. Stein, G.J. and Ozbal, R. 2006. A tale of two Oikumenai: variation in the expansionary dynamics of 'Ubaid' and Uruk Mesopotamia. Pp. 356–70 in Stone, E. C. (ed.) *Settlement and Society: Ecology, Urbanism, Trade and Technology in Mesopotamia and Beyond* (Robert McC. Adams Festschrift). Los Angeles, Cotsen Institute of Archaeology.

p. 159 'in the words of the archaeologist Gil Stein'. Stein, G.J. and Ozbal, R. 2006. A tale of two Oikumenai: variation in the expansionary dynamics of 'Ubaid' and Uruk Mesopotamia. Pp. 356–70 in: Stone, E. C. (ed.) *Settlement and Society: Ecology, Urbanism, Trade and Technology in Mesopotamia and Beyond* (Robert McC. Adams Festschrift). Los Angeles, Cotsen Institute of Archaeology.

p. 160 'The message those tablets tell is that the market came long before the other appurtenances of civilisation.' Basu, S., Dickhaut, J.W., Hecht, G., Towry, K.L. and Waymire, G.B. 2007. Recordkeeping alters economic history by promoting reciprocity. *PNAS* 106:1009–14.

p. 161 'Merchants and craftsmen make prosperity; chiefs, priests and thieves fritter it away.' Incidentally, I find it strange to recall that my education was utterly dominated by two stories: the Bible's and Rome's. Both were disappointing examples of history. One told the story of an obscure, violent and somewhat bigoted tribe and one of its later cults, who sat around gazing at their theological navels for a few thousand years while their fascinating neighbours – the Phoenicians, Philistines, Canaanites, Lydians and Greeks – invented respectively maritime trade, iron, the alphabet, coins and geometry. The other told the story of a barbarically violent people who founded one of the empires that institutionalised the plundering of its commercially minded neighbours, then went on to invent practically nothing in half a millennium and achieve an actual diminution in living standards for its citizens, very nearly extinguishing literacy as it died. I exaggerate, but there are more interesting figures in history than Jesus Christ or Julius Caesar.

p. 161 'Unlike hunter-gatherers or herders, farmers faced with taxes have to stay put and pay'. Carneiro, R.L. 1970. A theory of the origin of the state. *Science* 169: 733–8.

p. 161 'in the words of two modern historians'. Moore, K. and Lewis, D. 2000. Foundations of Corporate Empire. *Financial Times*/Prentice Hall.

p. 162 'As Sir Mortimer Wheeler wrote in his autobiography'. Quoted by Sally Greene in 1981, introduction to illustrated edition of *Man Makes Himself*. Childe, V. Gordon. 1956. Pitman Publishing.

p. 162 'the archaeologist Shereen Ratnagar concluded'. Ratnagar, S. 2004. *Trading Encounters: From the Euphrates to the Indus in the Bronze Age.* Oxford University Press India.

p. 162 'great wealth of the Indus cities was generated by trade'. Possehl, G.L.

2002. *The Indus Civilization: A Contemporary Perspective.* Rowman AltaMira.

p. 162 'the so-called Norte Chico civilisation'. Haas, J. and Creamer, W. 2006. Crucible of Andean civilization: The Peruvian coast from 3000 to 1800 BC. *Current Anthropology* 47:745–75.

pp. 163–4 'Intensification of trade came first'. The Chinese case remains unexplored here for the simple reason that the key moment in China, the Longshan culture, remains too poorly known, especially in terms of how much trade occurred.

p. 165 'silver-based prices, which fluctuated freely'. Aubet, M.E. 2001. *The Phoenicians and the West.* 2nd edition. Cambridge University Press.

p. 165 'the Uruk word for high priest is the same as the word for accountant'. Childe, V.G. 1956/1981. *Man Makes Himself.* Moonraker Press.

p. 165 'merchants from Ashur operated in "karum" enclaves'. Moore, K. and Lewis, D. 2000. *Foundations of Corporate Empire.* Pearson.

p. 165 'The profit margin was 100 per cent on tin and 200 per cent on textiles'. Chanda, N. 2007. *Bound Together: How Traders, Preachers, Adventurers and Warriors Shaped Globalisation.* Yale University Press.

p. 166 'Such merchants "did not devote themselves to trading in copper and wool because Assyria needed them, but because that trade was a means of obtaining more gold and silver".' Aubet, M.E. 2001. *The Phoenicians and the West.* 2nd edition. Cambridge University Press.

p. 167 'a different Phoenician invention, the bireme galley'. Holst, S. 2006. *Phoenicians: Lebanon's Epic Heritage.* Sierra Sunrise Publishing.

p. 168 '"Homer" displays a relentlessly negative attitude to Phoenician traders'. Aubet, M.E. 2001. *The Phoenicians and the West.* 2nd edition. Cambridge University Press.

p. 168 'Tyrian traders founded Gadir, present-day Cadiz, around 750 BC'. Aubet, M.E. 2001. *The Phoenicians and the West.* 2nd edition. Cambridge University Press.

p. 169 'said a Montagnais trapper to a French missionary in seventeenth-century Canada'. Brook, T. 2008. *Vermeer's Hat.* Profile Books.

p. 169 'When HMS *Dolphin*'s sailors found that a twenty-penny iron nail could buy a sexual encounter on Tahiti in 1767'. Bolyanatz, A. H. 2004. *Pacific Romanticism: Tahiti and the European Imagination.* Greenwood Publishing Group.

p. 170 'advanced by David Hume'. This argument goes back to David Hume's *History of Great Britain*, and has been pursued recently by Douglass North.

p. 170 'Miletus, the most successful of the Ionian Greek cities, sat "like a bloated spider" at the junction of four trade routes'. Cunliffe, B. 2001. *The Extraordinary Voyage of Pytheas the Greek.* Penguin.

p. 172 'Humanity's great battle over the last 10,000 years has been the battle against monopoly.' Kealey, T. 2008. *Sex, Science and Profits.* Random House.

p. 172 'The Mauryan empire in India'. Khanna, V. S. 2005. *The Economic History of the Corporate Form in Ancient India* (1 November 2005). Social Sciences Research Network.

p. 173 'without question the economic superpower of the day'. Maddison, A. 2006. *The World Economy*. OECD Publishing.

p. 173 'wrote Thomas Carney'. Carney, T.F. 1975. *The Shape of the Past*. Coronado Press.

p. 174 'Ostia was a trading city as surely as Hong Kong is today'. Moore, K. and Lewis, D. 2000. *Foundations of Corporate Empire*. Pearson.

p. 174 'Rome's continuing prosperity once the republic became an empire may be down at least partly to the "discovery" of India'. Chanda, N. 2007. *Bound Together: How Traders, Preachers, Adventurers and Warriors Shaped Globalisation*. Yale University Press.

p. 176 'The predatory expansion of the Carolingian Franks in the eighth century'. Kohn, M. 2008. How and why economies develop and grow: lessons from preindustrial Europe and China. Unpublished manuscript.

p. 177 'A dhow that sank off Belitung in Indonesia in AD 826'. Flecker, M. 2001. A 9th-century Arab or Indian shipwreck in Indonesian waters. *International Journal of Nautical Archaeology*.29:199–217.

p. 177 'Once the priesthood tightened its grip'. Norberg, J. 2006. *When Man Created the World*. Published in Swedish as *När människan skapade världen*. Timbro.

p. 178 'Maghribi traders developed their own rules of contract enforcement and punishment by ostracism'. Greif, A. 2006. *Institutions and the Path to the Modern Economy: Lessons from Medieval Trade*. Cambridge University Press.

p. 178 'a Pisan trader living in north Africa, Fibonacci'. Ferguson, N. 2008. *The Ascent of Money*. Allen Lane.

p. 178 'Genoa's trade with North Africa doubled after an agreement for the protection of merchants was reached'. Chanda, N. 2007. *Bound Together: How Traders, Preachers, Adventurers and Warriors Shaped Globalisation*. Yale University Press.

p. 179 'by 1500 Italy's GDP per capita was 60 per cent higher than the European average'. Maddison, A. 2006. *The World Economy*. OECD Publishing.

p. 179 'As late as 1600, European trade with Asia, dominated because of transport costs by luxuries such as spices, was only half the value of the inter-regional European trade in cattle alone'. Kohn, M. 2008. How and why economies develop and grow: lessons from preindustrial Europe and China. Unpublished manuscript.

p. 180 'According to Angus Maddison's estimates'. Maddison, A. 2006. *The World Economy*. OECD Publishing.

p. 181 'One of the paradoxical features of modern China is the weakness of a central, would-be authoritarian government.' Fukuyama, F. 2008. *Los Angeles Times*, 29 April 2008.

p. 181 'multi-spindle cotton wheels, hydraulic trip hammers, as well as umbrellas, matches, toothbrushes and playing cards'. Baumol, W. 2002. *The Free-market Innovation Machine.* Princeton University Press.

p. 181 'The Black Death'. Durand, J. 1960. The population statistics of China, A.D. 2–1953. *Population Studies* 13:209–56.

p. 182 'serfdom was effectively restored'. Findlay, R. and O'Rourke, K.H. 2007. *Power and Plenty: Trade, War and the World Economy.* Princeton University Press.

p. 182 'as Peter Turchin argues following the lead of the medieval geographer Ibn Khaldun'. Turchin, P. 2003. *Historical Dynamics.* Princeton University Press.

p. 182 'most clever people still call for government to run more things'. Note that this is also true of the financial crisis of 2008: government mismanagement of housing policy, interest rates and exchange rates bears just as big a responsibility as corporate mismanagement of risk. I wish there was space to expand upon this point, but see the writings of Northcote Parkinson, Mancur Olson, Gordon Tullock and Deepak Lal. It is strange to me that most people assume companies will be imperfect (as they are), but then assume that government agencies will be perfect, which they are not.

p. 182 'Not only did the Ming emperors nationalise much of industry and trade, creating state monopolies in salt, iron, tea, alcohol, foreign trade and education'. Landes, D. 1998. *The Wealth and Poverty of Nations.* Little, Brown.

p. 183 'As Etienne Balazs put it'. Balazs, E. quoted in Landes, D. 1998. *The Wealth and Poverty of Nations.* Little, Brown.

p. 183 'The behaviour of Hongwu, the first of the Ming emperors'. Brook, T. 1998. *The Confusions of Pleasure: Commerce and Culture in Ming China.* University of California Press.

p. 183 'a large Spanish galleon stuffed with silver'. Brook, T. 2008. *Vermeer's Hat.* Profile Books.

p. 184 'said Lactantius'. Quoted in Harper, F.A. 1955. Roots of economic understanding. *The Freeman* vol. 5, issue 11. http://www.thefreeman online.org/columns/roots-of-economic-understanding.

pp. 184–5 'The man in question, Johann Friedrich Bottger'. Gleason, J. 1998. *The Arcanum.* Bantam Press.

p. 185 'the Dutch so dominated European international trade that their merchant marine was bigger than that of France, England, Scotland, the Holy Roman Empire, Spain and Portugal – combined'. Blanning, T. 2007. *The Pursuit of Glory.* Penguin.

p. 186 'Both sides of the estuary of the River Plate became a vast slaughterhouse'. Edgerton, D. 2006. *The Shock of the Old: Technology and Global History since 1900.* Profile Books.

p. 186 'Yet in the aftermath of the First World War, one by one countries tried beggaring their neighbours in the twentieth century'. Findlay, R.

and O'Rourke, K.H. 2007. *Power and Plenty: Trade, War and the World Economy*. Princeton University Press.

p. 187 'China's Open Door policy, which cut import tariffs from 55 per cent to 10 per cent in twenty years, transformed it from one of the most protected to one of the most open markets in the world.' Lal, D. 2006. *Reviving the Invisible Hand*. Princeton University Press.

p. 188 'Farm subsidies and import tariffs on cotton, sugar, rice and other products cost Africa $500 billion a year in lost export opportunities'. Moyo, D. 2009. *Dead Aid*. Allen Lane.

p. 188 'Ford Madox Ford celebrated in his Edwardian novel *The Soul of London*'. Ford, F.M. 1905. *The Soul of London*. Alston Rivers.

p. 189 'says Suketa Mehta'. Mehta, S. Dirty, crowded, rich and wonderful. *International Herald Tribune*, 16 July 2007. Quoted in Williams, A. 2008. *The Enemies of Progress*. Societas.

p. 189 'writes Stewart Brand'. Brand, S. 2009. *Whole Earth Discipline*. Penguin.

p. 189 'says Deroi Kwesi Andrew, a teacher earning $4 a day in Accra'. Harris, R. 2007. Let's ditch this nostalgia for mud. *Spiked*, 4 December 2007.

p. 190 'people prefer to press into ever closer contact with each other in glass towers to do their exchanging'. Jacobs, J. 2000. *The Nature of Economies*. Random House.

p. 190 'As Edward Glaeser put it'. Glaeser, E. 2009. Green cities, brown suburbs. *City Journal* 19: http://www.city-journal.org/2009/19_1_green-cities.html.

p. 190 'the ecologist Paul Ehrlich had an epiphany'. Ehrlich, P. 1968. *The Population Bomb*. Ballantine Books.

## Chapter 6

p. 191 'The great question is now at issue'. Malthus, T. R. 1798. *Essay on Population*.

p. 191 Percentage increase in world population graph. United Nations Population Division.

p. 192 'The economist Vernon Smith, in his memoirs'. Smith, V.L. 2008. *Discovery – a Memoir*. Authorhouse.

p. 193 'The Malthusian crisis comes not as a result of population growth directly, but because of decreasing specialisation.' My argument here is part-way between the Malthusian one advanced by historians such as Greg Clark and the view that pre-industrial economies were always capable of greater productivity, but predation and other intrinsic factors prevented them – as advanced by George Grantham. See e.g. Grantham, G. 2008. Explaining the industrial transition: a non-Malthusian perspective. *European Review of Economic History* 12:155–65. See also Persson, K.-G. 2008. The Malthus delusion. *European Review of Economic History* 12: 165–73.

p. 193 'As Greg Clark puts it'. Clark, G. 2007. *A Farewell to Alms*. Princeton University Press.

p. 193 'Malthus'. Malthus, T.R. 1798. *Essay on Population*.

p. 193 'Ricardo'. Ricardo, D. 1817. *The Principles of Political Economy and Taxation*. (Adam Smith, looking at China, India and Holland, had thought the same.)

p. 194 'the Wiltshire village of Damerham'. Langdon, J. and Masschaele, J. 2006. Commercial activity and population growth in medieval England. *Past and Present* 190:35–81.

p. 195 'a miller in Feering in Essex'. Langdon, J. and Masschaele, J. 2006. Commercial activity and population growth in medieval England. *Past and Present* 190:35–81.

p. 195 'It came suddenly in the sodden summers of 1315 and 1317, when wheat yields more than halved all across the north of Europe.' Jordan, W.C. 1996. *The Great Famine: Northern Europe in the Early Fourteenth Century*. Princeton University Press.

p. 196 'neither the boom of the thirteenth century, nor the bust of the fourteenth, can be described in simplistic Ricardian and Malthusian terms'. See Meir Kohn's book *How and Why Economies Develop and Grow* at www.dartmouth.edu/~mkohn/Papers/ lessons%201r3.pdf.

pp. 196–7 'the site of a new windmill being constructed at Dover Castle in 1294'. Langdon, J. and Masschaele, J. 2006. Commercial activity and population growth in medieval England. *Past and Present* 190:35–81.

p. 197 'in Joel Mokyr's words'. Mokyr, J. 1990. *Lever of Riches*. Oxford University Press.

p. 197 'the Japanese had conquered Korea carrying tens of thousands of home-made arquebuses'. Noted in Perrin, N. 1988. *Giving Up the Gun: Japan's Reversion to the Sword*. Grodine.

p. 197–8 'As the traveller Isabella Bird remarked in 1880'. Macfarlane, A. and Harrison, S. 2000. Technological evolution and involution: a preliminary comparison of Europe and Japan. In Ziman, J. (ed.) *Technological Innovation as an Evolutionary Process*. Cambridge University Press.

p. 198 'Where Europeans used animal, water and wind power, the Japanese did the work themselves.' Macfarlane, A. and Harrison, S. 2000. Technological evolution and involution: a preliminary comparison of Europe and Japan. In Ziman, J. (ed.)*Technological Innovation as an Evolutionary Process*. Cambridge University Press.

p. 198 'They even gave up capital-intensive guns in favour of labour-intensive swords'. Perrin, N. 1988. *Giving Up the Gun: Japan's Reversion to the Sword*. Grodine.

p. 199 'Sir William Petty'. Petty, W. 1691. *Political Arithmetick*.

p. 199 'Adam Smith begged to differ'. *The Wealth of Nations*, quoted in Blanning, T. 2007. *The Pursuit of Glory*. Penguin.

p. 200 'By the 1800s, Denmark had become a country that was trapped by its own self-sufficiency.' Pomeranz, K. 2000. *The Great Divergence*. Princeton University Press.

p. 200 'On average a merchant in Britain who left £1,000 in his will had four surviving children, while a labourer who left £10 had only two'. Clark, G. 2007. *A Farewell to Alms*. Princeton University Press.

p. 203 'Johnson supposedly replied'. Epstein, H. 2008. The strange history of birth control. *New York Review of Books*, 18 August 2008.

p. 203 'Garrett Hardin, in his famous essay'. Hardin, G. 1968. The tragedy of the commons. *Science* 162:1243–8.

p. 203 'Hardin's view was nearly universal'. An exception was Barry Commoner, who argued at the UN conference on population in Stockholm in 1972 that the demographic transition would solve population growth without coercion.

p. 203 'wrote John Holdren (now President Obama's science adviser) and Paul and Anne Ehrlich in 1977'. Ehrlich, P., Ehrlich, A. and Holdren, J.F. 1977. *Eco-science*. W.H. Freeman.

p. 203 'Sanjay Gandhi, the son of the Indian prime minister, ran a vast campaign of rewards and coercion'. Connelly, M. 2008. *Fatal Misconception: the Struggle to Control World Population*. Harvard University Press.

p. 204 'Bangladesh had a birth rate of 6.8 children per woman'. The standard way of measuring the birth rate is the 'total fertility rate', which presumes to average the completed family size of each age cohort of the population. This is imperfect and confuses deferred reproduction with falling family size. But it is the best that is available and I have used it in this chapter for lack of a better measure.

p. 205 'As the environmentalist Stewart Brand puts it'. Brand, S. 2005. Environmental heresies. *Technology Review*, May 2005.

p. 206 'the entire world is experiencing the second half of a "demographic transition"'. Caldwell, J. 2006. *Demographic Transition Theory*. Springer.

p. 207 'a condescending blast by Paul Ehrlich and John Holdren'. Maddox's book was called *The Doomsday Syndrome* (1973, McGraw Hill) and Holdren's and Ehrlich's review is quoted by John Tierney at http://tierneylab.blogs.nytimes.com/2009/04/15/the-skeptical-prophet/.

p. 207 'demographic transition theory is a splendidly confused field.' Or to put it in academic-ese, 'the debate continues with a plethora of contending theoretical frameworks, none of which has gained wide adherence.' Hirschman, quoted in Bongaarts, J. and Watkins, S.C. 1996. Social interactions and contemporary fertility transitions. *Population and Development Review* 22:639–82.

p. 208 'Jeffrey Sachs recounts'. Sachs, J. 2008. *Common Wealth: Economics for a Crowded Planet*. Allen Lane.

p. 209 'Probably by far the best policy for reducing population is to encourage female education.' Connelly, M. 2008. *Fatal Misconception: the Struggle to Control World Population*. Harvard University Press.

p. 210 'A bold programme, driven by philanthropy or even government aid'. Sachs, J. 2008. *Common Wealth: Economics for a Crowded Planet*. Allen Lane.

p. 211 'Seth Norton found'. Norton, S. 2002. *Population Growth, Economic Freedom and the Rule of Law*. PERC Policy Series no. 24.

p. 211 'The Anabaptist sects in North America, the Hutterites and Amish, have largely resisted the demographic transition'. Richerson, P. and Boyd, R. 2005. *Not by Genes Alone*. Chicago University Press.

p. 211 'As Ron Bailey puts it'. Bailey, R. 2009. The invisible hand of population control. *Reason*, 16 June 2009. http://www.reason.com/news/show/134136.html.

p. 212 'Hans-Peter Kohler of the University of Pennsylvania'. Myrskylä, M., Kohler, H.-P. and Billari, F.C. 2009. Advances in development reverse fertility declines. *Nature*, 6 August 2009 (doi:10.1038/nature 08230).

## Chapter 7

p. 213 'With coal almost any feat is possible or easy; without it we are thrown back in the laborious poverty of earlier times'. Jevons, W.S. 1865. *The Coal Question: An Inquiry Concerning the Progress of the Nation, and the Probable Exhaustion of our Coal-mines*. Macmillan.

p. 213 Metal prices relative to US wages graph. Goklany, I. 2009. *Electronic Journal of Sustainable Development*. www.ejsd.org.

p. 214 'Writes the economist Don Boudreaux'. http://www.pittsburghlive.com/x/pittsburghtrib/opinion/columnists/boudreaux/s_304437.html.

p. 215 'In England, horses were 20 per cent of draught animals in 1086'. Fouquet, R. and Pearson, P.J.G. 1998. A thousand years of energy use in the United Kingdom. *Energy Journal* 19:1–41.

p. 215 'one for every fifty people in southern England'. Mokyr, J. 1990. *Lever of Riches*. Oxford University Press.

p. 215 'At Clairvaux'. The abbot of Clairvaux is quoted in Gimpel, J. 1976. *The Medieval Machine*. Penguin.

pp. 215–16 'peat gave the Dutch their chance'. De Zeeuw, J.W. 1978. Peat and the Dutch golden age. See http://www.peatandculture.org/documenten/Zeeuw.pdf.

p. 218 'In Gregory King's survey of the British population in 1688'. Kealey, T. 2008. *Sex, Science and Profits*. William Heinemann.

p. 218 'Even farm labourers' income rose during the industrial revolution'. Clark, G. 2007. *A Farewell to Alms*. Princeton University Press.

p. 218 'a patent for a hand-driven linen spinning machine from 1678'. Friedel, R. 2007. *A Culture of Improvement*. MIT Press.

p. 218 'The average Englishman's income, having apparently stagnated for three centuries, began to rise around 1800'. This is Clark's estimate. Others argue that because of the rapidly falling prices of goods like sugar, the purchasing power of average income was rising steadily in the 1700s. See Clark, G. 2007. *A Farewell to Alms*. Princeton University Press.

p. 219 'Here are three anecdotes'. The first case comes from an unpublished history of the village of Stannington written by my grandmother and others in the 1950s. The other two cases are cited in Rivoli, P. 2005. *The*

*Travels of a T-shirt in the Global Economy.* John Wiley.

p. 221 A famous print entitled 'The Distinguished Men of Science of Great Britain Living in the Year 1807-8'. The print was published alongside a book edited and published by William Walker, *Memoirs of the Distinguished Men of Science of Great Britain Living in the Year 1807–08.*

pp. 221–2 'like Gordon Moore and Robert Noyce, Steve Jobs and Sergey Brin, Herb Boyer and Leroy Hood'. Moore founded Intel, Noyce the microchip, Jobs Apple, Brin Google, Boyer Genentech, Hood Applied Biosystems.

p. 222 'explained one Hungarian liberal'. Gergely Berzeviczy, quoted in Blanning, T. 2007. *The Pursuit of Glory.* Penguin.

p. 222 'France, three times as populous as England, was "cut up by internal customs barriers into three major trade areas"'. Landes, D.S. 2003. *The Unbound Prometheus: Technological Change and Industrial Development in Western Europe from 1750 to the Present.* 2nd edition. Cambridge University Press.

p. 222 'Spain was "an archipelago, islands of local production and consumption, isolated from each other by centuries of internal tariffs".' John Lynch, quoted in Blanning, T. 2007. *The Pursuit of Glory.* Penguin.

p. 223 'a "glorious revolution" against James II's arbitrary government'. Jardine, L. 2008. *Going Dutch.* Harper.

p. 223 'this was not a bad place to start or expand a business in say 1700'. Baumol, W. 2002. *The Free-market Innovation Machine.* Princeton University Press.

p. 223 'says David Landes'. Landes, D.S. 2003. *The Unbound Prometheus: Technological Change and Industrial Development in Western Europe from 1750 to the Present.* 2nd edition. Cambridge University Press.

p. 224 'says Robert Friedel'. Friedel, R. 2007. *A Culture of Improvement.* MIT Press.

p. 224 'writes Neil McKendrick'. Quoted in Blanning, T. 2007. *The Pursuit of Glory.* Penguin.

p. 224 'Daniel Defoe, writing in 1728'. Quoted in Mokyr, J. 1990. *Lever of Riches.* Oxford University Press; Friedel, R. 2007. *A Culture of Improvement.* MIT Press.

p. 225 'it was by copying these Oriental imports that the industrialists got started'. Mokyr, J. 1990. *Lever of Riches.* Oxford University Press; Friedel, R. 2007. *A Culture of Improvement.* MIT Press.

p. 226 'the Calico Act'. Friedel, R. 2007. *A Culture of Improvement.* MIT Press; Rivoli, P. 2005. *The Travels of a T-shirt in the Global Economy.* John Wiley.

p. 226 'enclosure actually increased paid employment for farm labourers'. Here is how Landes puts it: 'For a long time, the most accepted view has been that propounded by Marx and repeated and embellished by generations of socialist and even non-socialist historians. This position explains the accomplishment of so enormous a social change – the creation of an industrial proletariat in the face of tenacious resistance –

by postulating an act of forcible expropriation: the enclosures uprooted the cottager and small peasant and drove them into the mills. Recent empirical research has invalidated this hypothesis; the data indicate that the agricultural revolution associated with the enclosures increased the demand for farm labour, and that indeed those rural areas that saw the most enclosure saw the largest increase in resident population. From 1750 to 1830 Britain's agricultural counties doubled their inhabitants. Whether objective evidence of this kind will suffice, however, to do away with what has become an article of faith is doubtful.' Landes, D.S. 2003. *The Unbound Prometheus: Technological Change and Industrial Development in Western Europe from 1750 to the Present.* 2nd edition. Cambridge University Press, pp. 114–15.

p. 227 'The historian Edward Baines noted in 1835'. Baines, E. 1835. *History of the Cotton Manufacture in Great Britain.* Quoted in Rivoli, P. 2005. *The Travels of a T-shirt in the Global Economy.* John Wiley.

p. 227 'reflected Joseph Schumpeter'. Schumpeter, J.A. 1943. *Capitalism, Socialism, and Democracy.* Allen & Unwin.

p. 227 'As the twentieth-century economist Colin Clark put it'. Clark, C. 1970. *Starvation or Plenty?* Secker and Warburg.

p. 228 'by 1800 the jenny was already obsolete'. Landes, D.S. 2003. *The Unbound Prometheus: Technological Change and Industrial Development in Western Europe from 1750 to the Present.* 2nd edition. Cambridge University Press.

p. 228 'price of a pound of fine-spun cotton yarn fell'. Friedel, R. 2007. *A Culture of Improvement.* MIT Press.

p. 228 'Cotton accounted for half of all American exports by value between 1815 and 1860.' Slavery delivered cheapness through increasing quantity of output, not by undercutting prices. Indian production did not decline in the nineteenth century: it expanded, but not as fast as American. Fogel, R.W. and Engerman, S.L. 1995. *Time on the Cross: The Economics of American Negro Slavery.* Reissue edition. W.W. Norton and Company.

p. 228 'As the economist Pietra Rivoli puts it'. Rivoli, P. 2005. *The Travels of a T-shirt in the Global Economy.* John Wiley.

p. 229 'There was never going to be enough wind, water or wood in England to power the factories, let alone in the right place.' Rolt, L.T.C. 1965. *Tools for the Job.* Batsford Press. Incidentally, coked coal was used to make iron (by Abraham Darby at Coalbrookdale in Shropshire) as early as 1709, but only inferior cast iron.

p. 230 'the country's demographic and economic centre of gravity shifted south to the Yangtze valley'. Pomeranz, K. 2000. *The Great Divergence.* Princeton University Press.

p. 230 'Coal's cost per tonne at the pithead in Newcastle rose slightly between the 1740s and 1860s'. Clark, G. and Jacks, D. 2006. *Coal and the Industrial Revolution, 1700–1869.* Working Paper #06-15, Department of Economics, University of California, Davis.

p. 231 'The wages of a coal hewer in the North-east of England were twice as high, and rising twice as fast, as those of a farm worker in the nineteenth century.' Clark, G. and Jacks, D. 2006. *Coal and the Industrial Revolution, 1700–1869.* Working Paper #06-15, Department of Economics, University of California, Davis. As one young English woman (my ancestor), the daughter of a judge, wrote to her mother after moving north from Bedfordshire to Northumberland in 1841: 'The more I see of the poor people about here the more I feel puzzled as to the possibility of doing them any good ... They all have immense wages and plenty of coal and are quite rich in comparison with our Millbrook people.' From Ridley, U. 1958/1990. *The Life and Letters of Cecilia Ridley 1819–1845.* Spredden Press.

p. 231 'As the historian Tony Wrigley has put it'. Wrigley, E.A. 1988. *Continuity, Chance and Change: the Character of the Industrial Revolution in England.* Cambridge University Press.

p. 232 'an Indian weaver could not compete with the operator of a steam-driven Manchester mule'. Clark, G. 2007. *A Farewell to Alms.* Princeton University Press.

p. 233 'Today most coal is used for generating electricity.' Fouquet, R. and Pearson, P.J.G. 1998. A thousand years of energy use in the United Kingdom. *Energy Journal* 19:1–41.

p. 234 'pulling ploughs by cable through a field at the Menier estate near Paris'. Rolt, L.T.C. 1967. *The Mechanicals.* Heinemann.

p. 234 'like the computer it took decades to show up in the productivity statistics'. David, P.A. 1990. The dynamo and the computer: an historical perspective on the modern productivity paradox. *American Economic Review* 80:355–61.

p. 234 'One recent study in the Philippines'. Barnes, D.F. (ed.). 2007. *The Challenge of Rural Electrification.* Resources for the Future Press.

p. 235 'Joule for joule, wood is less convenient than coal, which is less convenient than natural gas, which is less convenient than electricity, which is less convenient than the electricity currently trickling through my mobile telephone.' Huber, P.W. and Mills, M.P. 2005. *The Bottomless Well: the Twilight of Fuel, the Virtue of Waste, and Why We Will Never Run Out of Energy.* Basic Books.

p. 236 'in Adam Smith's words'. *The Wealth of Nations.*

p. 236 'the average person on the planet consumes power at the rate of about 2,500 watts'. A watt is a joule per second. A calorie is 4.184 joules. The figures of energy consumption in watts per capita come from the International Energy Agency. See http://en.wikipedia.org/wiki/Image:Energy_consumption_versus_GDP.png.

p. 236 'it would take 150 slaves'. By the way, twice as much energy is wasted turning grain into bicycle-cargo motion as is wasted turning oil into truck-cargo motion: or sixteen times as much if the grain goes into the cyclist via a chicken. Huber, P.W. and Mills, M.P. 2005. *The Bottomless*

*Well: the Twilight of Fuel, the Virtue of Waste, and Why We Will Never Run Out of Energy.* Basic Books.

p. 237 'an anxiety as old as fossil fuels themselves'. Jevons, W.S. 1865. *The Coal Question: An Inquiry Concerning the Progress of the Nation, and the Probable Exhaustion of our Coal-mines.* Macmillan.

p. 238 'If America were to grow all its own transport fuel as biofuel it would need 30 per cent more farmland'. Dennis Avery, cited in Bryce, R. 2008. *Gusher of Lies.* Perseus Books.

p. 239 'or hydroelectric dams with catchments one-third larger than all the continents put together'. The assumptions behind these calculations are optimistic, rather than conservative: that solar power can generate about 6 watts per square metre; wind about 1.2, hay-fed horses 0.8 (one horse needs 8 hectares of hay and pulls 700 watts, or one horse power); firewood 0.12; and hydro 0.012. America consumes 3,120 gigawatts. Spain covers 504,000 sq km; Kazakhstan 2.7m sq km; India and Pakistan 4m sq km; Russia and Canada 27m sq km; all the continents 148m sq km. All power density figures except horses are from Ausubel, J. 2007. Renewable and nuclear heresies. *International Journal of Nuclear Governance, Economy and Ecology* 1:229–43.

p. 239 'Just one wind farm at Altamont in California kills twenty-four golden eagles every year'. Bird risk behaviors and fatalities at the Altamont Pass wind resource area, by C.G. Thelander, K.S. Smallwood and L. Rugge of BioResource Consultants in Ojai, California, NREL/SR-500-33829, December 2003. To those who say far more birds are killed flying into windows – yes, but not golden eagles, which are both peculiarly rare and peculiarly vulnerable to wind turbines. When did a golden eagle last crash into your conservatory? As for the charge that an oil company would be prosecuted for causing such bird deaths, see Bryce, R. 2009. Windmills are killing our birds: one standard for oil companies, another for green energy sources. *Wall Street Journal*, 7 September 2009. http://online.wsj.com/article/SB1000142405297020370 6604574376543308399048.html?mod=googlenews_wsj.

p. 239 'Hundreds of orang-utans are killed a year because they get in the way of oil-palm biofuel plantations'. http://www.telegraph.co.uk/earth/main.jhtml?xml=/earth/2007/08/14/eaorang114.xml.

p. 239 'says the energy expert Jesse Ausubel'. Ausubel, J. 2007. Renewable and nuclear heresies. *International Journal of Nuclear Governance, Economy and Ecology* 1:229–43.

p. 241 'Between 2004 and 2007 the world maize harvest increased by fifty-one million tonnes'. Avery, D.T. 2008. *The Massive Food and Land Costs of US Corn Ethanol: an Update.* Competitive Enterprise Institute no. 144, 29 October 2008.

p. 241 'American car drivers were taking carbohydrates out of the mouths of the poor to fill their tanks'. Mitchell, D.A. 2008. *Note on Rising Food Prices.* World Bank Policy Research Working Paper no. 4682. Available at SSRN: http://ssrn.com/abstract=1233058.

p. 241 'So the question is: how much fuel does it take to grow fuel? Answer: about the same amount.' Bryce, R. 2008. *Gusher of Lies*. Perseus Books.

p. 242 'says Joseph Fargione of the Nature Conservancy'. Fargione, J. et al. 2008. Land clearing the biofuel carbon debt. *Science* 319:1235–8.

p. 242 'the biofuel industry is not just bad for the economy. It is bad for the planet, too.' Bryce, R. 2008. *Gusher of Lies*. Perseus Books.

p. 243 'to quote the ecologist E.O. Wilson'. Wilson. E.O. 1999. *The Diversity of Life*. Penguin.

p. 244 'as Peter Huber and Mark Mills put it'. Huber, P.W. and Mills, M.P. 2005. *The Bottomless Well: the Twilight of Fuel, the Virtue of Waste, and Why We Will Never Run Out of Energy*. Basic Books.

p. 244 'A modern combined-cycle'. A combined-cycle turbine uses burning gas itself to drive one turbine and then uses the heat to generate steam to drive another.

p. 245 'the Victorian economist Stanley Jevons'. Jevons, S. 1865. *The Coal Question: An Inquiry Concerning the Progress of the Nation, and the Probable Exhaustion of our Coal-mines*. Macmillan, p. 103.

p. 246 'Thomas Edison deserves the last word'. Edison in 1910, quoted in Collins, T. and Gitelman, L, *Thomas Edison and Modern America*. New York: Bedford/St Martin's, 2002, p. 60. Source: Bradley, R.J. 2004. *Energy: the Master Resource*. Kendall/Hunt.

## Chapter 8

p. 247 'He who receives an idea from me'. Thomas Jefferson letter to Isaac McPherson, 13 August 1813. http://www.let.rug.nl/usa/P/tj3/writings/brf/jefl220.htm.

p. 247 World product graph. Maddison, A. 2006. *The World Economy*. OECD Publishing.

p. 249 'said Ricardo'. Ricardo, D. 1817. *The Principles of Political Economy and Taxation*.

p. 249 'neo-classical economics gloomily forecast the end of growth'. Beinhocker, E. 2006. *The Origin of Wealth*. Random House.

p. 249 'As the economist Eamonn Butler puts it'. Butler, E. 2008. *The Best Book on the Market*. Capstone.

p. 250 'the failure of any particular market to match the perfect market no more constitutes "market failure"'. This point is made in Booth, P. 2008. Market failure: a failed paradigm. *Economic Affairs* 28:72–4.

p. 250 'the science of ecology has an enduring fallacy that in the natural world there is some perfect state of balance to which an ecosystem will return'. Kricher, J. 2009. *The Balance of Nature: Ecology's Enduring Myth*. Princeton University Press. 'As a result of research over the past several decades, ecologists have come to understand the reality of ecosystem dynamics, and have largely abandoned the notion that nature exists in some sort of meaningful natural balance.'

p. 251 'No country remains for long the leader in knowledge creation.' Indeed, so iron is the rule of ephemeral innovation that it has been given its own named law: Cardwell's Law. See Mokyr, J. 2003. *The Gifts of Athena*. Princeton University Press. That said, William Easterly has pointed out that since 1000 BC certain areas of the world have consistently stood at the forefront of technology and growth: Comin, D., Easterly, W. and Gong, E. 2006. *Was the Wealth of Nations Determined in 1000 BC?* NBER Working Paper no. 12657.

p. 252 'As Joel Mokyr puts it'. Mokyr, J. 2003. *The Gifts of Athena*. Princeton.

p. 253 'George Orwell was tired of the way the world appeared to be shrinking'. Orwell, G. 1944. *Tribune*, 12 May 1944.

p. 254 'when the credit card took off'. Nocera, J. 1994. *A Piece of the Action*. Simon and Schuster. (That said, there is little doubt that finance is one area of human activity in which too much innovation can be a bad thing. As Adair Turner has put it, whereas the loss of the knowledge of how to make a vaccine would harm human welfare, 'if the instructions for creating a CDO squared [a collateral debt obligation of collateral debt obligations] had somehow been mislaid, we will I think get along quite well without it.') See Turner, A. 2009. 'The Financial Crisis and the Future of Financial Regulation'. Inaugural Economist City Lecture, 21 January 2009. Financial Services Authority.

p. 254 'Lewis Mandell discovered'. Quoted in Nocera, J. 1994. *A Piece of the Action*. Simon and Schuster.

p. 254 'Michael Crichton once told me'. M. Crichton, email to the author, June 2007.

p. 254 'said William Petty in 1679'. Quoted in Mokyr, J. 2003. *The Gifts of Athena*. Princeton University Press.

p. 255 'in Alfred North Whitehead's words'. Whitehead, A.N. 1930. *Science and the Modern World*. Cambridge University Press.

p. 255 'As the scientist Terence Kealey has observed'. Kealey, T. 2007. *Sex, Science and Profits*. William Heinemann.

p. 256 'the biggest advances in the steam engine'. Kealey, T. 2008. *Sex, Science and Profits*. William Heinemann. Kealey argues that Watt vehemently denied any influence from Joseph Black. Joel Mokyr (in *The Gifts of Athena*) quotes Watt to the contrary.

p. 256 'efforts by eighteenth-century scientists to prove that Newcomen got his insights from Papin's theories have proved to be wholly without foundation'. Rolt, L.T.C. 1963. *Thomas Newcomen: the Prehistory of the Steam Engine*. David and Charles. Likewise, the establishment was so incredulous that the humble mine engineer George Stephenson could have invented a miner's safety lamp in 1815 without understanding the principle behind it, that they effectively accused him of stealing the idea from the scientist Sir Humphry Davy. The reverse accusation is more plausible: that Davy heard of Stephenson's experiments from the engineer John Buddle, who heard of them from the colliery doctor named Burnet, who had been told by Stephenson. See Rolt, L.T.C. 1960. *George and Robert Stephenson*. Longman.

p. 256 'the famous Lunar Society'. For more on the Lunar Society see Uglow, J. 2002. *The Lunar Men*. Faber and Faber.

p. 257 'a semi-directed, groping, bumbling process of trial and error by clever, dexterous professionals with a vague but gradually clearer notion of the processes at work'. Mokyr, J. 2003. *The Gifts of Athena*. Princeton.

p. 257 'It is a stretch to call most of this science'. Joel Mokyr has recently suggested (Mokyr, J. 2003. *The Gifts of Athena*. Princeton) that although the scientific revolution did not start the industrial, none the less the broadening of the epistemic base of knowledge – the sharing and generalisation of understanding – allowed a host of new applications of knowledge, which escaped diminishing returns and enabled the industrial revolution to continue indefinitely. I am not convinced. I think the prosperity generated by industry paid for an expansion of knowledge, which sporadically returned the favour. Even when, by the late nineteenth and early twentieth centuries, science appeared to make mighty contributions to new industries, the philosophers still played second fiddle to the engineers. Lord Kelvin's contributions to the physics of resistance and induction were driven more by practical problem-solving in the telegraph industry than esoteric rumination. And though it is true that the physics of James Clerk Maxwell produced an electrical revolution, the chemistry of Fritz Haber spawned an agricultural revolution, Leo Szilard's idea of a chain reaction of neutrons led to nuclear weapons and the biology of Francis Crick fathered biotechnology, it is none the less also true that these sages needed legions of engineers to turn their insights into things that could change living standards. Tinkering Thomas Edison, with his team of forty engineers, was more important to electrification than thinking Maxwell; practical Carl Bosch mattered more than esoteric Haber; administrative Leslie Groves than dreamy Szilard; practical Fred Sanger than theoretical Crick.

p. 258 'One of Britain's advantages in the eighteenth century'. Hicks, J.R. 1969. *A Theory of Economic History*. Clarendon Press.

p. 259 'By contrast in France'. Ferguson, N. 2008. *The Ascent of Money*. Allen Lane.

p. 259 'fully one-third of successful start-ups in California between 1980 and 2000 had Indian- or Chinese-born founders'. Baumol, W.J., Litan, R.E. and Schramm, C.J. 2007. *Good Capitalism, Bad Capitalism*. Yale University Press.

p. 259 'A telling anecdote about glass repeated by several Roman authors'. Moses Finley, cited in Baumol, W. 2002. *The Free-market Innovation Machine*. Princeton University Press.

p. 260 'A Christian missionary in Ming China wrote'. Quoted in Rivoli, P. 2005. *The Travels of a T-shirt in the Global Economy*. John Wiley.

p. 260 'The proportion of GDP spent by firms on research and development in America has more than doubled'. Kealey, T. 2007. *Sex, Science and Profits*. William Heinemann.

p. 261 'The pioneer venture capitalist Georges Doriot said'. Quoted in Evans, H. 2004. *They Made America*. Little, Brown.

p. 262 'as Don Tapscott and Anthony Williams call it'. Tapscott, D. and Williams, A. 2007. *Wikinomics*. Atlantic.

p. 263 'The dye industry relied mostly on secrecy till the 1860s'. See Moser, P. 2009. Why don't inventors patent? http://ssrn.com/abstracts= 930241.

p. 264 'Emmanuelle Fauchart discovered by interviewing ten *chefs de cuisine*'. Fauchart, E. and Hippel, E. von. 2006. *Norm-based Intellectual Property Systems: the Case of French Chefs*. MIT Sloan School of Management working paper 4576-06. http://web.mit.edu/evhippel/www /papers/vonhippelfauchart2006.pdf.

p. 264 'Yet there is little evidence that patents are really what drive inventors to invent.' There is a lively debate going on about whether James Watt's aggressive enforcement of his broadly worded patents on steam engines in 1769 and 1775 actually shut down innovation in the steam industry. See Rolt, L.T.C. 1960. *George and Robert Stephenson*. Longman. ('With coal so readily available, the north country colliery owners preferred to forgo the superior economy of the Watt engine rather than pay the dues demanded by Messrs. Boulton and Watt.'); also www.thefreemanonline. org/featured/do-patents-encourage-or-hinder-innovation-the-case-of-the-steam-engine/; Boldrin, M. and Levine, D.K. 2009. Against intellectual monopoly. Available online: http://www.micheleboldrin. com/research/aim.html; and Von Hippel, E. 2005. *Democratizing Innovation*. MIT Press. The contrary view, that Watt's patent did little to hinder innovation and that without it he would never have attracted Boulton's backing, is put by George Selgin and John Turner: Selgin, G. and Turner, J.L. 2006. James Watt as intellectual monopolist: comment on Boldrin and Levine. *International Economic Review* 47:1341–8; and Selgin, G. and Turner, J.L. 2009. Watt, again? Boldrin and Levine still exaggerate the adverse effect of patents on the progress of steam power. 18 August 2009, prepared for the Center for Law, Innovation and Economic Growth conference, Washington University School of Law, April 2009.

p. 264 'the list of significant twentieth-century inventions that were never patented is a long one'. Cited in Shermer, M. 2007. *The Mind of the Market*. Times Books.

p. 264 'the Wright brothers effectively grounded the nascent aircraft industry'. Heller, M. 2008. *The Gridlock Economy*. Basic Books.

p. 264 'a logjam in the manufacture of radios caused by the blocking patents held by four firms'. Benkler, Y. 2006. *The Wealth of Networks*. Yale University Press.

p. 265 'the biggest generators of new patents in the US system are "patent trolls" – firms that buy up weak patent applications'. I am indebted to R. Litan for this information.

p. 265 'Research in Motion, the Canadian company that manufactures BlackBerries'. Baumol, W.J., Litan, R.E. and Schramm, C.J. 2007. *Good Capitalism, Bad Capitalism.* Yale University Press.

p. 265 'Michael Heller's analogy for the patent trolls is to the state of the river Rhine between the decay of Holy Roman imperial power and the emergence of modern states'. Heller, M. 2008. *The Gridlock Economy.* Basic Books.

p. 266 'In one survey of 650 R&D executives from 130 different industries'. Von Hippel, E. 2005. *Democratizing Innovation.* MIT Press.

p. 266 'most of the money goes towards me-too drugs for diseases of Westerners'. Boldrin, M. and Levine, D.K. 2009. Against intellectual monopoly. Available online: http://www.micheleboldrin.com/research/aim.html.

p. 267 'only one country had allowed the copyrighting of music'. Boldrin, M. and Levine, D.K. 2009. Against intellectual monopoly. Available online: http://www.micheleboldrin.com/research/aim.html.

p. 267 'Just as newspapers have derived little of their income from licensing copyrights'. Benkler, Y. 2006. *The Wealth of Networks: How Social Production Transforms Markets and Freedom.* Yale University Press. (Benkler's book, true to his argument, is available free online.)

p. 268 'chronically entrepreneurial'. Audretsch, D.B. 2007. *The Entrepreneurial Society.* Oxford University Press.

p. 268 'laying the foundations for their global dominance at the expense of precisely the big companies dirigistes admired'. Postrel, V. 1998. *The Future and Its Enemies.* Free Press.

p. 269 'A large study by the OECD'. Quoted in Kealey, T. 2007. *Sex, Science and Profits.* William Heinemann.

p. 270 'a recent survey of forty-six major inventions'. Agarwal, R. and Gort, M. 2001. First mover advantage and the speed of competitive entry: 1887–1986. *Journal of Law and Economics* 44:161–78.

p. 270 'sired by the bicycle out of the horse carriage'. Rolt, L.T.C. 1967. *The Mechanicals.* Heinemann.

p. 271 'cross-fertilisation … does … happen between species of bacteria, 80 per cent of whose genes have been borrowed from other species'. Dagan, T., Artzy-Randrup, Y. and Martin, W. 2008. Modular networks and cumulative impact of lateral transfer in prokaryote genome evolution. *PNAS* 105:10039–44: 'At least 81 +- 15% of the genes in each genome studied were involved in lateral gene transfer at some point in their history.'

p. 271 'able to produce only sterile offspring'. The sterility of hybrids was a problem that greatly exercised Charles Darwin, chiefly because it was being claimed by some American anthropologists that black people were a separately created species, which justified slavery, and even that hybrids between blacks and whites were sterile. See Desmond, A. and Moore, J. 2009. *Darwin's Sacred Cause*. Penguin.

p. 271 'Technologies emerge from the coming together of existing technologies into wholes that are greater than the sum of their parts.' Arthur, B. and Polak, W. 2004. *The Evolution of Technology within a Simple Computer Model*. Santa Fe working paper 2004-12-042.

p. 271 'Henry Ford once candidly admitted'. Evans, H. 2004. *They Made America*. Little, Brown.

p. 272 'in the historian George Basalla's words'. Basalla, G. 1988. *The Evolution of Technology*. Cambridge University Press.

p. 272 'an invention looking for a job'. http://laserstars.org/history/ruby.html.

p. 273 'Eric von Hippel, incidentally, practises what he preaches'. Von Hippel, E. 2005. *Democratizing Innovation*. MIT Press.

p. 274 'as Geoffrey Miller reminds us'. Miller, G. 2009. *Spent*. Heinemann.

p. 275 'full-scale trebuchets capable of tossing pianos more than 150 yards'. *Wall Street Journal*, 15 January 1992.

p. 276 'It was Paul Romer's great achievement in the 1990s to rescue the discipline of economics from the century-long cul-de-sac into which it had driven by failing to incorporate innovation.' Warsh, D. 2006. *Knowledge and the Wealth of Nations*. W.W. Norton.

p. 276 'As Paul Romer puts it'. Romer, P. 1995. *Beyond the Knowledge Worker*. Wordlink.

## Chapter 9

p. 279 'I have observed that not the man who hopes when others despair'. Speech by John Stuart Mill to the London Debating Society on 'perfectibility', 2 May 1828.

p. 279 US air pollutant emissions graph. US Environmental Protection Agency.

p. 280 'the economist Julian Simon tried it in the 1990s'. Simon, J. 1996. *The Ultimate Resource 2*. Princeton University Press.

p. 280 'Bjørn Lomborg tried it in the 2000s'. Lomborg, B. 2001. *The Sceptical Environmentalist*. Cambridge University Press.

p. 280 'said Hayek'. Hayek, F.A. 1960. *The Constitution of Liberty*. Routledge.

p. 280 'As Warren Meyer has put it'. http://www.coyoteblog.com/coyote_blog/2005/02/in_praise_of_ro.html.

p. 280 'The environmentalist Lester Brown, writing in 2008'. Brown, L. 2008. *Plan B 3.0: Mobilizing to Save Civilisation*. Earth Policy Institute.

p. 283 'Pessimists have always been ubiquitous and have always been feted'. Herman, A. 1997. *The Idea of Decline in Western History*. The Free Press.

p. 283 'wrote Adam Smith at the start of the industrial revolution'. Smith, A. 1776. *The Wealth of Nations*.

pp. 283–4 'cried the *Quarterly Review*'. Smiles, S. 1857. *The Life of George Stephenson, Railways Engineer*. John Murray.

p. 284 'Dr Arnold was more enlightened'. Quoted in Williams, A. 2008. *The Enemies of Progress*. Societas.

p. 284 'Robert Southey had just published a book'. Southey, R. 1829. *Sir Thomas More: Or, Colloquies on the Progress and Prospects of Society*. John Murray.

p. 285 'the modern philosopher John Gray'. Quoted in Postrel, V. 1998. *The Future and Its Enemies*. Free Press.

p. 285 'Thomas Babington Macaulay'. Macaulay, T.B. 1830. Review of Southey's Colloquies on Society. *Edinburgh Review*, January 1830.

p. 286 'in his *History of England*'. Macaulay, T.B. 1848. *History of England from the Accession of James the Second*.

p. 287 'said Macaulay in 1830'. Macaulay, T.B. 1830. Review of Southey's Colloquies on Society. *Edinburgh Review*, January 1830.

p. 288 'a book called *Degeneration*, by the German Max Nordau'. Quoted in Leadbetter, C. 2002. *Up the Down Escalator: Why the Global Pessimists Are Wrong*. Viking.

p. 288 'said Winston Churchill in a memo to the prime minister'. Asquith papers, December 1910, quoted in Addison, P. 1992. *Churchill on the Home Front 1900–1955*. Jonathan Cape.

p. 288 'Theodore Roosevelt was even more explicit'. *The Works of Theodore Roosevelt*, National Edition, XII, p. 201.

p.288 'as Isaiah Berlin put it'. Quoted in Byatt, I. 2008. Weighing the present against the future: the choice and use of discount rates in the analysis of climate change. In *Climate Change Policy: Challenging the Activists*. Institute of Economic Affairs.

p. 289 'Oswald Spengler in 1923 in his bestselling polemic *The Decline of the West*'. Spengler, O. 1923. *The Decline of the West*. George Allen & Unwin.

p. 290 'the opening words of Agenda 21'. Preamble to Agenda 21, 1992.

p. 290 'in the words of Charles Leadbetter'. Leadbetter, C. 2002. *Up the Down Escalator: Why the Global Pessimists Are Wrong*. Viking.

p. 291 'groaned the wealthy environmentalist Edward Goldsmith'. Quoted in Postrel, V. 1998. *The Future and Its Enemies*. Free Press.

p. 291 'in the words of the Prince of Wales'. HRH Prince of Wales 2000. The civilized society. *Temenos Academy Review*. See http://www.prince ofwales.gov.uk/speechesandarticles/an_article_by_hrh_the_prince_of_wales_titled_the_civilised_s_93.html000.

p. 291 'says a professor of psychology'. Barry Schwartz, quoted in Easterbrook, G. 2003. *The Progress Paradox*. Random House.

p. 291 'This notion dates from Herbert Marcuse'. Saunders, P. 2007. Why capitalism is good for the soul. *Policy Magazine* 23:3–9.

p. 291 'the poet Hesiod was nostalgic for a lost golden age'. Hesiod, *Works and Days* II.

p. 292 'Plato, who deplored writing as a destroyer of memorising'. Barron, D. 2009. *A Better Pencil*. Oxford University Press.

p. 292 'scalded and defoliated by a kind of cognitive Agent Orange'. John Cornwell. Is technology ruining our children? *The Times*, 27 April 2008.

p. 292 'The psychoanalyst Adam Phillips'. Phillips, A. and Taylor, B. 2009. *On Kindness*. Hamish Hamilton. Excerpted in *The Guardian*, 3 January 2009.

p. 293 'Bill McKibben's best-selling dirge of 1989'. McKibben, W. 1989. *The End of Nature*. Random House.

p. 293 'Robert Kaplan told the world in 1994'. www.theatlantic.com/doc/199402/anarchy.

p. 293 'our stolen future'. Colburn, T., Dumanoski, D. and Myers, J.P. 1996. *Our Stolen Future*. Dutton. See Breithaupt, H. 2004. *A Cause without a Disease*. EMBO Reports 5:16–18.

p. 293 'Jared Diamond fell under the spell of fashionable pessimism'. Diamond, J. 1995. *The Rise and Fall of the Third Chimpanzee*. Radius.

p. 294 'Martin Rees in his book'. Rees, M. 2003. *Our Final Century*. Heinemann.

p. 294 'what Greg Easterbrook calls'. Easterbrook, G. 2003. *The Progress Paradox*. Random House.

p. 294 'People … tend to assume that they will live longer, stay married longer and travel more than they do'. Gilbert, D. 2007. *Stumbling on Happiness*. Harper Press.

p. 294 'Dane Stangler calls this'. Stangler, D., personal communication.

p. 294 'people much more viscerally dislike losing a sum of money than they like winning the same sum'. McDermott, R., Fowler, J.H. and Smirnov, O. 2008. On the evolutionary origin of prospect theory preferences. *The Journal of Politics* 70:335–50.

p. 294 'pessimism genes might quite literally be commoner than optimism genes'. Fox, E., Ridgewell, A. and Ashwin, C. 2009. Looking on the bright side: biased attention and the human serotonin transporter gene. *Proceedings of the Royal Society B* (doi:10.1098/rspb.2008.1788).

p. 295 'the 7-repeat version of the DRD4 gene accounts for 20 per cent of financial risk taking in men'. Dreber, A. et al. 2009. The 7R polymorphism in the dopamine receptor D4 gene (DRD4)is associated with financial risk taking in men. *Evolution and Human Behavior* (in press).

p. 295 'The day I was writing a first draft of this paragraph, the BBC reported'. 1 May 2008.

p. 295 'how the *New York Times* reported the reassuring news in 2009 that world temperature had not risen for a decade'. *New York Times*, 23 September 2009.

p. 296 'a new theory suggests that cosmic rays are a bigger cause of the Antarctic ozone hole than chlorine is'. Lu, Q.-B. 2009. Correlation between cosmic rays and ozone depletion. *Physical Review Letters* 102:118501–9400.

p. 297 'Rachel Carson, influenced by Hueper, set out in her book *Silent Spring* (1962) to terrify her readers'. Carson, R. 1962. *Silent Spring*. Houghton Mifflin.

p. 297 'other causes of childhood death were declining faster'. Bailey, R. 2002. Silent Spring at 40. *Reason*, June 2002. http://www.reason.com/news/show/34823.html.

p. 298 'wrote the environmentalist Paul Ehrlich in 1971'. Ehrlich, P. 1970. *The Population Bomb.* 2nd edition. Buccaneer Press.

p. 298 'Later he was more specific'. Special Earthday edition of *Ramparts* magazine, 1970.

p. 298 'both cancer incidence and death rate from cancer fell steadily'. Ames, B.N. and Gold, L.S. 1997. Environmental pollution, pesticides and the prevention of cancer: misconceptions. *FASEB Journal* 11:1041–52.

p. 298 'Richard Doll and Richard Peto had concluded that age-adjusted cancer rates were falling'. Doll, R. and Peto, R. 1981. The causes of cancer: quantitative estimates of avoidable risks of cancer in the United States today. *Journal of the National Cancer Institute* 66:1193–1308.

p. 298 'As Bruce Ames famously demonstrated in the late 1990s'. Ames, B.N. and Gold, L.S. 1997. Environmental pollution, pesticides, and the prevention of cancer: misconceptions. *FASEB Journal* 11:1041–52.

p. 299 'Ames says'. Bruce Ames, personal communication.

p. 299 'sparing, targeted use of DDT against malarial mosquitoes can be done without any such threat to wildlife'. http://www.nationalreview.com/comment/bate200406030904.asp; http://www.prospect-magazine.co.uk/article_details.php?id=10176.

p. 300 'says Greg Easterbrook'. Easterbrook, G. 2003. *The Progress Paradox.* Random House.

p. 300 'Lester Brown predicted'. Various sources for these Brown quotes, including Smil, V. 2000. *Feeding the World.* MIT Press, and Bailey, R. 2009. Never right, but never in doubt: famine-monger Lester Brown still gets it wrong after all these years. *Reason* magazine, 12 May 2009: http://reason.com/archives/2009/05/05/never-right-but-never-in-doubt. See also Brown, L. 2008. *Plan B 3.0: Mobilizing to Save Civilization.* Earth Policy Institute.

p. 301 'by William and Paul Paddock'. Paddock, W. and Paddock, P. 1967. *Famine, 1975! America's Decision: Who Will Survive?* Little, Brown.

p. 301 'William Paddock was calling for a moratorium'. Paddock, William C. Address to the American Phytopathological Society, Houston, Texas 12 August 1975.

p. 301 '*The Population Bomb*'. Ehrlich, P. 1971. *The Population Bomb.* 2nd edition. Buccaneer.

p. 301 '*The Dominant Animal*'. Ehrlich, P. and Ehrlich, A. 2008. *The Dominant Animal.* Island Press.

p. 302 'wrote the economist Joseph Schumpeter in 1943'. Schumpeter, J.A. 1943. *Capitalism, Socialism, and Democracy.* Allen & Unwin.

p. 302 '*Limits to Growth*'. It should be noted that the authors of *Limits to Growth* have argued since that they only wished to illustrate what might happen if exponential use continued and no new reserves of these minerals were discovered, which they realised was unlikely. But this is a highly generous reading of both their mathematics and their prose. 'There will be a desperate arable land shortage before the year 2000' and 'The world population will be 7 billion in 2000' sound like predictions to me. Even in more recent updates, the main prognosis remains that civilisation will – or should – collapse for lack of resources in the current century: 'Humanity must draw back, ease down, and heal if it wants to continue to live.' See Meadows, D.H., Meadows, D.L. and Randers, J. 1992. *Beyond the Limits*. Chelsea Green Publishing; and Meadows, D.H., Randers, J. and Meadows, D. 2004. *Limits to Growth: The 30-Year Update*. Chelsea Green Publishing.

p. 303 'school textbooks soon parroting its predictions minus the caveats'. See Bailey, R. 2004. science and public policy. *Reason*: http://www.reason.com/news/show/34758.html.

p. 303 'In 1990 the economist Julian Simon won $576.07 in settlement of a wager'. Simon, J. 1996. *The Ultimate Resource 2*. Princeton University Press.

p. 304 '*Life* magazine promised its readers'. Quoted in http://www.ihatethemedia.com/earth-day-predictions-of-1970-the-reason-you-should-not-believe-earth-day-predictions-of-2009.

p. 304 'Professor Bernd Ulrich said it was already too late for Germany's forests'. Mauch, C. 2004. *Nature in German History*. Berghahn Books.

p. 305 'The *New York Times* declared "a scientific consensus"'. Easterbrook, G. 1995. *A Moment on the Earth*. Penguin. See also *Fortune* magazine, April 1986.

p. 305 'When asked if he had been pressured to be optimistic, one of the authors said the reverse was true'. Mathiesen, M. 2004. *Global Warming in a Politically Correct Climate*. Universe Star.

p. 306 'The activist Jeremy Rifkin said'. Miller, H.I. 2009. The human cost of anti-science activism. *Policy Review*, April/May 2009. http://www.hoover.org/publications/policyreview/41839562.html.

p. 307 'Ebola outbreaks'. Colebunders, R. 2000. Ebola haemorrhagic fever – a review. *Journal of Infection* 40:16–20.

pp. 307–8 'The proportion of the population infected with HIV is falling'. http://data.unaids.org/pub/GlobalReport/2008/JC1511_GR08_Executive Summary_en.pdf.

p. 308 'Hugh Pennington'. http://news.bbc.co.uk/1/hi/sci/tech/573919.stm.

p. 308 'the number of deaths has reached 166'. See http://www.cjd.ed.ac.uk/figures.htm

p. 308 'no extra birth defects at all'. Little, J. 1993. The Chernobyl accident, congenital anomalies and other reproductive outcomes. *Paediatric Perinatal Epidemiology* 7:121–51. The World Health Organisation concluded in 2006 that: 'A modest but steady increase in reported congenital malformations in both contaminated and uncontaminated areas of Belarus appears related to better reporting, not radiation.' See

http://www.iaea.org/NewsCenter/Focus/Chernobyl/pdfs/pr.pdf.

p. 308 'The evacuation of the area has caused wildlife to flourish there to an extraordinary degree'. Brand, S. 2009. *Whole Earth Discipline*. Penguin.

p. 309 'As one commentator concluded'. Fumento, M. 2006. The Chicken Littles were wrong: bird flu threat flew the coop. *The Standard*, 25 December 2006.

pp. 309–10 'you are far more likely to get the flu from a person who is well enough to go to work than one who is ill enough to stay at home'. Wendy Orent. Swine flu poses a risk, but no reason to panic. *Los Angeles Times*, 29 April 2009. http://articles.latimes.com/2009/apr/29/opinion/oe-orent29.

p. 311 'in the words of President Obama's science adviser John Holdren'. Holdren, J., Ehrlich, A. and Ehrlich, P. 1973. *Human Ecology: Problems and Solutions*. W.H. Freeman and Company, p.279.

p. 311 'in the words of Maurice Strong, first executive director of the United Nations Environment Programme (UNEP)'. http://www.spiked-online.com/index.php/site/article/7314.

p. 311 'in the words of the journalist George Monbiot'. *The Guardian*, 18 August 2009.

p. 311 'When they speak of retreat'. See http://www.climate-resistance.org/2009/08/folie-a-deux.html.

**Chapter 10**

p. 313 'It is possible to believe that all the past is but the beginning of a beginning'. Wells, H.G. 'The Discovery of the Future' Lecture at the Royal Institution, 24 January 1902, published in *Nature* 65:326–31. Reproduced with the permission of AP Watt Ltd on behalf of the Literary Executors of the Estate of H.G. Wells.

p. 313 Greenland ice cap temperature graph. NCDC. See ncdc.noaa.gov.

p. 314 'says the environmentalist Jonathan Porritt'. Ecologist Online April 2007. See www.optimumpopulation.org/ecologist.j.porritt.April07.doc.

p. 315 'Paul Collier's phrase'. Collier, P. 2007. *The Bottom Billion*. Oxford University Press.

p. 316 'life expectancy is rising rapidly'. As of this writing, life expectancy is still falling in South Africa, Mozambique, and of course Zimbabwe.

pp. 316–7 'Paul Collier and his colleagues at the World Bank encountered a storm of protest'. Collier, P. 2007. *The Bottom Billion*. Oxford University Press.

p. 317 'would by now have given Zambians the income per head of the Portuguese'. Moyo, D. 2009. *Dead Aid*. Allen Lane.

p. 317 'these conclusions were later dashed by Raghuram Rajan and Arvind Subramanian of the International Monetary Fund'. Rajan, R.G. and Subramanian, A. 2005. *Aid and Growth: What Does the Cross-Country Evidence Really Show?* NBER Working Papers 11513, National Bureau of Economic Research.

p. 318 'the recommendations of the Zambian economist Dambisa Moyo'. Moyo, D. 2009. *Dead Aid*. Allen Lane.

p. 318 'As William Easterly puts it'. Easterly, W. 2006. *The White Man's Burden: Why the West's Efforts to Aid the Rest Have Done So Much Ill and So Little Good*. Oxford University Press.

p. 318 'the example of insecticide-treated mosquito bed nets'. Easterly, W. 2006. *The White Man's Burden: Why the West's Efforts to Aid the Rest Have Done So Much Ill and So Little Good*. Oxford University Press.

p. 321 'consistently the most successful economy in the world in recent decades'. Acemoglu, D., Johnson, S.H. and Robinson, J.A. 2001. *An African Success Story: Botswana*. MIT Department of Economics Working Paper no. 01-37.

p. 323 'developers leave the poor to build their own slums'. Boudreaux, K. 2008. Urbanisation and informality in Africa's housing markets. *Economic Affairs*, June 2008: 17–24.

p. 323 'a Cairo house owner will build up to three illegal storeys on top of his house'. De Soto, H. 2000. *The Mystery of Capital*. Bantam Press.

p. 323 'the end of a long, exhausting and bitter struggle between elitist law and a new order brought about by massive migration and the needs of an open and sustainable society'. De Soto, H. 2000. *The Mystery of Capital*. Bantam Press.

p. 324 'Bart Wilson and his colleagues set up a land of three virtual villages inhabited by real undergraduates'. Kimbrough, E.O., Smith, V.L. and Wilson, B.J. 2008. Historical property rights, sociality, and the emergence of impersonal exchange in long-distance trade. *American Economic Review* 98:1009–39.

p. 324 'well crafted property rights are also the key to wildlife and nature conservation'. Anderson, T. and Huggins, L. 2008. *Greener Than Thou*. Hoover Institution Press.

p. 324 'fish off Iceland'. Costello, C., Gaines, S.D. and Lynham, J. 2008. Can catch shares prevent fisheries collapse? *Science* 321:1678–80. (doi: 0.1126/ science.1159478).

p. 325 'De Soto's assistants found that to do the same in Tanzania would take 379 days'. Institute of Liberty and Democracy. 2005. Tanzania: the diagnosis. http://www.ild.org.pe/en/wnatwedo/diagnosis/tanzania.

p. 326 'Bamako in Mali could build upon its strong musical traditions'. Schulz, M. and van Gelder, A. 2008. *Nashville in Africa: Culture, Institutions, Entrepreneurship and Development*. Trade, Technology and Development discussion paper no. 2, International Policy Network.

p. 326 'Micro-finance banking, mobile telephony and the internet are now merging'. Talbot, D. 2008. Upwardly mobile. *Technology Review*, November/December 2008: 48–54.

p. 326 'opportunities to the poor of Africa that were not available to the poor of Asia a generation ago'. Rodrik, D. (ed.). 2003. *In Search of Prosperity*. Princeton University Press.

p. 327 'a study of the sardine fishermen of Kerala in southern India'. Jensen, Robert T. 2007. The digital provide: information (technology), market performance and welfare in the South Indian fisheries sector. *Quarterly Journal of Economics* 122: 879–924.

p. 328 'demographic dividend'. Bloom, D.E. et al. 2007. *Realising the Demographic Dividend: Is Africa Any Different?* PGDA Working Paper no. 23, Harvard University.

p. 328 'charter city in Africa'. www.chartercities.com.

p. 329 'The weather is always capricious'. *Newsweek*, 22 January 1996. On the web at http://www.newsweek.com/id/101296/page/1.

p. 329 'Meteorologists disagree about the cause and extent'. *Newsweek*, 28 April 1975. On the web at http://www.denisdutton.com/cooling_world.htm.

p. 329 'that the last three decades of relatively slow average temperature changes are more compatible with a low-sensitivity than a high-sensitivity model of greenhouse warming'. Lindzen, R.S. and Choi, Y.S. 2009. On the determination of climate feedbacks from ERBE data. *Geophysical Research Letters*. In press. Schwartz, S.E., R.J. Charlson, and H. Rhode, 2007: Quantifying climate change – too rosy a picture? Nature Reports Climate Change 2:23-24, and Schwartz S. E. 2008. Reply to comments by G. Foster et al., R. Knutti et al., and N. Scafetta on Heat capacity, time constant, and sensitivity of Earth's climate system. J. Geophys. Res. 113, D15105. (doi:10.1029/2008JD009872).

p. 329 'that clouds may slow the warming as much as water vapour may amplify it'. Paltridge, G., Arking, and Pook, M. 2009. Trends in middle- and upper-level tropospheric humidity from NCEP reanalysis data. *Theoretical and Applied Climatology*. (doi: 10.1007/ s00704-009-0117-x).

p. 329 'that the increase in methane has been (erratically) decelerating for twenty years'. M.A.K. Khalil, C.L. Butenhoff and R.A. Rasmussen, Atmospheric methane: trends and cycles of sources and sinks, *Environmental Science & Technology* 41:2131–7.

p. 329 'that there were warmer periods in earth's history in medieval times and about 6,000 years ago yet no accelerations or 'tipping points' were reached'. Loehle, C. 2007. A 2000-year global temperature reconstruction based on non-treering proxies. *Energy & Environment* 18: 1049-58; and Moberg, A., D.M. Sonechkin, K. Holmgren, N.M. Datsenko, and W. Karlén, 2005. Highly variable Northern Hemisphere temperatures reconstructed from low- and high-resolution proxy data. *Nature* 433:613-7.

p. 330 'the Intergovernmental Panel on Climate Change (IPCC)'. The full IPCC reports are available at www.ipcc.ch.

p. 331 'the Dutch economist Richard Tol'. www.ff.org/centers/csspp/pdf/ 20061031_tol.pdf.

p. 331 'With a higher discount rate, Stern's argument collapses'. See Weitzman, M. 2007. Review of the Stern Review on the economics of climate change. *Journal of Economic Literature* 45 (3): 'The present discounted value of a given global-warming loss from a century hence at the non-Stern annual interest rate of 6 per cent is one-hundredth of the value of the same loss at Stern's centuries-long discount rate of 1.4 per cent.'

p. 331 'Nigel Lawson asks, reasonably enough'. Lawson, N. 2008. *An Appeal to Reason.* Duckworth.

p. 331 'all six of the IPCC's scenarios assume that the world will experience so much economic growth that the people alive in 2100 will be on average 4–18 times as wealthy as we are today'. http://www.ipcc.ch/ipcc reports/sres/emission/014.htm.

p. 332 'In the hottest scenario, income rises from $1,000 per head in poor countries today to more than $66,000 in 2100 (adjusted for inflation)'. Goklany, I. 2009. Is climate change 'the defining challenge of our age'? *Energy and Environment* 20: 279–302.

p. 332 'Note that this is true even if climate change itself cuts wealth by Stern's 20 per cent by 2100: that would mean the world becoming 'only' 2–10 times as rich.' See http://sciencepolicy.colorado.edu/prometheus/archives/climate_change/001165a_ comment_on_ipcc_wo.html.

p. 332 'the Prince of Wales said in 2009'. http://www.spectator.co.uk/politics/all/5186108/the-spectators-notes.thtml.

p. 332 'All the futures use market exchange rates instead of purchasing power parities for GDP, further exaggerating warming.' Castles, I. and Henderson, D. 2003. Economics, emissions scenarios and the work of the IPCC. *Energy and Environment* 14:422–3. See also Maddison. A. 2007. *Contours of the World Economy.* Oxford University Press.

pp. 332–3 'The trouble with this reasoning is that it applies to all risks, not just climate change.' http://cowles.econ.yale.edu/P/cd/d16b/d1686.pdf; and http://www.economics.harvard.edu/faculty/weitzman/files/ReactionsCritique.pdf.

p. 334 'some countries will continue to gain more land from siltation than they lose to erosion'. Despite this, the journalist George Monbiot incites murder: 'Every time someone dies as a result of floods in Bangladesh, an airline executive should be dragged out of his office and drowned.' (*Guardian*, 5 December 2006); and James Hansen demands trials for crimes against humanity for having an outlying view: 'James Hansen, one of the world's leading climate scientists, will today call for the chief executives of large fossil fuel companies to be put on trial for high crimes against humanity and nature, accusing them of actively spreading doubt about global warming' (*Guardian*, 23 June 2008).

p. 334 'even the highest estimates of Greenland's melting'. Luthke, S.B. et al. 2006. Recent Greenland ice mass loss from drainage system from satellite gravity observations. *Science* 314:1286–9. If anything the rate of melting in Greenland is slowing: van de Wal, R.S.W., et al. 2008. Large and rapid melt-induced velocity changes in the ablation zone of the Greenland ice sheet. *Science* 321:111.

p. 334 'warming will itself reduce the total population at risk for water shortage'. Arnell, N.W., 2004. Climate change and global water resources: SRES emissions and socio-economic scenarios. *Global Environmental Change* 14: 31–52. Commenting on how the IPCC's summary for policymakers misreported this paper by omitting all mention of the positive effects caused by more rain falling on populated areas, Indur Goklany writes: 'To summarize, with respect to water resources, Figure SPM.2 – and its clones – don't make any false statements, but by withholding information that might place climate change in a positive light, they have perpetrated a fraud on the readers.' See http://wattsupwiththat.com/2008/09/18/how-the-ipcc-portrayed-a-net-positive-impact-of-climate-change-as-a-negative/#more-3138.

p. 334 'previous warm episodes'. The famous 'hockey stick' graph that seemed to prove that the Medieval Warm Period never happened has since been comprehensively discredited. It relied far too heavily on two sets of samples from bristlecone pine trees and Siberian larch trees that have since been shown to be highly unreliable; it spliced together proxies and real thermometer data in a selective way, obscuring the fact that the proxies did not mirror modern temperatures, and it used statistical techniques that made a hockey stick out of red noise. Subsequent non-tree-ring proxies have emphatically reinstated the Medieval Warm Period as warmer than today. See Holland, D. 2007. Bias and concealment in the IPCC process: the 'hockey-stick' affair and its implications. *Energy and Environment* 18:951–83; http://republicans. energycommerce.house.gov/108/home/07142006_Wegman_Report.pdf; www.climateaudit.org/?p=4866#more-4866; http://wattsupwiththat. com/2009/03/18/steve-mcintyres-iccc09-presentation-with-notes/#more-6315; http://www.climateaudit.org/?p=7168. See also Loehle, C. 2007. A 2000-year global temperature reconstruction based on non-tree ring proxies. *Energy and Environment* 18:1049–58; and Moberg, A., Sonechkin, D.M., Holmgren, K., Datsenko, N.M. and Karlén, W, 2005. Highly variable Northern Hemisphere temperatures reconstructed from low- and high-resolution proxy data. *Nature* 433:613–17. For papers on the Holocene warm period, between 8,000 and 5,000 years ago, see http://climatesanity.wordpress.com/2008/10/15/dont-panic-the-arctic-has-survived-warmer-temperatures-in-the-past/; http://adsabs.harvard. edu/abs/2007AGUFMPP11A0203F; and http://adsabs.harvard.edu/abs/2007AGUFMPP11A0203F; and http://nsidc.org/ arcticseaicenews/faq. html#summer_ice.

p. 334 'the net population at risk of water shortage by 2100 falls under all their scenarios'. Goklany, I. 2009. Is climate change the defining challenge of our age? *Energy and Environment* 20:279–302.

p. 335 'no increase in either the number or the maximum wind speed of Atlantic hurricanes making landfall'. Pielke, R.A., Jr., Gratz, J., Landsea, C.W., Collins, D., Saunders, M.A. and Muslin, R, 2008: Normalized hurricane damage in the United States: 1900–2005. *Natural Hazard Review* 9:29–42.

p. 335 'death rate from weather-related natural disasters has declined by a remarkable 99 per cent'. Goklany, I. 2007. Deaths and death rates due to extreme weather events. *Civil Society Report on Climate Change.* International Policy Network.

p. 335 'cold weather continues to exceed the number of excess deaths during heatwaves by a large margin'. Lomborg, B. 2007. *Cool It.* Marshall Cavendish.

p. 336 'malaria is not limited by climate'. Reiter, P. 2008. Global warming and malaria: knowing the horse before hitching the cart. *Malaria Journal* 7 (supplement 1):S3.

p. 336 'says Paul Reiter, a malaria expert'. Reiter, P. 2007. Human ecology and human behavior. *Civil Society Report on Climate Change.* International Policy Network.

p. 336 'the possibility that global warming might increase that number by 30,000'. Goklany, I. 2004. Climate change and malaria. *Science* 306:56–7. The treatment of Paul Reiter, an expert on malaria, by the IPCC is a strange tale: 'The IPCC rejected Professor Reiter's nomination to write the malaria segment of the health chapter of its 2007 Climate Assessment Report first by pretending he had not been nominated and then by pretending that it had not received the four copies of the nomination papers that he had sent to separate officials. The two lead authors of that segment, unlike Professor Reiter, were not experts on malaria, and had published only one paper on the subject between them. One was not a scientist but an environmental campaigner.' From http://scienceand publicpolicy.org/images/stories/papers/scarewatch/scarewatch_agw_ spread_malaria.pdf.

p. 336 'a jump in tick-borne disease in eastern Europe around 1990'. Randolph, S.E. 2008. Tick-borne encephalitis in Central and Eastern Europe: consequences of political transition. *Microbes and Infection* 10:209–16.

p. 337 'Kofi Annan's Global Humanitarian Forum doubled the number of climate deaths to 315,000 a year'. For a good discussion of this issue see http://cstpr.colorado.edu/prometheus/?p=5410; also http://www.climate-resistance.org/2009/06/the-age-of-the-age-of-stupid.html; also the *Wall Street Journal*: http://online.wsj.com/article/SB124424567009790525. html.

p. 337 'Wheat, for example, grows 15–40 per cent faster in 600 parts per million of carbon dioxide'. Pinter, P.J., Jr., Kimball, B.A., Garcia, R.L., Wall, G.W., Hunsaker, D.J. and LaMorte, R.L. 1996. Free-air $CO_2$ enrichment: Responses of cotton and wheat crops. In Koch, G.W. and Mooney, H.A. (eds). 1996. *Carbon Dioxide and Terrestrial Ecosystems.* Academic Press.

pp. 337–8 'leaving only 5 per cent of the world under the plough in 2100, compared with 11.6 per cent today'. Goklany, I. cited in Bailey, R. 2009. What planetary emergency? *Reason*, 10 March 2009. See http://www.reason.com/news/show/132145.html.

p. 338 'The richest and warmest version of the future will have the least hunger'. Parry, M.L., Rosenzweig, C., Iglesias, A., Livermore, M. and Fischer, G, 2004: Effects of climate change on global food production under SRES emissions and socio-economic scenarios. *Global Environmental Change* 14:53–67.

p. 338 'will have ploughed the least extra land to feed itself'. Levy, P.E. et al. 2004. Modelling the impact of future changes in climate, $CO_2$ concentration and future land use on natural ecosystems and the terrestrial carbon sink. *Global Environmental Change* 14:21–30.

p. 338 'hunger, dirty water, indoor smoke and malaria, which kill respectively about seven, three, three and two people per minute'. UN estimates: 3.7m deaths from hunger each year; 1.7m from dirty water, 1.6m from indoor smoke; 1.1m from malaria.

p. 338 'Economists estimate that a dollar spent on mitigating climate change brings 90 cents of benefits'. Lomborg, B. 2008. How to get the biggest bang for 10 billion bucks. *Wall Street Journal*, 28 July 2008.

p. 338 'The polar bear, still thriving today (eleven of thirteen populations are growing or steady)'. http://www.sciencedaily.com/releases/2008/10/0810 20095850.htm. See also Dyck, M.G., Soon, W., Baydack, R.K., Legates, D.R., Baliunas, S., Ball, T.F. and Hancock, L.O. 2007. Polar bears of western Hudson Bay and climate change: Are warming spring air temperatures the 'ultimate' survival control factor? *Ecological Complexity* 4:73–84. See also Dr Mitchell Taylor's presentation at http://www.you tube.com/watch?v=I63Dl14Pemc.

pp. 339–40 'Charlie Veron, an Australian marine biologist ... Alex Rogers of the Zoological Society of London'. Both quoted in the *Guardian*, 2 September 2009. http://www.guardian.co.uk/environment/2009/sep/02/coral-catastrophic-future.

p. 340 'not even in the Persian Gulf where water temperatures reach 35C'. This is what a Canadian biologist wrote on a blog in August 2008: 'I just got back from Iranian side of the Persian Gulf – the Asaluyeh/Nyband Bay region. Air temps 40, sea temps 35. (Email me privately if you want comments on the joys of doing field work under those conditions.)We observed corals at depths from 4–15m. No corals, at any depths, were bleached. Gives perhaps some relevance to the term "resilience." BTW, those mostly undescribed reefs had coral cover of approx 30% – higher than the Florida Keys.' http://coral.aoml.noaa.gov/pipermail/coral-list/2008-August /037881.html.

p. 340 'corals become more resilient the more they experience sudden warmings'. Oliver, T.A. and Palumbi, S.R. 2009. Distributions of stress-resistant coral symbionts match environmental patterns at local but not regional scales. *Marine Ecology* Progress Series 378:93–103. See also

Baker, A.C. et al. 2004. Coral reefs: Corals' adaptive response to climate change. *Nature* 430:741, who say: 'The adaptive shift in symbiont communities indicates that these devastated reefs could be more resistant to future thermal stress, resulting in significantly longer extinction times for surviving corals than had been previously assumed.'

p. 340 'Some reefs may yet die if the world warms rapidly in the twenty-first century, but others in cooler regions may expand.' Kleypas, J.A., Danabasoglu, G. and Lough. J.M. 2008. Potential role of the ocean thermostat in determining regional differences in coral reef bleaching events, *Geophysical Research* Letters 35: L03613. (doi:10.1029/2007GL03 2257).

p. 341 'a rash of empirical studies showing that increased carbonic acid either has no effect or actually increases the growth of calcareous plankton'. Iglesias-Rodriguez, M.D. et al. 2008. Phytoplankton calcification in a high-$CO_2$ world. *Science* 320:336–40. Other studies of the carbonate issue are summarised by Idso, C. 2009. *$CO_2$, Global Warming and Coral Reefs*. Vales Lake Publishing.

p. 341 'said Bill Clinton once'. Speech to the US National Academy of Sciences, 15 July 1998.

p. 341 'As Indur Goklany puts it'. Goklany, I. 2008. *The Improving State of the World*. Cato Institute.

p. 341 'The results of thirteen economic analyses of climate change'. Summarised in Tol, R. S. J. 2009. The Economic Effects of Climate Change. *Journal of Economic Perspectives*, 23:29–51. http://www.aeaweb.org/articles.php? doi=10.1257/jep.23.2.29. See also the essay by Jerry Taylor at http://www.masterresource.org/2009/11/the-economics-of-climate-change-essential-knowledge.

p. 342 'quoting from the IPCC's 2007 report'. IPCC AR4, Working Group III, p. 204.

p. 342 'says the physicist David MacKay'. MacKay, D. 2009. *Sustainable Energy – without the Hot Air*. UIT, Cambridge.

p. 343 '125 kilowatt-hours per day per person of work that give Britons their standard of living'. Numbers in this paragraph recalculated from MacKay, D. 2009. *Sustainable Energy – without the Hot Air*. UIT, Cambridge. Compare this number (125 kWh per person per day) with the number given in chapter 7 from a different source: England consumes 250 gigawatts (250 gigajoules per second) in total, or 5,000 joules per person per second, assuming the population of England is 50m. There are 3.6m joules in a kilowatt hour and 86,400 seconds in a day so 5,000 x 86,400 = 432m joules per person per day. 432/3.6 = 120 kWh per person per day.

p. 344 'a Spanish study confirms that wind power subsidies destroy jobs'. Donald Hertzmark, 6 April 2009 at http://masterresource.org/?p=1625. See also http://www.juandemariana.org/pdf/090327-employment-public-aid-renewable.pdf, and http://masterresource.org/?p=5046#more-5046.

p. 344 'writes Peter Huber'. Huber, P. 2009. Bound to burn. *City Journal*, spring 2009.

p. 344 'quite soon engineers will be able to use sunlight to make hydrogen directly from water with ruthenium dye as a catalyst'. Bullis, K. 2008. Sun + water = fuel. *Technology Review*, November/December, 56–61.

pp. 344–5 'Once solar panels can be mass-produced at $200 per square metre and with an efficiency of 12 per cent, they could generate the equivalent of a barrel of oil for about $30'. Ian Pearson, 8.9.08: http://www.futurizon.net/blog.htm.

p. 345 'human energy use over the past 150 years as it migrated from wood to coal to oil to gas'. Ausubel, J.H. 2003. 'Decarbonisation: the Next 100 Years'. Lecture at Oak Ridge National Laboratory, June 2003. http://phe.rockefeller.edu/PDF_FILES/oakridge.pdf.

p. 346 'Jesse Ausubel predicts'. Ausubel, J.H. and Waggoner, P.E. 2008. Dematerialization: variety, caution and persistence. *PNAS* 105:12774–9. See also: http://www.nytimes.com/2009/04/21/science/earth/21tier.html.

p. 346 'carbon-rich oceanic organisms called salps'. Lebrato, M. and Jones, D.O.B. 2009. Mass deposition event of Pyrosoma atlanticum carcasses off Ivory Coast (West Africa). *Limnology and Oceanography* 54:1197–1209.

## Chapter 11

p. 349 'I hear babies cry, I watch them grow'. Thiele, B. and Weiss, G. D. 1967. 'What a Wonderful World'. Range Road Music, Inc., Bug Music – Quartet Music, Inc. and Abilene Music, Inc., USA. Copyright renewed. All rights reserved. Reproduced with permission of Carlin Music Corp., London.

p. 349 IPCC projections for world GDP graph. Intergovernmental Panel on Climate Change, 4th Assessment Report 2007.

p. 352 'said H.G. Wells'. Wells, H.G. 'The Discovery of the Future' Lecture at the Royal Institution, 24 January 1902, published in *Nature* 65:326–31. Reproduced with the permission of AP Watt Ltd on behalf of the Literary Executors of the Estate of H.G. Wells.

p. 354 'As Paul Romer puts it'. Quotes are from Romer, P. 'Economic growth' in the *Concise Encyclopedia of Economics (edited by David R Henderson, published by Liberty Fund)*; and Romer, P. 1994. New goods, old theory, and the welfare costs of trade restrictions. *Journal of Development Economics* 43:5–38.

p. 355 'the world economy will be doubling in months or even weeks'. Hanson, R. 2008. Economics of the Singularity. *IEEE Spectrum* (June 2008) 45:45–50.

p. 355 'a technological "singularity"'. This notion has been explored by Vernor Vinge and Ray Kurzweil. See Kurzweil, R. 2005. *The Singularity Is Near*. Penguin.

p. 355 'says Stephen Levy.' Levy, S. 2009. Googlenomics. *Wired*, June 2009.

p. 356 'says the author Clay Shirky'. Shirky, C. 2008. *Here Comes Everybody*. Penguin.

p. 356 'Says Kevin Kelly'. Kelly, K. 2009. The new socialism. *Wired*, June 2009.

p. 358 'The wrong kind of chiefs, priests and thieves could yet snuff out future prosperity on earth.' Meir Kohn has written eloquently on this point. See www.dartmouth.edu/~mkohn/Papers/lessons% 201r3.pdf.

p. 359 'Said Lord Macaulay'. Macaulay, T.B. 1830. Southey's Colloquies on Society. *Edinburgh Review*, January 1830.

p. 359 'In Thornton Wilder's play *The Skin of Our Teeth*'. Wilder, T. 1943. *The Skin of Our Teeth*. HarperCollins.

# INDEX

Index

internet: access to 253, 268;
blogging 257; and charitable
giving 318–19, 356; cyber-crime
99–100, 357; development of 263,
268, 270, 356; email 292; free
exchange 105, 272–3, 356; packet
switching 263; problem-solving
applications 261–2; search engines
245, 256, 267; shopping 37, 99,
107, 261; social networking
websites 262, 268, 356; speed of
252, 253; trust among users
99–100, 356; World Wide Web
273, 356
Inuits 44, 61, 64, 126
IPCC (Intergovernmental Panel on
Climate Change) 330, 331, 332,
333–4, 338, 342, 347, 349, 419,
420, 421, 422
IQ levels 19
Iran 162
Iraq 31, 158, 161
Ireland 24, 129, 199, 227
iron 166, 167, 169, 181, 184, 223,
229, 230, 302, 404
irradiated food 150–51
irrigation 136, 147–8, 159, 161, 163,
198, 242, 281
Isaac, Glyn 64
Isaiah 102, 168
Islam 176, 357, 358
Israel 53, 69, 124, 148
Israelites 168
Italy: birth rate 208; city states
178–9, 181, 196; fascism 289;
Greek settlements 170–71, 173–4;
infant mortality 15; innovations
196, 251; mercantilism 89, 103,
178–9, 180, 196; prehistoric 69
ivory 70, 71, 73, 167

Jacob, François 7
Jacobs, Jane 128
Jamaica 149
James II, King 223

Japan: agriculture 197–8; birth rates
212; dictatorship 109; economic
development 103, 322, 332;
economic and technological
regression 193, 197–9, 202;
education 16; happiness 27;
industrialisation 219; life
expectancy 17, 31; trade 31, 183,
184, 187, 197
Jarawa tribe 67
Java 187
jealousy 2, 351
Jebel Sahaba cemeteries, Egypt 44, 45
Jefferson, Thomas 247, 249, 269
Jenner, Edward 221
Jensen, Robert 327
Jericho 127, 138
Jevons, Stanley 213, 237, 245
Jews 89, 108, 177–8, 184
Jigme Singye Wangchuck, King of
Bhutan 25–6
Jobs, Steve 221, 264, 403
John, King of England 118
Johnson, Lyndon 202–3
Jones, Rhys 79
Jordan 148, 167
Jordan river 127
Joyce, James 289
justice 19–20, 116, 320, 358

Kalahari desert 44, 61, 76
Kalkadoon aborigines 91
Kanesh, Anatolia 165
Kangaroo Island 81
kangaroos 62, 63, 69–70, 84, 127
Kant, Immanuel 96
Kaplan, Robert 293
Kay, John 184, 227
Kazakhstan 206
Kealey, Terence 172, 255, 408
Kelly, Kevin 356
Kelvin, William Thomson, 1st
Baron 409
Kenya 42, 87, 155, 209, 316, 326,
336, 353

# About the author

# Read on

Insights,
Interviews
& More...

# Meet Matt Ridley

*by Natasha Loder*

© 2000 by Jerry Bauer

WHEN WE LOOK AT ANY HUMAN, it is natural to wonder what made them the way they are. On meeting Matt Ridley, one finds a tall, well-bred, thoughtful man who cares about the details of the world around him. Even slightly trivial questions are carefully considered. So was it nature or nurture that made Matt Ridley the man he is today? Ridley insists that it was a bit of both. Not surprising, really, since this is a fundamental message of his book *Nature Via Nurture* (now titled *The Agile Gene*).

Ridley was born in England in 1958. He grew up outside Newcastle on an idyllic dairy and wheat farm that has belonged to his family for three

> 66 Was it nature or nurture that made Matt Ridley the man he is today? 99

centuries. It doesn't get much nicer than that, he observes wistfully. Educated at Eton, he went on to study zoology at Oxford. He had an incredibly privileged background, he admits, with both the best nature and the best nurture.

One factor above all others appears to have steered the course of Ridley's life: during boyhood he had a "complete and dominating obsession" with bird-watching. His father was also a keen bird-watcher. Was it in the genes, then? No, says Ridley, "my grandfather was an engineer. I inherited the personality that led me to be interested in these things, and I learned the habit that was then reinforced by practice."

Bird-watching led to an interest in natural history, which in turn led to a degree in zoology, scientific research on pheasants, and a PhD from Oxford in 1983. While working on his thesis, though, he discovered that he enjoyed writing more than scientific research. So he left academia and joined *The Economist*.

Ridley worked for nine years at *The Economist*, first as a science correspondent, next as the magazine's science editor, then finally as its Washington correspondent. In 1996 he became the founding chairman of the international Centre for Life in Newcastle (a £70,000,000 science park and education center devoted to research in genetics)—a role of which he is proud. He has also been a columnist for the *Daily Telegraph* and *Sunday Telegraph*, and has written a number of highly acclaimed books on popular science, ▶

66 While working on his thesis . . . he discovered that he enjoyed writing more than scientific research. 99

including *The Red Queen: Sex and the Evolution of Human Nature* (1994), *Genome*, which was an international bestseller first published in 2000, and *The Rational Optimist.* Ridley is also a regular columnist for the *Wall Street Journal.*

Married with two children, Ridley lives near Newcastle with his wife, Anya Hurlbert. She is, he says, a "real scientist"—a neuroscientist trying to understand how the brain interprets what the eye tells it. The integration of his work and home life seems a touch uncanny. As Ridley himself notes: "It is perhaps no total coincidence that I wrote a book about mating soon after I got married and have written about nature and nurture soon after I've had children."

He claims to have been "mugged by the realities of parenthood," and had expected to exercise much greater control over the development of his children. "It's amazing how they seem to come into the world not just with their own personalities, but with fully formed sets of behaviors that are extremely resistant to you." It is said, he adds, that people believe in nurture when they have one child, and nature when they have two. Parent of two children, Ridley manages to see how incredibly different they are despite having been brought up in similar environments.

If he hadn't learned about the effect of genes, he might have gone on to feel that

> He claims to have been 'mugged by the realities of parenthood.'

his children's behavior resulted almost purely from their parents' input. Parenting certainly matters, he now recognizes, but it does not fine-tune the personality. Children seem to be much more aware of peer pressure and relationships, and adjust to these factors more than to the will of their parents. He thinks children calibrate themselves against siblings and peers in order to judge what sort of person they are and what they are good at. Then there is a feedback effect. If they are good at something like tennis, they enjoy playing it, spend more time doing it, and therefore get even better. But if they determine they are bad at something they will give up.

Ridley enjoys fly-fishing, seeks inspiration on the Internet, and wishes he had written the Tom Stoppard play *Arcadia*. He never leaves home without his compass, which he uses "on emerging from tube stations." ∾

# Have You Read?
## More by Matt Ridley

### THE AGILE GENE:
### HOW NATURE TURNS ON NURTURE

Armed with extraordinary new discoveries about our genes, acclaimed science writer Matt Ridley turns his attention to the nature versus nurture debate and brings us a stunning book about the roots of human behavior.

In February 2001 it was announced that the human genome contained not 100,000 genes, as was originally expected, but only 30,000. This startling revision led some scientists to conclude that there are simply not enough human genes to account for all the different ways people behave; we must be made by nurture, not nature. Ridley argues that the emerging truth is far more interesting than this myth. Nurture depends on genes, too, and genes need nurture. Genes not only predetermine the broad structure of the brain, they also absorb formative experiences, react to social cues, and even manage memory. They are consequences as well as causes of the will.

Ridley recounts the hundred years' war between partisans of nature and nurture to explain how this paradoxical creature, the human being, can simultaneously exercise free will and be motivated by instinct and culture. *The Agile Gene* is an enthralling account of how genes build brains to absorb experience.

"Matt Ridley's marvelous new book . . . [is] a sweepingly ambitious work that tackles, in lucid and often poetic prose, many of the biggest questions in biology and beyond. . . . Ridley is a superb writer whose exquisite, often moving descriptions of life's designs remind me of the best work of the late Lewis Thomas. Scorning hackneyed, imprecise metaphors (a gene, he insists, is nothing like a blueprint), he crafts some of the clearest explanations of complex biological processes that I have encountered. What's more, he captures their slippery beauty."
—Susan Okie, *Washington Post*

**THE RED QUEEN: SEX AND THE EVOLUTION OF HUMAN NATURE**

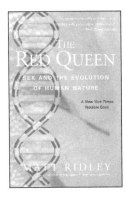

Dandelions don't do it; topminnows don't do it; and the tiny freshwater creature called a bdelloid rotifer definitely doesn't do it (there are no males in the species). But human beings come in two sexes, which mate in spite of all the trouble that goes along with it. This fascinating, delightfully literate work of evolutionary theory explains why. Referencing Lewis Carroll's Red Queen (*Through the Looking Glass*), who has to keep running to stay in the same place, Matt Ridley demonstrates why sex is humanity's best strategy for outwitting its constantly mutating internal predators. *The Red Queen* answers dozens of other riddles of human nature and culture—including why men propose marriage, the method behind our maddening notions of beauty, and the disquieting fact that a woman is more likely to conceive a child by an adulterous lover than by her husband. The result is a brilliantly written book that compels us to rethink everything from the persistence of sexism to the endurance of romantic love.

"[L]iterary . . . witty . . . humane."
—*Boston Globe*

"[A] dazzling display of creativity and wit." —*The Independent* (London)

### THE BEST AMERICAN SCIENCE WRITING 2002
(Matt Ridley, editor)

*The Best American Science Writing 2002* gathers top writers and scientists covering the latest developments in the fastest-changing, farthest-reaching scientific fields, such as medicine, genetics, computer technology, evolutionary psychology, cutting-edge physics, and the environment. The book's twenty-one essays include: "The Made-to-Order Savior," in which Lisa Belkin spotlights two desperate families seeking an unprecedented cure by a medically and ethically unprecedented means—creating a genetically matched child; "Rethinking the Brain," in which Michael Specter reports on the shock waves rippling through the field of neuroscience following the revolutionary discovery that adult brain cells might in fact regenerate; and "I Love My Glow Bunny," in which Christopher Dickey recounts with sly humor a peculiar episode in which genetic engineering and artistic culture collide.

"Superb brain candy."   —*Kirkus Reviews*

### GENOME

Arguably the most significant scientific discovery of the new century, the mapping of the twenty-three pairs of chromosomes that make up the human genome raises almost as many questions as it answers. Questions that will profoundly impact the way we think about disease, about longevity, and about free will. Questions that will affect the rest of your life.

*Genome* offers extraordinary insight into the ramifications of this incredible breakthrough. By picking one newly discovered gene from each pair of chromosomes and telling its story, Matt Ridley recounts the history of our species and its ancestors from the dawn of life to the brink of future medicine. From Huntington's disease to cancer, from the applications of gene therapy to the horrors of eugenics, Matt Ridley probes the scientific, philosophical, and moral issues arising as a result of the mapping of the genome. It will help you understand what this scientific milestone means for you, for your children, and for humankind.

Don't miss the next book by your favorite author. Sign up now for AuthorTracker by visiting www.AuthorTracker.com.